Computer Science and Data Analysis Series

Design and Modeling for Computer Experiments

Chapman & Hall/CRC
Computer Science and Data Analysis Series

The interface between the computer and statistical sciences is increasing, as each discipline seeks to harness the power and resources of the other. This series aims to foster the integration between the computer sciences and statistical, numerical, and probabilistic methods by publishing a broad range of reference works, textbooks, and handbooks.

SERIES EDITORS
John Lafferty, Carnegie Mellon University
David Madigan, Rutgers University
Fionn Murtagh, Royal Holloway, University of London
Padhraic Smyth, University of California, Irvine

Proposals for the series should be sent directly to one of the series editors above, or submitted to:

Chapman & Hall/CRC
23-25 Blades Court
London SW15 2NU
UK

Published Titles

Bayesian Artificial Intelligence
Kevin B. Korb and Ann E. Nicholson

Pattern Recognition Algorithms for Data Mining
Sankar K. Pal and Pabitra Mitra

Exploratory Data Analysis with MATLAB®
Wendy L. Martinez and Angel R. Martinez

Clustering for Data Mining: A Data Recovery Approach
Boris Mirkin

Correspondence Analysis and Data Coding with Java and R
Fionn Murtagh

R Graphics
Paul Murrell

Design and Modeling for Computer Experiments
Kai-Tai Fang, Runze Li, and Agus Sudjianto

Computer Science and Data Analysis Series

Design and Modeling for Computer Experiments

Kai-Tai Fang
Hong Kong Baptist University
Hong Kong, P.R. China

Runze Li
The Pennsylvania State University
University Park, PA, U.S.A.

Agus Sudjianto
Bank of America
Charlotte, NC, U.S.A.

CRC Press
Taylor & Francis Group
Boca Raton London New York

CRC Press is an imprint of the
Taylor & Francis Group, an **informa** business

Published in 2006 by
Chapman & Hall/CRC
Taylor & Francis Group
6000 Broken Sound Parkway NW, Suite 300
Boca Raton, FL 33487-2742

First issued in paperback 2020

ISBN 13: 978-0-367-57800-8 (pbk)
ISBN 13: 978-1-58488-546-7 (hbk)

This book contains information obtained from authentic and highly regarded sources. Reasonable efforts have been made to publish reliable data and information, but the author and publisher cannot assume responsibility for the validity of all materials or the consequences of their use. The authors and publishers have attempted to trace the copyright holders of all material reproduced in this publication and apologize to copyright holders if permission to publish in this form has not been obtained. If any copyright material has not been acknowledged please write and let us know so we may rectify in any future reprint.

**Visit the Taylor & Francis Web site at
http://www.taylorandfrancis.com**

**and the CRC Press Web site at
http://www.crcpress.com**

Library of Congress Cataloging-in-Publication Data

Fang, Kaitai.
  Design and modeling for computer experiments / Kai-Tai Fang, Run-ze Li, and Agus Sudjianto.
      p. cm. -- (Computer science and data analysis series ; 6)
  Includes bibliographical references and index.
  ISBN 1-58488-546-7 (alk. paper)
  1. Computer simulation. 2. Experimental design. I. Li, Run-ze. II. Sudjianto, Agus. III. Title. IV.
Series.

QA76.9.C65F36 2005
003'.3--dc22   2005051751

Preface

With the advent of computing technology and numerical methods, engineers and scientists frequently use computer simulations to study actual or theoretical physical systems. To simulate a physical system, one needs to create mathematical models to represent physical behaviors. Models can take many forms and different levels of granularity of representation of the physical system. The models are often very complicated and constructed with different levels of fidelity such as the detailed physics-based model as well as more abstract and higher level models with less detailed representation. A physics-based model may be represented by a set of equations including linear, nonlinear, ordinary, and partial differential equations. A less detailed and a more abstract model may employ stochastic representations such as those commonly used to simulate processes based on empirical observations (e.g., discrete events simulations for queuing networks). Because of the complexity of real physical systems, there is usually no simple analytic formula to sufficiently describe the phenomena, and systematic experiments to interrogate the models are required. In particular, systematic experiments are useful in the following situations: 1) the model is very complex where the underlying relationships are nonlinear with many variables interacting with one another; 2) the model exhibits stochastic behavior; or 3) the model produces multiple and complicated outputs such as functional response or even 3D animation. In such situations, it is often difficult and even impossible to study the behavior of computer models using traditional methods of experiments. One approach to studying complex input-output relationships exhibited by the simulation model is by constructing an approximation model (also known as a metamodel) based on a set of limited observation data acquired by running the simulation model at carefully selected design points. This approximation model has an analytic form and is simpler to use and apply.

The availability of a metamodel as a surrogate to the original complex simulation model is particularly useful when applying the original model is a resource-intensive process (e.g., in terms of model preparation, computation, and output analysis), as is typically the case in complex simulation models. The need for a "cheap-to-compute" metamodel is even greater when one would like to use the model for more computationally demanding tasks such as sensitivity analysis, optimization, or probabilistic analysis. To perform the aforementioned tasks successfully, a high quality metamodel developed from a limited sample size is a very important prerequisite. To find a high quality metamodel, choosing a good set of "training" data becomes an important issue for computer simulation. The so-called space-filling design provides ways to generate efficient training data. By efficient, we mean able to generate a set of sample points that capture maximum information between the input-output relationships. There are many space-filling designs such as Latin hypercube sampling and its modifications, optimal Latin hypercube samplings and uniform designs. How to find the best suited metamodel is another key issue in computer experiments. Various modeling techniques (e.g., Kriging models, (orthogonal) polynomial regression models, local polynomial regression, multivariate spline and wavelets, Bayesian methods, and neural networks) have been proposed and applied by many practitioners and researchers. Therefore, **design** and **modeling** are two key issues in computer experiments. Part II of this book introduces various space-filling designs and their constructions while Part III presents most useful modeling techniques. In the last chapter, we introduce functional response modeling, which is becoming increasingly more common in modern computer experiments. This approach addresses cases where response data are collected over an interval of some index with an intensive sampling rate.

As design and modeling of computer experiments have become popular topics in recent years, hundreds of papers have been published in both statistics and engineering communities. Several comprehensive review papers have emphasized the importance and challenge of this relatively new area. In summer 2002, the first two authors of the book were invited by the third author, who was a manager at Ford Motor Company, to visit Ford for a technical exchange and a series of lectures to explore the synergy between statistical theory and the practice of computer experiments. In recent years, Ford has intensively applied computer experiments (including uniform design) to support new product development, in particular, automotive engine design, with tremendous practical success. The pervasive use of computer experiments at Ford to support early stages of product design before the availability of physical prototypes is a cornerstone in its "Design for Six Sigma (DFSS)" practice. After careful discussion, we decided to collaborate on a book on design and modeling for computer experiments. Our motivation to write the book was, at that time, strengthened by the lack of a single comprehensive resource on the subject as well as our desire, to share our experience of blending modern and sound statistical approach with extensive practical engineering applications. The collaboration has been challenging because of the geographical barriers as well as busy schedules. In summer 2003, we learned about the new book "_The Design and Analysis of Computer Experiments_" by Santner, Williams and Notz (2003), which provides a good reference and complementary material to our book.

Because of our motivation to provide useful techniques to practicing engineers and scientists who are interested in applying the methodology, this book has been structured to emphasize practical methodology with applications, and it purposefully omits the corresponding theoretical details. This has been a challenge for us because, at the same time, we would like to provide readers with sufficient sound statistical justifications. Various examples are provided in the book to clarify the methods and their practical implementations. The book presents the most useful design and modeling methods for computer experiments, including some original contributions from the authors. The book can be used as a textbook for undergraduate and postgraduate students and as a reference for engineers.

We thank Professors C. R. Rao, Y. Wang, R. Mukerjee, Fred J. Hickernell and D. K. J. Lin for their encouragement and/or valuable comments. Fang, the first author of the book, is grateful to his collaborators L. Y. Chan, G. Ge, E. Liski, M. Q. Liu, X. Lu, C. X. Ma, P. Winker, and J. X. Yin, for their successful collaboration and their contributions, and to his former postgraduate students, W. M. Ho, Y. K. Lo, J. X. Pan, H. Qin, Y. Tang, G. L. Tian, M. Y. Xie, and A. J. Zhang, for significant contributions to this area. Fang is very grateful to Professor C.F. Ng, the President of Hong Kong Baptist University, for his support and encouragement. Li, the second author, would like to thank John Dziak, Yiyun Zhang, and Zhe Zhang for enthusiastic support of his work. Sudjianto would like to thank many of his former colleagues at Ford Motor Company, the company where he spent his 14-year engineering career, for their generous collaboration and support, in particular, Thomas McCarthy, Georg Festag, Radwan Hazime, Feng Shen, Richard Chen, Ren-Jye Yang, Nathan Soderborg, Jovan Zagajac, Joseph Stout, Frank Fsadni, John Koszewnik, Steve Wang, Liem Ferryanto, Mahesh Vora and many others. He also would like to thank his collaborators Wei Chen, Ruichen Jin, and Xiaoping Du. All three authors thank Miss P. Kuan, Mr. P.H. Cheung, and Mr. K. Yeung for their excellent technical assistance.

Fang, the first author, acknowledges financial support from Hong Kong RGC grant RGC/HKBU 200804, Hong Kong Baptist University grant FRG/03-04/II-711, and the Statistics Research and Consultancy Centre, Hong Kong Baptist University. Li, the second author, is grateful for financial support from the National Science Foundation and the National Institute on Drug Abuse.

Kai-Tai Fang, Hong Kong Baptist University
and Chinese Academy of Sciences
Runze Li, The Pennsylvania State University
Agus Sudjianto, Bank of America

Contents

Part I An Overview

This part provides a brief introduction to the design and modeling of computer experiments, including such experimental design concepts as orthogonal arrays and design optimality. It also presents some interesting and heuristic examples of computer experiments that will serve as illustrations for methods presented later in the book. An illustrative case study is also provided, and guidelines for reading the rest of the book are suggested.

1

Introduction

1.1 Experiments and Their Statistical Designs

The study of *experimental design* is extremely important in modern industry, science, and engineering. Nowadays experiments are performed almost everywhere as a tool for studying and optimizing processes and systems.

The purpose of an experiment in industrial engineering is

- to improve process yields;

- to improve the quality of products, such as to reduce variability and increase reliability;

- to reduce the development time; and

- to reduce the overall costs.

A good experimental design should minimize the number of runs needed to acquire as much information as possible. Experimental design, a branch of statistics, has enjoyed a long history of theoretical development as well as applications. Comprehensive reviews for various kinds of designs can be found in *Handbook of Statistics, Vol. 13*, edited by S. Ghosh and C.R. Rao as well as in *Handbook of Statistics, Vol. 22: Statistics in Industry*, Edited by R. Khattree and C.R. Rao.

Experiments can be classified as
- physical experiments, or
- computer (or simulation) experiments.

Traditionally, an experiment is implemented in a laboratory, a factory, or an agricultural field. This is called *physical experiment* or *actual experiment*, where the experimenter physically carries out the experiment. There always exist random errors in physical experiments so that we might obtain different outputs under the identical experimental setting. Existence of random errors creates complexity in data analysis and modeling. Therefore, the experimenter may choose one or few factors in the experiment so that it is easy to explore the relationship between the output and input; alternatively, the experimenter may use some powerful statistical experimental designs.

The statistical approach to the design of experiments is usually based on a statistical model. A good design is one which is optimal with respect to a statistical model under our consideration. There are many statistical models for physical experiments, among which the *fractional factorial design* – based on an ANOVA model – and the *optimum design* – based on a regression model – are the most widely used in practice. These models involve unknown parameters such as main effects, interactions, regression coefficients, and variance of random error. Good designs (e.g., orthogonal arrays and various optimal designs) may provide unbiased estimators of the parameters with smaller or even the smallest variance-covariance matrix under a certain sense. Useful concepts for design of (physical) experiments and basic knowledge of the orthogonal arrays and the optimal designs will be given in the next section.

Many physical experiments can be expensive and time consuming because "physical processes are often difficult or even impossible to study by conventional experimental methods. As computing power rapidly increasing and accessible, it has become possible to model some of these processes by sophisticated computer code" (Santner, Williams and Notz (2003)). In the past decades *computer experiments* or *computer-based simulation* have become topics in statistics and engineering that have received a lot of attention from both practitioners and the academic community. The underlying model in a computer experiment is deterministic and given, but it is often too complicated to manage and to analyze. One of the goals of computer experiments is to find an approximate model ("metamodel" for simplicity) that is much simpler than the true one. Simulation experiments study the underlying process by simulating the behavior of the process on a computer. The true model is deterministic and given as in a computer experiment, but errors on the inputs are considered and assumed. The simulation of the random process is conducted by incorporating random inputs into the deterministic model.

For detailed introductions for the above three kinds of experiments, we need related concepts, which will be given in the next section.

1.2   Some Concepts in Experimental Design

This section introduces some basic concepts in *factorial design, optimal design,* and *computer experiments.* For details the reader can refer to Box et al. (1978) and Dey and Mukerjee (1999) for factorial designs, to Atkinson and Donev (1992) for optimal designs, and to Santner, Williams and Notz (2003) for computer experiments.

Factor: A factor is a controllable variable that is of interest in the experiment. A factor may be *quantitative* or *qualitative*. A *quantitative factor* is one whose values can be measured on a numerical scale and that fall in an interval, e.g., temperature, pressure, ratio of two raw materials, reaction time length, etc. A *qualitative factor* is one whose values are categories such as different equipment of the same type, different operators, several kinds of a material, etc. The qualitative factor is also called a *categorical factor* or an *indicator factor*. In computer experiments, factors are usually called *input variables*. The variables that are not studied in the experiment are not considered as factors, and they are set to fixed values. The variables that cannot be controlled at all are treated as the *random error*. In *robust design* study, some uncontrollable factors during regular operation are purposefully controlled and included during the experiment so that the product can be designed to perform well in varying operating environments. These are known as *noise factors*.

Experimental domain, **level**, and **level-combination**: Experimental domain is the space where the factors (input variables) take values. In computer experiments the experimental domain is also called *input variable space*. A factor may be chosen to have a few specific values in the experimental domain, at which the factor is tested. These selected values are called *levels* of the factor. Levels are used in the ANOVA model where the experimenter wants to test whether the response y has a significant difference among the levels. A *level-combination* is one of the possible combinations of levels of the factors. A level-combination is also called a *treatment combination*. A level-combination can be considered a point in input variable space and called an *experimental point*. In computer experiments the concept of level loses the original statistical meaning defined in factorial plans, but we retain this concept to ease the construction of designs (see Chapter 2).

Run, trial: A run or trial is the implementation of a level-combination in the experimental environment. A run might be a physical or computer experiment. Because a computer experiment is deterministic, multiple trials will produce identical outputs; thus, a trial (in the sense of one of several runs at the same experimental point) is only meaningful in physical experiments.

Response: Response is the result of a run based on the purpose of the experiment. The response can be numerical value or qualitative or categorical, and can be a function that is called *functional response*. Chapter 7 in this book will discuss computer experiments with functional response. In computer experiments responses are also called *outputs*.

Random error: In any industrial or laboratory experiment there exist random errors. We might obtain different results in two runs of the identical environment due to the random error. The random error can often be assumed to be distributed as a normal distribution $N(0, \sigma^2)$ in most experiments. The variance σ^2 measures magnitude of the random error. In computer experiments, however, *there is no random error*. Therefore, the statistical theory and methods that have been constructed to address random errors cannot be directly applied to analyze data from computer experiments. Nevertheless, the ideas and principles of many statistical methods can be extended to those cases without random errors. The sensitivity analysis in Chapter 6 is such an example.

Factorial design: A factorial design is a set of level-combinations with main purpose of estimating main effects and some interactions of the factors. A factorial design is called *symmetric* if all factors have the same number of levels; otherwise, it is called *asymmetric*.

Full factorial design: A design where all the level-combinations of the factors appear equally often is called a *full factorial design* or a *full design*. Clearly, the number of runs in a full factorial design should be $n = k \prod_{j=1}^{s} q_j$, where q_j is the number of levels of the factor j and k is the number replications for all the level-combinations. When all the factors have the same number of levels, q, $n = kq^s$. The number of runs of a full factorial design increases exponentially with the number of factors. Therefore, we consider implementing a subset of all the level-combinations that have a good representation of the complete combinations.

Fractional factorial design: A fraction of a full factorial design (FFD) is a subset of all level-combinations of the factors (Dey and Mukerjee (1999)). A FFD can be expressed as an $n \times s$ matrix, where n is the number of runs and s the number of factors, and the jth column of the matrix has q_j levels (entries). For example, the design in Example 1 below can be expressed as

$$
\mathbf{D} = \begin{bmatrix}
80 & 90 & 5 & a \\
80 & 120 & 7 & b \\
80 & 150 & 9 & c \\
90 & 90 & 7 & c \\
90 & 120 & 9 & a \\
90 & 150 & 5 & b \\
100 & 90 & 9 & b \\
100 & 120 & 5 & c \\
100 & 150 & 7 & a
\end{bmatrix}
\tag{1.1}
$$

where there are four factors, each having three levels. Totally, there are $81 = 3^4$ possible level-combinations, but the design $\mathbf{D}$ chooses only nine of them. How to choose a good subset is the most important issue in FFD. A carefully selected combination known as the orthogonal array is recommended in the literature and has been widely used in practice.

Orthogonal array: An orthogonal array (OA) of strength t with n runs and s factors, denoted by $OA(n, s, q, r)$, is a FFD where any subdesign of n runs and m ($m \leq r$) factors is a full design. Strength two orthogonal arrays are extensively used for planning experiments in various fields and are often expressed as orthogonal design tables. The reader can refer to Hedayat, Sloane and Stufken (1999) for the details.

Orthogonal design tables: An orthogonal design table, denoted by $L_n(q_1 \times \cdots \times q_s)$, is an $n \times s$ matrix with entries $1, 2, \cdots, q_j$ at the jth column such that:

1) each entry in each column appears equally often;
2) each entry-combination in any two columns appears equally often.

When all the factors have the same number of levels, it is denoted by $L_n(q^s)$, which is an orthogonal array $OA(n, s, q, 2)$. The first five columns of Table 1.1 give an orthogonal table $L_9(3^4)$, where the design serves for an experiment of 9 runs and 4 factors, each having three levels: '1,' '2,' and '3.' In fact, we can use other three symbols, such as 'α,' 'β,' and 'γ' to replace '1,' '2,' and '3,' in $L_9(3^4)$.

Example 1 Suppose that there are four factors in a chemical engineering experiment and each chooses three levels as follows:
  A (the temperature of the reaction): 80°C, 90°C, 100°C;
  B (the time length of the reaction): 90 min., 120 min., 150 min.;
  C (the alkali percentage): 5%, 7%, 9%;
  D (operator): a, b, c.
There are $3^4 = 81$ level-combinations. If we use $L_9(3^4)$ to arrange a fractional factorial design, the columns of $L_9(3^4)$ can be filled for factors A, B, C, and D by any choices of $4! = 24$ ones. Suppose that the four factors are put in columns in a natural order. Then, replace the 'abstract' three levels of the four columns by the factor's levels and obtain a fractional factorial design on the right-hand side of Table 1.1. There are nine level-combinations chosen. The first one is (80°C, 90 min., 5 %, a), and the last one is (100°C, 150 min., 7 %, a). Note that the factor D (operator) is not of interest, but we have to consider its effect as the operator may influence the output of the experiment. Such a factor is called a *noise factors*.

TABLE 1.1
Orthogonal Design $L_9(3^4)$ and Related Design

No	1	2	3	4	A	B	C	D
1	1	1	1	1	80°C	90 min.	5%	a
2	1	2	2	2	80°C	120 min.	7%	b
3	1	3	3	3	80°C	150 min.	9%	c
4	2	1	2	3	90°C	90 min.	7%	c
5	2	2	3	1	90°C	120 min.	9%	a
6	2	3	1	2	90°C	150 min.	5%	b
7	3	1	3	2	100°C	90 min.	9%	b
8	3	2	1	3	100°C	120 min.	5%	c
9	3	3	2	1	100°C	150 min.	7%	a

Isomorphism: Exchanging columns and rows of an orthogonal design table still gives an orthogonal design table. Two orthogonal designs are called *isomorphic* if one can be obtained from the another by exchanging rows and columns, and permuting levels of one or more columns of the design matrix. In the traditional factorial design two isomorphic factorial designs are considered to be equivalent.

ANOVA models: A factorial plan is based on a statistical model. For one factor experiments with levels $x_1, \cdots, x_q$, the model can be expressed as

$$y_{ij} = \mu + \alpha_j + \varepsilon_{ij} = \mu_j + \varepsilon_{ij}, \quad j = 1, \cdots, q, i = 1, \cdots, n_j, \quad (1.2)$$

where μ is the overall mean of y, μ_j is the true value of the response y at x_j, and ε_{ij} is the random error in the ith replication at the jth level x_j. All the ε_{ij}'s are assumed to be independently identically distributed according to $N(0, \sigma^2)$. Then the mean μ_j can be decomposed into $\mu_j = \mu + \alpha_j$, where α_j is called the *main effect* of y at x_j and they satisfy $\alpha_1 + \cdots + \alpha_q = 0$. The number of runs in this experiment is $n = n_1 + \cdots + n_q$. The model (1.2) is called an ANOVA model.

An ANOVA model for a two factor experiment, factor A and factor B, can be expressed as

$$y_{ijk} = \mu + \alpha_i + \beta_j + (\alpha\beta)_{ij} + \varepsilon_{ijk},$$
$$i = 1, \cdots, p; \quad j = 1, \cdots, q; \quad k = 1, \cdots, K, \quad (1.3)$$

where
μ = overall mean;
α_i = main effect of factor A at level α_i;
β_j = main effect of factor B at level β_j;
ε_{ijk} = random error in the kth trial at level-combination $\alpha_i\beta_j$;
$(\alpha\beta)_{ij}$ = interaction between A and B at level-combination $\alpha_i\beta_j$ under the restrictions:

$$\sum_{i=1}^{p} \alpha_i = 0; \quad \sum_{j=1}^{q} \beta_j = 0;$$

$$\sum_{i=1}^{p} (\alpha\beta)_{ij} = \sum_{j=1}^{q} (\alpha\beta)_{ij} = 0.$$

There are $p-1$ independent parameters α_is, $q-1$ β_js, and $(p-1)(q-1)$ $(\alpha\beta)_{ij}$s. The total number of these parameters is $pq - 1$. For an s factor factorial experiment with $q_1, \cdots, q_s$ levels, we may consider interactions among three factors, four factors, etc., and the total number of main effects and interactions becomes $(\prod_{j=1}^{s} q_j - 1)$, which exponentially increases as s increases. Therefore, the *sparsity principle* and *hierarchical ordering principle* are proposed.

Sparsity principle: The number of relatively important effects and interactions in a factorial design is small.

Hierarchical ordering principle: Lower order effects are more likely to be important than higher order effects; main effects are more likely to be important than interactions; and effects of the same order are equally likely to be important.

Supersaturated design: Supersaturated designs are fractional factorials in which the number of estimated (main or interaction) effects is greater than the number of runs. In industrial and scientific experiments, especially in their preliminary stages, very often there are a large number of factors to be studied and the run size is limited because of high costs. However, in many situations only a few factors are believed to have significant effects. Under the effect sparsity assumption, supersaturated designs can be effectively used to identify the dominant factors. Consider a design of n runs and s factors each having q levels. The design is called *unsaturated* if $n - 1 > s(q - 1)$; *saturated* if $n - 1 = s(q - 1)$; and *supersaturated* if $n - 1 < s(q - 1)$.

Optimal designs: Suppose the underlying relationship between the response y_k and the input factors $\mathbf{x}_k = (x_{k1}, \cdots, x_{ks})$ can be expressed in a regression model:

$$y_k = \sum_{j=1}^{m} \beta_j g_j(x_{k1}, \cdots, x_{ks}) + \varepsilon_k = \sum_{j=1}^{m} \beta_j g_j(\mathbf{x}_k) + \varepsilon_k, \quad k = 1, \cdots, n, \quad (1.4)$$

where $g_j(\cdot)$ are pre-specified or known functions, ε is random error with $E(\varepsilon) = 0$, and $\mathrm{Var}(\varepsilon) = \sigma^2$. The model (1.4) involves many useful models, such as, the linear model,

$$y = \beta_0 + \beta_1 x_1 + \cdots + \beta_s x_s + \varepsilon,$$

and the quadratic model,

$$y = \beta_0 + \sum_{i=1}^{s} \beta_i x_i + \sum_{1 \leq i \leq j \leq s} \beta_{ij} x_i x_j + \varepsilon.$$

Note that functions g_j can be nonlinear in $\mathbf{x}$, such as $\exp(-x_i), \log(x_j), 1/(a + x_i x_j)$, etc. The model (1.4) can be expressed in terms of vector and matrix as

$$\mathbf{y} = \mathbf{G}\boldsymbol{\beta} + \boldsymbol{\epsilon}, \quad (1.5)$$

where

$$\mathbf{G} = \begin{bmatrix} g_1(\mathbf{x}_1) & \cdots & g_m(\mathbf{x}_1) \\ \vdots & & \vdots \\ g_1(\mathbf{x}_n) & \cdots & g_m(\mathbf{x}_n) \end{bmatrix}, \quad \boldsymbol{\beta} = \begin{bmatrix} \beta_1 \\ \vdots \\ \beta_m \end{bmatrix}. \quad (1.6)$$

The matrix $\mathbf{G}$, called the *design matrix*, represents both the data and the model. $\mathbf{M} = \frac{1}{n}\mathbf{G}'\mathbf{G}$ is called the *information matrix*, where $\mathbf{G}'$ is the transpose of $\mathbf{G}$; it can also be written as $\mathbf{M}(D_n)$ to emphasize that it is a function of the design $D_n = \{\mathbf{x}_k = (\mathbf{x}_1, \cdots, \mathbf{x}_s)\}$. The covariance matrix of the least squares estimator is given by

$$\mathrm{Cov}(\hat{\boldsymbol{\beta}}) = \frac{\sigma^2}{n}\mathbf{M}^{-1}. \quad (1.7)$$

Clearly, we wish $\text{Cov}(\hat{\beta})$ to be as small as possible in a certain sense. That suggests maximizing $\mathbf{M}(D_n)$ with respect to D_n. However, $\mathbf{M}$ is an $m \times m$ matrix, so we have to find a scale criterion of maximization of $\mathbf{M}$, denoted by $\phi(\mathbf{M}(D_n))$, and we can then find a ϕ-optimal design that maximizes $\phi(\mathbf{M})$ over the design space. Many authors have proposed many criteria, such as

- D-**optimality**: maximize the determinant of $\mathbf{M}$;
- A-**optimality**: minimize the trace of $\mathbf{M}^{-1}$;
- E-**optimality**: minimize the largest eigenvalue of $\mathbf{M}^{-1}$.

The concepts of 'determinant', 'trace', and 'eigenvalue' of a matrix are reviewed in Section A.1. In multivariate statistical analysis the determinant of the covariance matrix is called the *generalized variance*. D-optimality is equivalent to minimizing the generalized variance and also is equivalent to minimizing the volume of the confidence ellipsoid of $(\beta - \hat{\beta})'\mathbf{M}(\beta - \hat{\beta}) \leq a^2$ for any $a^2 > 0$. The A-optimality is equivalent to minimizing the sum of the variances of $\hat{\beta}_1, \cdots, \hat{\beta}_m$. The reader can find more optimalities and related theory in Atkinson and Donev (1992), Pukelsheim (1993), etc.

When the true model is known, optimal designs are the best under the given optimality. Optimal designs have many attractive properties, but they lack robustness to model misspecification. When the underlying model is unknown, optimal designs may have a poor performance. If one can combine efficiency of the optimal design and robustness of the uniform design, the new design may have a good performance. Yue and Hickernell (1999) provide some discussion on this direction.

1.3   Computer Experiments

1.3.1   Motivations

Much scientific research involves modeling complicated physical phenomena using a mathematical model

$$y = f(x_1, \cdots, x_s) = f(\mathbf{x}), \quad \mathbf{x} = (x_1, \cdots, x_s)' \in T, \qquad (1.8)$$

where $\mathbf{x}$ consists of input variables, y is an output variable, the function f may or may not have an analytic formula, and T is the input variable space. Model (1.8) may be regarded as a solution of a set of equations, including linear, nonlinear, ordinary, and/or partial differential equations, and it is often impossible to obtain an analytic solution for the equations. Engineers or scientists use models to make decisions by implementing the model to predict behavior of systems under different input variable settings. Hence, the model is a vital element in scientific investigation and engineering design. The availability of a model is often crucial in many situations because physical experiments to fully understand the relationship between response y and inputs x_j's can be too expensive or time consuming to conduct. Thus, computer models become very important for investigating complicated physical phenomena. Scientists and engineers make use of computer simulations to explore the complex relationship between inputs and outputs. One of the goals of computer experiments is first to find an approximate model that is much simpler than the true (but complicated) model (cf. Figure 1.1) by simulating the behavior of the device/process on a computer. The distinct nature of computer experiments poses a unique challenge, which requires special design of experiment approach.

Design of computer experiments has received a great deal of attention in the past decades. Hundreds of papers have been published on this topic. Among them, Sacks, Welch, Mitchell and Wynn (1989), Bates, Buck, Riccomagno and Wynn (1996), and Koehler and Owen (1996) give comprehensive reviews on this rapidly developing area and provide references therein. It is worthwhile to note the unique characteristics of computer experiments compared to physical experiments, such as:

- Computer experiments often involve larger numbers of variables compared to those of typical physical experiments.

- Larger experiment domain or design space is frequently employed to explore complicated nonlinear functions.

- Computer experiments are deterministic. That is, samples with the same input setting will produce identical outputs.

To be better suited to the aforementioned characteristics, computer experiments necessitate different approaches from those of physical experiments.

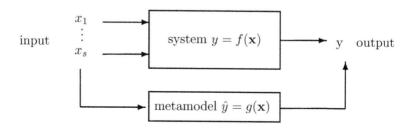

FIGURE 1.1
Computer experiments.

1.3.2   Metamodels

As previously discussed, one of the goals of computer experiment is to seek a
suitable approximate model

$$y = g(x_1, \cdots, x_s), \quad \mathbf{x} \in T, \tag{1.9}$$

which is close to the real one but faster to run, and yields insight into the
relationships between y and $\mathbf{x}$. Such an approximate model is also called a
"model of the model" or *"metamodel"* (Kleijnen (1987)). As the metamodel
g is easier to compute and has an analytic formula, many applications can be
done based on g; for example, a metamodel can provide:

1. **Preliminary study and visualization**

   Graphical tools are very useful for visualizing the behavior of the function g. Since g has an expressive form, we may easily construct various
   plots of y against each input factor. This will help us examining the
   direction of main effects of each factor. Contour plots of the output
   variable against each pair of the factors can be used to examine whether
   there is interaction between the input factors. 3-D plots or animation
   can intuitively show us how many local minimum/maximum of the true
   model may exist, and give us a better understanding of the relationship
   between the input and output variables.

2. **Prediction and Optimization**

   Given a sample set from design of experiment, one may be interested in
   predicting the value of y at untried inputs. For example, one may want
   to compute the overall mean of y over a region T,

   $$E(y|T) = \frac{1}{\text{Vol}(T)} \int_T f(\mathbf{x}) d\mathbf{x}, \tag{1.10}$$

   where $\text{Vol}(T)$ is the volume of T, but a direct computation cannot be
   easily done. In this case, a metamodel can be used as an estimator,

   $$I(g|T) = \frac{1}{\text{Vol}(T)} \int_T g(\mathbf{x}) d\mathbf{x}. \tag{1.11}$$

   Very often, global minimum/maximum of the response is required. The
   solution can be approximately found by finding a point $\mathbf{x}^* = (x_1^*, \cdots, x_s^*)$
   $\in T$ such that

   $$g(x_1^*, \cdots, x_s^*) = \min_{\mathbf{x} \in T} g(x_1, \cdots, x_s).$$

   The minimum/maximum point $\mathbf{x}^*$ can be found analytically (for instance, the metamodel g has continuous derivatives) or numerically by
   some computer programs. The above optimization frequently involves

multiple responses to be optimized and constraints to be satisfied. This kind of constrained optimization often cannot be done by using response surface methodology alone. But a combination of nonlinear programming techniques and metamodel can handle this problem easily.

3. **Sensitivity Analysis**

Sensitivity analysis, which attempts to quantify and rank the effects of input variables, is one of the main interests in decision-making processes such as engineering design. Here, one would like to quantify the leverage of each input variable to the output. Sensitivity calculation such as quantification of the proportion of variation of outputs, y, explained by variations of input variables x_i (McKay (1995))

$$CR_i = \frac{\text{Var}\{E(y|x_i)\}}{\text{Var}(y)}$$

for $i = 1, \cdots, s$, can be easily done using the metamodel. Analysis of variance (ANOVA) and its generalization for nonlinear models such as global sensitivity index frequently require integration in high dimensional space using Monte Carlo techniques. When direct computation of y is expensive, the availability of a cheap-to-compute metamodel becomes an attractive numerical choice.

4. **Probabilistic Analysis**

In many applications, one often is interested in studying the effect of input uncertainty on the variability of the output variable. In particular, reliability and risk assessment applications often exercise computer models subjected to assumed stochastic inputs to assess the probability that output will meet a certain criterion (Haldar and Mahadevan (2000)). In probabilistic analysis, the input $\mathbf{x}$ is assumed to follow random distributions with an interest to predict the probability of y to exceed a certain target value, y_0,

$$R = P(y > y_0) = \int_{y_0}^{\infty} p(y)dy,$$

or

$$R = P\{f(\mathbf{x}) > y_0\} = \int_{f(\mathbf{x}) > y_0} p(\mathbf{x})d\mathbf{x},$$

where $\mathbf{x}$ are stochastic input variables, $f(\mathbf{x})$ is a deterministic function of the system model, $p(y)$ is the probability density function of y, and $p(\mathbf{x})$ is the joint density of x.

Since the number of random variables in practical applications is usually large, multidimensional integration is involved. Due to the complexities of the model, $f(\mathbf{x})$, there is rarely a closed form solution and it is also often difficult to evaluate the probability with numerical integration methods. Therefore, Monte Carlo integration (Melchers (1999)) and its approximations such as the first order reliability method (Hasofer and Lind (1974)), Second Order Reliability Method (Breitung (1984)), or Saddlepoint Approximation (Du and Sudjianto (2004)) are frequently applied. When function evaluations, $f(\mathbf{x})$, are computationally expensive, the metamodel becomes a popular choice for estimating probability, e.g.,

$$\hat{R} = P\{g(\mathbf{x}) > y_0\} = \int_{g(\mathbf{x})>y_0} p(\mathbf{x})d\mathbf{x}$$

(Sudjianto, Juneja, Agrawal and Vora (1998), Jin, Du and Chen (2003), Zou, Mahadevan, Mourelatos and Meernik (2002)).

5. Robust Design and Reliability-Based Design

Robust design is a method for improving the quality of a product through minimizing the effect of input variation without eliminating the source of variation, i.e., the so-called *noise factors* (Phadke (1989), Taguchi (1986), Taguchi (1993), Wang, Lin and Fang (1995), Nair, Taam and Ye (2002), Wu and Hamada (2000)). It emphasizes the achievement of performance robustness (i.e., reducing the variation of a performance) by optimizing the so-called *controllable input variables*. In robust design, the input variables, $\mathbf{x}$, consist of three types of variables:

- **Control factors**, $\mathbf{x}_c \in T$, deterministic variables for which the values are optimized to minimize the variability of response variable, y.
- **Noise factors**, $\mathbf{x}_n \in T$ stochastic variables representing uncontrollable sources of variations such as manufacturing variation, environment, and usage.
- **Input signals**, $\mathbf{x}_s \in T$, deterministic variable for which the response must be optimized within a specified range of these variables (Nair et al. (2002)).

Robust design problems can be formulated as a constrained variance minimization problem to find $\mathbf{x}_c^* = (x_1^*, \cdots, x_s^*) \in T$ such that (Chen, Allen, Mistree and Tsui (1996), Parkinson, Sorensen and Pourhassan (1993))

$$Var\{f(\mathbf{x}_c^*, \mathbf{x}_n, \mathbf{x}_s)\} = \min_{\mathbf{x}_c \in T} Var\{f(\mathbf{x}_c, \mathbf{x}_n, \mathbf{x}_s)\}$$

subject to $E\{f(\mathbf{x}_c^*, \mathbf{x}_n, \mathbf{x}_s)\} = Y.$

Related to robust design is an approach called *reliability-based design* (RBD). The RBD approach focuses on searching $\mathbf{x}^* = (x_1^*, \cdots, x_s^*) \in T$ such that the response variable $y_{obj} = f_{obj}(\mathbf{x}) \in R$ is optimized, while maintaining other constraints, $y_i, i \in 1, \cdots, d$, to satisfy specified minimum probabilistic levels, α_i (Tu, Choi and Young (1999), Wang, Grandhi and Hopkins (1995), Wu and Wang (1998)).

$$f_{obj}(\mathbf{x}*) = \min_{\mathbf{x} \in T} f_{obj}(\mathbf{x})$$

subject to $Pr\{f_i(x)\} \geq \alpha_i, i \in 1, \cdots, d.$

Du, Sudjianto and Chen (2004) presented a unified framework for both robust design and reliability-based design for optimization under uncertainty using inverse probability approach.

Note that both robust design and RBD require repetitive evaluations of system performances, $f(\mathbf{x})$, for which the availability of metamodels, $g(\mathbf{x})$, becomes crucial when the original models are computationally expensive.

In summary, the statistical approach for computer experiments involves two parts:

1. **Design:** We wish to find a set of n points, i.e., a design matrix, denoted by D_n, in the input space T so that an approximate model can be 'best' constructed by modeling techniques based on the data set that is formed by D_n and the output on D_n. Considering the previously mentioned requirements of computer experiments, a natural idea is to put points of D_n to be uniformly scattered on T. Such a design is called a *space-filling design*, or *uniform design* in the literature. Part II of this book provide in-depth discussion of many space-filling designs best suited for computer experiments.

2. **Modeling:** Fitting highly adaptive models by various modeling techniques. Because of the space-filling nature of the experimental design, an adaptive model which can represent non-linearity as well as provide good prediction capability at untried points is very important. Because of the complexity of highly adaptive models, which typically have the form of "model-free" non-parametric regression (i.e., no preconceived rigid structural model is assumed during the model building), straightforward model interpretations are often unavailable. To alleviate this problem, one may need to employ a more sophisticated ANOVA-like global sensitivity analysis to interpret the metamodel. Part III of this book introduces a variety of modeling techniques popular for computer experiments as well as sensitivity analysis techniques.

Figure 1.2 demonstrates the situation of computer experiments. There is a surface that is too complicated for applications. We want to choose a set of experimental points that are presented in the figure and to calculate the corresponding output.

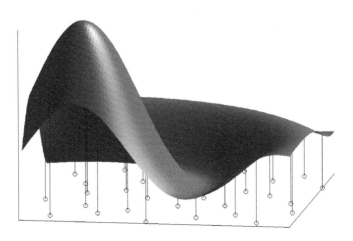

FIGURE 1.2
Space-filling design.

1.3.3   Computer experiments in engineering

The extensive use of models in engineering design practice is often necessitated by the fact that data cannot be obtained from physical tests of the system prior to its construction, yet the system cannot be built until the decisions are made. In a complex product development environment, reliance on models is crucial when a series of design decisions must be made upfront prior to the availability of a physical prototype, because they determine the architecture of the product and affect other downstream design decisions (e.g., manufacturing facilities, design of interfacing systems). Reversing a decision at a later time can have very expensive consequences both financially as well as in loss of valuable product development time. For example, in internal combustion engine design, the crankshaft and engine block designs (see Figure 1.3) determine the architecture of an engine which other components designs are dependent upon. The decisions about the crankshaft and engine block are also major determinants of the expense and development time of manufacturing processes. Thus, some decisions about good design need to be made before the effects of these decisions can be tested. In this situation, the engine designer frequently uses a crank and block interaction dynamic model (Loibnegger et al. (1997)) to make structural decisions to optimize durability and vibration characteristics prior to the availability of a physical prototype.

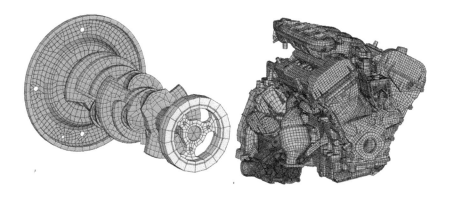

FIGURE 1.3
Engine crankshaft and block interaction finite element and dynamic model.

In other cases, the reliance on the model is necessitated because physical testing is expensive, time consuming, harmful, or in some situations, is even prohibitive. For example, Yang, Gu, Tho and Sobieszczanski-Sobieski (2001) performed vehicle crash design optimization using a finite element model to simulate frontal crashes, roof crush, and noise vibration and harshness to help design vehicle structures to absorb crash energy through structure deformation and impact force attenuation.

Models are often used by engineers to gain some insight into certain phenomena which may be lacking from physical experiments due to measurement system limitations and practicality. For example, a computational fluid dynamics (CFD) model (see Figure 1.5) is used to study the oil mist separator system in the internal combustion engine (Satoh, Kawai, Ishikawa and Matsuoka (2000)) to improve its efficiency by reducing engine oil consumption and emission effects. Using this model, engineers study the relationships of various parameters, such as gas flow rate, oil mist mass flow rate, oil droplet size distribution, to the oil mist velocity, pressure loss, and droplet accumulation, in a running engine under various engine speeds and loads. By understanding

FIGURE 1.4
Vehicle crash model.

the behavior of the systems, engineers create oil separation design concepts and optimize their design parameters to achieve maximum oil separation efficiency.

| | Pressure loss | Oil droplet |
| Velocity map | distribution | accumulation |

FIGURE 1.5
Oil separator CFD model.

While sophisticated computer models have become ubiquitous tools to investigate complicated physical phenomena, their effectiveness to support timely design decisions in fast-paced product development is often hindered by their excessive requirements for model preparation, computation, and output post-processing. The computational requirement increases dramatically when the simulation models are used for the purpose of design optimization or probabilistic analysis, since various design configurations must be iteratively evaluated. Recently, robust design and probabilistic-based design optimization approaches for engineering design have been proposed (see, for example, Wu and Wang (1998); Du and Chen (2004); Kalagnanam and Diwekar (1997); Du, Sudjianto and Chen (2004)) and have gained significant adoption in industrial practices (e.g., Hoffman, Sudjianto, Du and Stout (2003); Yang et al. (2001); Simpson, Booker, Ghosh, Giunta, Koch and Yang (2000); Du et al. (2004)). For this purpose, the computation demand becomes even greater because both probabilistic computation (e.g., Monte Carlo simulation) and optimization must be performed simultaneously, for which a "double loop" procedure may be required, as illustrated in Figure 1.6.

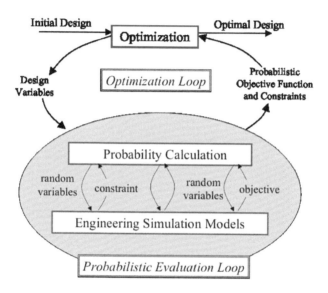

FIGURE 1.6
Double loop procedure in probabilistic design, adapted from Li and Sudjianto (2005).

The outer loop is the optimization process, while the inner loop is probability calculations for design objective and design constraints. This computational challenge makes obvious the importance of a cheap-to-compute metamodel (Kleijnen (1987)) as a surrogate to a computationally expensive engineering simulation model. This cheap-to-compute metamodel approach has become a popular choice in many engineering applications (e.g., Booker, Dennis, Frank, Serafini, Torczon and Trosset (1999), Hoffman et al. (2003), and Du et al. (2004)). In this approach, an approximation model that is much simpler and faster to compute than the true model (see Figure 1.1) is employed during engineering analysis and optimization. Because of the computational requirements and sophistication of the true models and the need for accurate metamodels, statistical techniques such as design of experiment play very important roles in the creation of metamodels such as to enable a minimum number of samples (i.e., number of runs in design of experiment) and at the same time capture maximum information (i.e., nonlinearity in the true model). Accordingly, sophisticated statistical modeling technique is required to analyze and represent the sample data into accurate and fast-to-compute metamodels.

In many industrial experiments, for example, the *factorial design* has been widely used (Montgomery (2001), Wu and Hamada (2000)). The most basic experimental design is a *full factorial design* (see Section 1.2), by which the experimenter can estimate main effects and interactions of the factors. The number of runs of a full factorial experiment increases exponentially with the number of factors, therefore, *fractional factorial design* (see Section 1.2) is used when many factors are considered and experiments are costly in terms of budget or time (Dey and Mukerjee (1999)). When there are many factors, the *sparsity of effects principle*, introduced in Section 1.2, can be invoked whereby the response is dominated by few effects and interactions (Montgomery (2001), Wu and Hamada (2000)). However, the sparsity of effects principle is not always valid. Furthermore, the experimenter may not know the underlying model well and may have difficulty guessing which interactions are significant and which interactions can be ignored. A similar difficulty will be encountered in the use of optimal designs (Kiefer (1959) and Atkinson and Donev (1992)). In this case, a space-filling design is a useful alternative and various modeling techniques can be evaluated to get the best metamodel. The study of the flow rate of water discussed in Section 1.8 will give an illustration. Therefore, the design and modeling techniques in computer experiments can also be applied in many factorial plans where the underlying model is unknown or partially unknown.

1.4   Examples of Computer Experiments

Computer experiments are required in many fields. In this section we present several examples discussed in the literature. Some of them will be used as examples throughout the book.

Example 2 **Engine block and head joint sealing assembly** *(Chen, Zwick, Tripathy and Novak (2002))* The head and block joint sealing assembly is one of the most crucial and fundamental structural designs in the automotive internal combustion engine. The design of the engine block and head joint sealing assembly is very complex, involving multiple components (block and head structures, gasket, and fasteners) with complicated geometry to maintain proper sealing of combustion, high pressure oil, oil drain, and coolant. The selection of design parameter settings in this assembly (block and head structures, gasket, and fasteners) cannot be analyzed separately because of strong assembly interaction effects. Design decisions for this system must be made upfront in the product development stage prior to the availability of physical prototype. To support the decision, computer simulation is commonly used in the design process. To best simulate the engine assembly process and its operation, a finite element model is used to capture the effects of three-dimensional part geometry, the compliance in the components, non-linear gasket material properties, and contact interface among the block, gasket, head, and fasteners. An example of a finite element model for this system is shown in Figure 1.7.

The computer model simulates the assembly process (e.g., head bolt run down) as well as various engine operating conditions (e.g., thermal and cylinder pressure cyclical loads due to the combustion process). The factors investigated in the computer simulation model consist of several design and manufacturing variables. The manufacturing variables include process variations and are referred to as noise factors. The number of experimental runs must be small due to simulation setup complexity and excessive computing requirements. The objective of the design in this example is to optimize the design variables to minimize the "gap lift" of the assembly. Additionally, one is also interested in finding an optimal design setting that minimizes the gap lift sensitivity to manufacturing variation. We will continue this example in Section 5.2 in which a list of design factors and noise factors will be given.

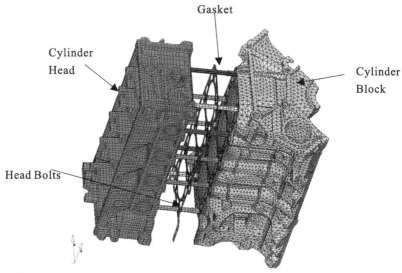

FIGURE 1.7
Finite element model of head and block joint sealing assembly.

Example 3 **Robot arm** *(An and Owen (2001))* The movement trajectory of a robot arm is frequently used as an illustrative example in neural network literature. Consider a robot arm with m segments. The shoulder of the arm is fixed at the origin in the (u, v)-plane. The segments of this arm have lengths $L_j, j = 1, \cdots, m$. The first segment is at angle θ_1 with respect to the horizontal coordinate axis of the plane. For $k = 2, \cdots, m$, segment k makes angle θ_k with segment $k - 1$. The end of the robot arm is at

$$\begin{cases} u = \sum_{j=1}^{m} L_j \cos(\sum_{k=1}^{j} \theta_k), \\ v = \sum_{j=1}^{m} L_j \sin(\sum_{k=1}^{j} \theta_k), \end{cases} \tag{1.12}$$

and the response y is the distance $y = \sqrt{u^2 + v^2}$ from the end of the arm to the origin expressed as a function of $2m$ variables $\theta_j \in [0, 2\pi]$ and $L_j \in [0, 1]$. Ho (2001) gave an approximation model $y = g(\theta_1, \theta_2, \theta_3, L_1, L_2, L_3)$ for the robot arm with 3 segments.

Example 4 **Chemical kinetics** *(Miller and Frenklach (1983) and Sacks, Schiller and Welch (1989))* In this example, chemical kinetics is modeled by a linear system of 11 differential equations:

$$h_j(x, t) = g_j(\eta, x, t), \; j = 1, \cdots, 11, \tag{1.13}$$

where x is a set of rate constants, the inputs to the system. A solution to (1.13) can be obtained numerically for any input x by solving 11 differential equations, yielding concentrations of five chemical species at a reaction time of 7×10^{-4} seconds. One might be interested in finding a closed form for an approximate model which would be much simpler than the original one (1.13).

Another case study of a computer experiment on the kinetics of a reversible chemical reaction was discussed by Xu, Liang and Fang (2000).

Example 5 **Dynamical system** *(Fang, Lin, Winker and Zhang (2000))* A one-off ship-based flight detector launch can be modeled as a dynamical system. The launching parameters are determined by a combination of several factors: the motion of wave and wind, the rock of the ship, the relative motion between the detector and launching support, the direction of launching support relative to the base system, plus other factors. For this problem, a series of coordinate systems are set up and involve many characteristics: 1) six degrees of freedom for ship rock; 2) two degrees of freedom for the circumgyration launching system; 3) the guide launching; 4) choice-permitted launching position; 5) two-period launching procedure; and 6) large-range direction of the launching support composed with the no-limitation azimuth and wide-range elevation angle. Because of the high cost of the experiment, the experimenter considered only two key factors: azimuth angle and pitching angle. Two responses, ω_1 and ω_2, are of interest. For a given set of values of the above six groups of parameters, the corresponding ω_1 and ω_2 can be calculated by solving a set of differential equations. Due to the complexity of the system, the experimenter wishes to find a simple model to replace solving the set of differential equations. Fang, Lin, Winker and Zhang (2000) applied uniform design (see Chapter 3) to this problem and gave a detailed discussion on finding a suitable polynomial metamodel.

Example 6 **Integrated circuit** *(Sacks, Schiller and Welch (1989) and Lo, Zhang and Han (2000))* The design of analog integrated circuit behaviors is commonly done using computer simulation. Here $\mathbf{x}$ denotes various circuit parameters, such as transistor characteristics, and y is the measurement of the circuit performance, such as output voltage.

Figure 1.8 shows the circuit of a night detector, where R_1, R_2, R_3, and R_4 are resistors inside the circuit, and y is the output current. The aim of night detector design is that the circuit current remains stable around a target value (180 mA) via adjustment of the resistances $R_1, \cdots, R_4$. These adjustable value resistors are called controllable factors. Once the values of the controllable factors are set, one would like to study the variation of the output current due to the effect of external and internal noises (V_{be}, V_{cc}, respectively) and the biases $R_1', \cdots, R_4'$ of $R_1, \cdots, R_4$. The output current y is given by

$$y = \frac{R_1 V_{cc}}{(R_1 + R_2)R_3} - \frac{R_3 + 2R_4}{2R_3 R_4} V_{bc}. \tag{1.14}$$

In a robust design study, one wishes to find the best combination of $R_1, \cdots, R_4$ such that the output response variability is minimized under various levels of variability due to tolerance variations of $R_1', \cdots, R_4', V_{be}$, and V_{cc}.

Example 7 **Environmental experiment** *(Fang and Wang (1994) and Li (2002))* To study how environmental pollutants affect human health, an experiment was conducted. Environmentalists believe that contents of some metal elements in water would directly affect human health. An experiment using uniform design was conducted at the Department of Biology, Hong Kong Baptist University. The purpose of this study was to understand the association between the mortality of some kind of cell of mice and contents of six metals: Cadmium (Cd), Copper (Cu), Zinc (Zn), Nickel (Ni), Chromium (Cr), and Lead (Pb). To address this issue, the investigator took 17 levels for the content of each metal: 0.01, 0.05, 0.1, 0.2, 0.4, 0.8, 1, 2, 5, 8, 10, 12, 16, 18, 20 (ppm). The experimenters considered the mortality of cells due to high density of the chemical elements. To fit the obtained data, we consider a regression model

$$y = f(\mathbf{x}) + \varepsilon, \tag{1.15}$$

where ε is random error with mean zero and an unknown variance. In Section A.3, we will use this example to illustrate the strategy of linear regression analysis. This example also shows applications of design and modeling techniques for computer experiments to industrial experiments with an unknown model.

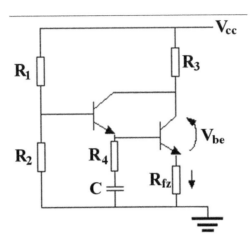

FIGURE 1.8
Night detector.

1.5   Space-Filling Designs

Without loss of generality, consider an experimental region in an s-dimensional unit cube $C^s = [0,1]^s$. For a given number of runs, n, one is interested in how to find a good design $D_n = \{\mathbf{x}_1, \cdots, \mathbf{x}_n\}$, where $\mathbf{x}_i \in C^s$, such that the deviation

$$\text{Dev}(\mathbf{x}; f, g) = f(\mathbf{x}) - g(\mathbf{x}), \tag{1.16}$$

is as small as possible for all $\mathbf{x} \in C^s$, where $y = f(\mathbf{x})$ is the true model and $y = g(\mathbf{x})$ is a metamodel. Because there are many metamodel alternatives, considering all the deviations in (1.16) becomes too difficult. Therefore, the *overall mean model*, which aims to find the best estimator of the overall mean, has been adopted by many authors to search for the best sampling (i.e., design of experiment) scenario. The most preliminary aim of the design is to obtain the best estimator of the overall mean of y

$$E(y) = \int_{C^s} f(\mathbf{x}) d\mathbf{x}. \tag{1.17}$$

The sample mean method suggests using

$$\bar{y}(D_n) = \frac{1}{n} \sum_{i=1}^{n} f(\mathbf{x}_i) \tag{1.18}$$

as an estimator of the overall mean $E(y)$. One wants to find a design such that the estimator $\bar{y}(D_n)$ is optimal in a certain sense. There are two kinds of approaches to assessing a design D_n:

(A) *Stochastic Approach:* From the statistical point of view we want to find a design D_n such that the sample mean $\bar{y}(D_n)$ is an unbiased or asymptotically unbiased estimator of $E(y)$ and has the smallest possible variance. Let $\mathbf{x}_1, \cdots, \mathbf{x}_n$ be independent samples of the uniform distribution on C^s, denoted by $U(C^s)$. It is easy to see that in this case the sample mean is unbiased with variance $\text{Var}(f(\mathbf{x}))/n$, where $\mathbf{x}$ is uniformly distributed on C^s, denoted by $\mathbf{x} \sim U(C^s)$. From the central limit theorem with an approximate probability 0.95 we have

$$\text{diff-mean} = |E(y) - \bar{y}(D_n)| \leq 1.96\sqrt{\text{Var}(f(\mathbf{x}))/n}.$$

Unfortunately, the variance $\text{Var}(f(\mathbf{x}))/n$ is too large for many case studies. Thus, many authors proposed different ways to lower the variance of $\bar{y}(D_n)$. McKay, Beckman and Conover (1979) proposed the so-called *Latin hypercube sampling* (LHS), that is, to randomly choose $\mathbf{x}_1, \cdots, \mathbf{x}_n$ such that they are dependent and have the same marginal distribution. In this case

$$\text{Var}(\bar{y}(D_n)) = \frac{1}{n}\,\text{Var}(f(\mathbf{x})) + \frac{n-1}{n}\,\text{Cov}(f(\mathbf{x}_1), f(\mathbf{x}_2)). \qquad (1.19)$$

Obviously, $\text{Var}(\bar{y}(D_n)) < \text{Var}(f(\mathbf{x}))/n$ if and only if the covariance of $f(x_1)$ and $f(x_2)$ is negative. McKay et al. (1979) showed that this covariance is negative whenever $f(x_1, \cdots, x_s)$ is monotonic in each variable. LHS has been widely used in computer experiments and many modifications of LHS have been proposed. We shall introduce LHS and its versions in Chapter 2.

(B) *Deterministic Approach:* From a deterministic point of view, one wishes to find a sampling scenario so that the difference

$$\text{diff-mean} = |E(y) - \bar{y}(D_n)| \qquad (1.20)$$

will be as small as possible. The *Koksma-Hlawka inequality* in *quasi-Monte Carlo methods* gives an upper bound of the diff-mean as follows

$$\text{diff-mean} = |E(y) - \bar{y}(D_n)| \leq V(f)D(D_n), \qquad (1.21)$$

where $D(D_n)$ is the *star discrepancy* of D_n, not depending on g, and $V(f)$ is the *total variation of the function f* in the sense of Hardy and Krause (see Niederreiter (1992) and Hua and Wang (1981)). The star discrepancy is a *measure of uniformity* used in quasi-Monte Carlo methods and is just the *Kolmogorov-Smirnov statistic* in goodness-of-fit tests. The lower the star discrepancy, the better uniformity the set of points has. The details of the above concepts will be further explained in Section 3.2.1. The Koksma-Hlawka inequality suggests minimizing the star discrepancy $D(D_n)$ over all designs of n runs on C^s, that is, to find a set of n points uniformly scattered on C^s. The latter is called a *uniform design* (UD for short). It was proposed by Fang and Wang (Fang (1980) and Wang and Fang (1981)) based on the three big projects in system engineering. The Koksma-Hlawka inequality also indicates that the uniform design is *robust* against model specification. For example, if two models $y = g_1(\mathbf{x})$ and $y = g_2(\mathbf{x})$ have the same variation $V(g_1) = V(g_2)$, a uniform design may have the same upper bound of the diff-mean for these two models. Many authors have discussed model robust designs, for example, Sack and Ylvisaker (1985), Sacks and Ylvisaker (1984), and Chang and Notz (1996).

The LHS and UD were motivated by the overall mean model. Of course, the overall mean model is far not enough in practice, but it provides a simple way to develop methodology and theory. Surprisingly, the LHS, UD and their modifications are good space-filling designs. Chapters 2 and 3 will give a detailed introduction to these designs and their constructions. It has been demonstrated that these space-filling designs have a good performance not only for estimation of the overall mean, but also for finding a good approximate model. The argument for the latter will be further discussed in Section 3.4.

Koehler and Owen (1996) proposed a different way to classify approaches to computer experiments: "There are two main statistical approaches to computer experiments, one based on Bayesian statistics and a frequentist one based on sampling techniques." Both LHS and UD belong to frequentist experimental designs, and optimal Latin hypercube designs are Bayesian designs according to their philosophy.

1.6   Modeling Techniques

This section presents a brief overview of popular metamodeling methods in the literature. Chapter 5 is devoted to introducing various modeling approaches in details accompanied by illustrations using case study examples to add clarity and practical applications. The construction of an accurate and simple metamodel $y = g(\mathbf{x})$ to approximate the true model, $y = f(\mathbf{x})$, plays a crucial role in analysis of computer experiments. For a given true model $y = f(\mathbf{x})$ there are many possible metamodels. Very often it is difficult to say which metamodel is the best. The purpose of the modeling is to find some useful metamodels. Good modeling practice requires that the modeler provides an evaluation of the confidence in the model, possibly assessing the uncertainties associated with the modeling process and with the outcome of the model itself (see Saltelli et al. (2000)).

Most metamodels can be represented using a linear combination of a set of specific bases. Let $\{B_1(\mathbf{x}), B_2(\mathbf{x}), \cdots, \}$ be a set of basis functions defined on the experimental domain T. A metamodel g usually can be written in the following general form:

$$g(\mathbf{x}) = \beta_1 B_1(\mathbf{x}) + \beta_2 B_2(\mathbf{x}) + \cdots, \qquad (1.22)$$

where β_j's are unknown coefficients to be estimated.

Polynomial models have traditionally enjoyed popularity in many modeling contexts including computer experiments. The models employ a polynomial basis $x_1^{r_1} \cdots x_s^{r_s}$ where $r_1, \ldots, r_s$ are non-negative integers. The number of polynomial basis functions dramatically increases with the number of input variables and the degree of polynomial. Low-order polynomials such as the second-order polynomial model,

$$g(\mathbf{x}) = \beta_0 + \sum_{i=1}^{s} \beta_i x_i + \sum_{i=1}^{s} \sum_{j=i}^{s} \beta_{ij} x_i x_j, \qquad (1.23)$$

are the most popular polynomial metamodels for computer experiment modeling. The term *response surfaces* (Myers and Montgomery (1995) and Morris and Mitchell (1995)) in the literature refers to this second-order polynomial model. To make numerical computation stable, one usually centralizes the input variables before building polynomial models. For example, the *centered quadratic model*,

$$g(\mathbf{x}) = \beta_0 + \sum_{i=1}^{s} \beta_i (x_i - \bar{x}_i) + \sum_{i=1}^{s} \sum_{j=i}^{s} \beta_{ij} (x_i - \bar{x}_i)(x_j - \bar{x}_j), \qquad (1.24)$$

where $\bar{x}_i$ is the sample mean of the i-component of $\mathbf{x}$, has been recommended in the literature. In the presence of high-order polynomials, the polynomial basis may cause a collinearity problem, i.e., there are high correlations among regressors. In such situations, *orthogonal polynomial models* are recommended to overcome this difficulty of multi-collinearity. See Section 5.2 for construction of orthogonal polynomial bases.

Polynomial metamodels are usually used to discover the overall trend of the true model. When the domain T is large and the true model is more complicated, for example, there are many local minimums/maximums, so one needs high-degree polynomials to approximate the true model. In such cases the collinearity becomes a serious problem, so splines can be employed to construct a set of basis for expansion in (1.22). In the literature, *spline* approaches mainly include smoothing splines (Wahba (1990)) and regression splines (see, for example, Stone, Hansen, Kooperberg and Truong (1997)). For a univariate x-variable, the power spline basis has the following general form:

$$1, x, x^2, \cdots, x^p, (x - \kappa_1)_+^p, \cdots, (x - \kappa_K)_+^p, \qquad (1.25)$$

where $\kappa_1, \cdots, \kappa_K$ are a set of selected knots (locations at which the $(p+1)$-th derivative of g is allowed to be discontinuous), and a_+ stands for the positive part of a, i.e., $a_+ = aI(a > 0)$. Multivariate spline basis may be constructed from the univariate spline basis using the tensor product approach. Section 5.3 explains the spline model in more detail.

As an alternative, the *Fourier basis* is commonly used to model responses with a periodic function. For example, the true model $f(\mathbf{x})$ in Example 6 is appropriately modeled using Fourier regression. Multivariate Fourier bases usually are constructed from univariate Fourier basis:

$$1, \cos(2\pi x), \sin(2\pi x), \cdots, \cos(2k\pi x), \sin(2k\pi x), \cdots,$$

by using the tensor product method (see Chapter 5). Similar to polynomial basis, the number of terms in full multivariate Fourier bases dramatically increases as the dimension of $\mathbf{x}$ increases. In practice, people often use the following Fourier metamodel (see Riccomango, Schwabe and Wynn (1997)):

$$g(\mathbf{x}) = \beta_0 + \sum_{i=1}^{s} \{[\alpha_i \cos(2\pi x_i) + \beta_i \sin(2\pi x_i)] + [\alpha_i^{(2)} \cos(4\pi x_i) + \beta_i^{(2)} \sin(4\pi x_i)]$$
$$+ \cdots + [\alpha_i^{(m)} \cos(2m\pi x_i) + \beta_i^{(m)} \sin(2m\pi x_i)]\}. \tag{1.26}$$

Wavelets bases have also been proposed in the literature to improve the Fourier approximation, especially in cases where the function being approximated may not be smooth. See Chui (1992), Daubechies (1992), and Antoniadis and Oppenheim (1995) for systematic introductions to wavelets.

Metamodels built on polynomial bases, spline bases, Fourier bases, and wavelets bases are powerful when the input variable is one-dimensional. They also may be useful when input variables are low-dimensional and the function can be adequately represented using low-order basis functions. The tensor product method has been widely employed to extend these basis functions from univariate inputs to multivariate inputs. In practice, however, these extensions may be difficult to implement because the number of terms in a set of multivariate basis functions exponentially increases as the dimension of input variables increases. Thus, techniques such as Kriging models, neural networks, and local polynomial models may provide a more natural construct to deal with multivariate basis functions.

A **Kriging model** assumes that

$$y(\mathbf{x}) = \mu + z(\mathbf{x}), \tag{1.27}$$

where μ is the overall mean of $y(\mathbf{x})$, and $z(\mathbf{x})$ is a Gaussian process with mean $E\{z(\mathbf{x})\} = 0$ and covariance function

$$\text{Cov}(z(\mathbf{x}_i), z(\mathbf{x}_j)) = \sigma^2 R(\mathbf{x}_i, \mathbf{x}_j), \tag{1.28}$$

where σ^2 is the unknown variance of $z(\mathbf{x})$, and R is a *correlation function* with a pre-specified functional form and some unknown parameters. A typical choice for the correlation function is

$$r(\mathbf{x}_1, \mathbf{x}_2) = \exp\{-\sum_{k=1}^{s} \theta_i (x_{k1} - x_{k2})^2\},\qquad(1.29)$$

where θ_i's are unknown. Let $\{\mathbf{x}_1, \cdots, \mathbf{x}_n\}$ consist of a design, and $y_1, \cdots, y_n$ are their associated outputs. The resulting metamodels of the Kriging approach can be written as

$$g(\mathbf{x}) = \sum_{i=1}^{n} \beta_i r(\mathbf{x}, \mathbf{x}_i)\qquad(1.30)$$

which is of the form (1.22) regarding $r(\mathbf{x}, \mathbf{x}_i)$ to be a basis function. Chapter 5 gives a thorough discussion of the Kriging approach. One of the advantages of the Kriging approach is that it constructs the basis directly using the correlation function. Under certain conditions, it can be shown that the resulting metamodel of the Kriging approach is the best linear unbiased predictor (see Section 5.4.1). Furthermore, the Gaussian Kriging approach admits a Bayesian interpretation.

Bayesian interpolation was proposed to model computer experiments by Currin, Mitchell, Morris and Ylvisaker (1991) and Morris, Mitchell and Ylvisaker (1993). Bayesian interpolation can be beneficial in that it easily incorporates auxiliary information in some situations. For instance, Morris et al. (1993) demonstrated how to use a Bayesian Kriging method to create computer models that can provide both the response and its first partial derivatives. (See Section 5.5 for further discussions.)

The **multilayer perceptron network** (MLP) is the most ubiquitous neural network model. The network consists of input, hidden, and output layers with nonlinear input transformation as follows:

$$\hat{y} = \sum_{j=1}^{d} z_j(v_j)\beta_j + \beta_0,$$

where β_j is the "regression coefficient" of the j-th basis functions $z_j(v_j)$, in the following forms,

$$z_j(v_j) = \frac{1}{1 + \exp(-\lambda v_j)} \quad \text{or} \quad z_j(v_j) = \tanh(\lambda v_j),$$

where v_j's are linear combinations of the input variables

$$v_j = \sum_{i=1}^{p} w_{ji} x_i + w_{j0}.$$

The "weight" parameters, w_{ji}, in the linear combination and the regression coefficient, β_j, of the basis functions are estimated using "least square criteria" between the model fit and the true values of the response variables

$$E = \frac{1}{2} \sum_{k=1}^{n} (y_k - \hat{y}_k)^2.$$

The network "training" process (i.e., parameter estimation) involves nonlinear optimization for the above least square objective function. There are numerous training algorithms in the literature (e.g., Haykin (1998), Hassoun (1995)). Among them is a learning algorithm known as the "Backpropagation" (Rumelhart, Hinton and Williams (1986)), and this approach is often called a back-propagation network. The application of this model for computer experiment will be discussed in Section 5.6.

Radial basis function methods have been used for neural network modeling and are closely related to the Kriging approach. Let $\{\mathbf{x}_1, \cdots, \mathbf{x}_n\}$ consist of a design, and $y_1, \cdots, y_n$ are their associated outputs. In general, a radial basis function has the following form:

$$K(\|\mathbf{x} - \mathbf{x}_i\|/\theta), i = 1, \cdots, n,$$

where $K(\cdot)$ is a kernel function and θ is a smoothing parameter. The resulting metamodel is as follows:

$$g(\mathbf{x}) = \sum_{i=1}^{n} \beta_i K(\|\mathbf{x} - \mathbf{x}_i\|/\theta).$$

Taking the kernel function to be the Guassian kernel function, i.e., the density function of normal distribution, the resulting metamodel has the same form as that in (1.30) with the correlation function (1.29) where $\theta_i = \theta$. The reader can refer to Section 5.6 for a systematic discussion on neural network modeling.

Local polynomial modeling has been applied for nonparametric regression in statistical literature for many years. Fan (1992) and Fan and Gijbels (1996) give a systematic account in the context of nonparametric regression. The idea of local polynomial regression is that a datum point closer to $\mathbf{x}$ carries more information about the value of $f(\mathbf{x})$ than one which is far away. Therefore, an intuitive estimator for the regression function $f(x)$ is the running local average. An improved version of the local average is the locally weighted average:

$$g(\mathbf{x}) = \sum_{i=1}^{n} w_i(\mathbf{x}) y_i, \tag{1.31}$$

where $w_i(\mathbf{x}), i = 1, \cdots, n$ are weights with $\sum_{i=1}^{n} w_i(\mathbf{x}) = 1$. Of course, the weight function depends on $\mathbf{x}_1, \cdots, \mathbf{x}_n$. It further depends on a smoothing parameter which relates to how we define "close" or "far." The resulting metamodel by local polynomial models can be represented in the form of (1.22). In Section 5.7, we will introduce the fundamental ideas of local modeling and demonstrate the connections among the Kriging method, the radial basis function approach, and local polynomial modeling. More modeling techniques, like the Bayesian approach and neural network, will be introduced in Chapter 5.

1.7 Sensitivity Analysis

Sensitivity analysis studies the relationships between information flowing in and out of the model. The metamodel here is intended to give insight into the relationships of input variables with the response variable. Understanding the relationships between inputs and output can be motivated by several reasons. First, one may want to do a sanity check on the model. This is particularly important when one is dealing with a very complicated model with numerous input variables. By quantifying the sensitivity, one can cross check the results produced by the model and compare them with known general physical laws. The other reason, especially in engineering design, is that one would like to make design decisions by knowing which input variables have the most influence on the output. These needs lead us to consider the following questions: how to rank the importance of variables and how to interpret the resulting metamodel. To address these questions, the so-called *sensitivity analysis* (SA) was proposed. Many authors, such as Sobol' (1993, 2001) and Saltelli, Chan and Scott (2000), have comprehensively studied this topic. Saltelli et al. (2000) stated: "Sensitivity analysis (SA) is the study of how the variation in the output of a model (numerical or otherwise) can be apportioned, quantitatively, to different sources of variation, and of how the given model depends upon the information fed into it. ... As a whole, SA is used to increase the confidence in the model and its predictions, by providing an understanding of how the model response variables respond to changes in the inputs, be they data used to calibrate it, model structures, or factors, i.e. the model-independent variables."

When a low-order polynomial model is used to fit the computer experiment data, traditional decomposition of sum of squares or traditional analysis of variance (ANOVA) decomposition can be directly employed to rank the importance of variables and interactions, and the magnitude of the scaled regression coefficient also explains the effect of each term. However, it would be difficult to directly understand the metamodels if sophisticated basis functions, such as Kriging or neural networks, are used.

Let D be the total variation of the output $y = f(\mathbf{x}) = f(x_1, \cdots, x_s)$. We can split the total variation D into

$$D = \sum_{k=1}^{s} \sum_{t_1 < \cdots < t_k}^{s} D_{t_1 \cdots t_k}, \tag{1.32}$$

where D_i gives the contribution of x_i to the total variation D; $D_{i,j}$ the contribution of interaction between x_i and x_j; and so on. The Sobol' functional ANOVA method suggests

$$D = \int (f(\mathbf{x}) - f_0(\mathbf{x}))^2 d\mathbf{x} = \int f^2 d\mathbf{x} - f_0^2,$$

where

$$f_0 = \int f(\mathbf{x}) d\mathbf{x}.$$

And

$$D_{t_1 \cdots t_k} = \int f_{t_1 \cdots t_k}^2 dx_{t_1} \cdots dx_{t_k},$$

where

$$f_i(x_i) = \int f(\mathbf{x}) \prod_{k \neq i} dx_k - f_0,$$

$$f_{ij}(x_i, x_j) = \int f(\mathbf{x}) \prod_{k \neq i,j} dx_k - f_0 - f_i(x_i) - f_j(x_j),$$

and so on. Therefore, the true model has the decomposition

$$f(\mathbf{x}) = f_0 + \sum_{i=1}^{s} f_i(x_i) + \sum_{1=i<j}^{s} f_{ij}(x_i, x_j) + \cdots + f_{12\cdots n}(x_1, x_2, \cdots, x_s). \quad (1.33)$$

The ratios

$$S_{t_1 \cdots t_k} = \frac{D_{t_1 \cdots t_k}}{D}$$

are called *Sobol' global sensitivity indices*. If one uses a metamodel g to replace the true one f, we can carry out the above process by the use of g to replace f. In Section 6.3, we will illustrate how to use the Sobol' indices to rank the importance of input variables and their interactions and will further introduce the correlation ratio, and FAST in a systematical way. The Sobol' indices, correlation ratios and Fourier amplitude sensitivity test (FAST) are three popular measures to quantify the importance of input variables in the literature.

The aim of sensitivity analysis is to investigate how a given metamodel responds to variations in its inputs. The investigator may be concerned with

- whether the metamodel resembles the system or the true model,
- which input variables and which interactions give the greatest contribution to the output variability,
- whether we can remove insignificant input factors of some items in the metamodel to improve the efficiency of the metamodel, or
- which level-combination of the input variables can reach the maximum or minimum of the output y.

In the literature there are many different approaches for SA. These can generally be divided into three classes:

- Screening methods: i.e., factor screening, where the aim is to identify influential factors in the true model;
- Local SA: this emphasizes the local impact of the input variables on the model, the local SA helps to compute partial derivatives of the metamodel with respect to the input variables;
- Global SA: this emphasizes apportioning the output uncertainty to the uncertainty in the input variables. It typically takes a sampling approach to distributions for each input variable.

Chapter 6 will provide details on SA and the application of SA to some examples in the previous chapters of the book.

1.8 Strategies for Computer Experiments and an Illustration Case Study

This section gives an illustrative example of computer experiment where a space-filling design and related modeling will be discussed. The following are the typical steps in a computer experiment:

Factor screening: There are many possible related factors according to the purpose of the experiment. The experimenter has to screen out the inactive factors so that one can focus on a few active factors. Choose a suitable factor space, i.e., experimental domain.

Construct a design for the experiment: Selecting a design with an appropriate number of runs and levels for each variable to ensure sufficient design space coverage is one of the important early steps in design and modeling computer experiment. There are many high quality space-filling designs in the literature. See, for example,

http://www.math.hkbu.edu.hk/UniformDesign

The number of runs and levels depends on the complexity of the true model. Some models such as finite element analysis and computational fluid dynamics are very expensive to run. In addition, the aforementioned types of models often are also time consuming to construct and, similar to physical experiments, it is often difficult to incorporate a large number of levels. Therefore, an optimized design which has a limited number of runs and levels yet still has sufficient space-filling property may be needed (see Chapter 4).

Metamodel searching: This step includes the selection of appropriate modeling techniques as well as estimation of the model parameters. One may want to consider several modeling technique alternatives for a given data set (see Chapter 5 for a detailed discussion on modeling techniques).

Verify the metamodel: One may consider several alternatives of modeling methods to find the most suitable model. Conducting comparisons among these models may be necessary to select the 'best' one. Readers should refer to Section 5.1.1 for a detailed discussion of model selection. Selecting the 'best' model can be a difficult task. For the purpose of generalization capability, i.e., satisfactory predictions at untried points, we can verify the model using the *mean square error* (MSE) of prediction at untried points. The smaller the MSE value, the better the metamodel. The procedure is as follows: Randomly choose N points $\mathbf{x}_k, i = 1, \cdots, N$ on the domain T and calculate their output y-values, $y(\mathbf{x}_k)$, and estimated values $\hat{y}(\mathbf{x}_k)$ using the recommended metamodel. The MSE is given by

$$MSE = \frac{1}{N} \sum_{k=1}^{N} (y(\mathbf{x}_k) - \hat{y}(\mathbf{x}_k))^2. \tag{1.34}$$

The MSE or the square root of MSE is a departure of the metamodel from the true model. In general, when the computer model is computationally expensive, users will not have the luxury of performing cross validation with a large sample size. In this case, verification will be performed using limited samples or in some cases using approximation methods such as k-fold cross validation (see Section 5.1.1 for details).

Interpret the model and perform sensitivity analysis: Interpretation of the chosen metamodel as discussed in Section 1.7 is an important step in computer experiments especially for complicated metamodels such as Kriging models. The ability to interpret the metamodel to understand the relationship between inputs and output is often as important as getting the best prediction accuracy from a metamodel. Chapter 6 gives a detailed discussion on this issue.

Further study: If one cannot find a satisfactory metamodel based on the data set, it may be because of either poor design selection or metamodel choice. If the latter is the problem, one may consider other more suitable modeling techniques discussed in Chapter 5. However, when the poor result is due to the former, one may want to consider either augmenting the existing data set or creating a new experiment with better design structure to ensure a sufficient number of levels (to capture nonlinearity) and a sufficient number of runs to improve the space filling of the design.

Example 8 (**Demo Example**) For illustration of design and modeling for computer experiments, the following example gives the reader a brief demonstration. A case study was investigated by many authors such as Worley (1987), An and Owen (2001), Morris et al. (1993), Ho and Xu (2000), and Fang and Lin (2003). A brief introduction of design and modeling for this case study is given as follows:

1. Choose factors and experimental domain.

In the following example, we are using a simple engineering example of flow rate of water through a borehole from an upper aquifer to a lower aquifer separated by an impermeable rock layer. Although this is a simple example and does not represent a typical case of computer experiment in complex engineering design practice, it is a good demonstrative example for computer experiments. The flow rate through the borehole can be studied using an analytical model derived from Bernoulli's law using an assumption that the flow is steady-state laminar and isothermal. The response variable y, the flow rate through the borehole in m^3/yr, is determined by

$$y = \frac{2\pi T_u[H_u - H_l]}{\ln(\frac{r}{r_w})\left[1 + \frac{2LT_u}{\ln(r/r_w)r_w^2 K_w} + \frac{T_u}{T_l}\right]}, \tag{1.35}$$

where the 8 input variables are as follows:

$r_w(m)$ = radius of borehole
$r(m)$ = radius of influence
$T_u(m^2/yr)$ = transmissivity of upper aquifer
$T_l(m^2/yr)$ = transmissivity of lower aquifer
$H_u(m)$ = potentiometric head of upper aquifer
$H_l(m)$ = potentiometric head of lower aquifer
$L(m)$ = length of borehole
$K_w(m/yr)$ = hydraulic conductivity of borehole

and the domain T is given by

$r_w \in [0.05, 0.15]$, $r \in [100, 50000]$, $T_u \in [63070, 115600]$, $T_l \in [63.1, 116]$, $H_u \in [990, 1110]$, $H_l \in [700, 820]$, $L \in [1120, 1680]$, $K_w \in [9855, 12045]$. The input variables and the corresponding output are denoted by $\mathbf{x} = (x_1, \cdots, x_8)$ and $y(\mathbf{x})$, respectively.

2. Construct design of experiment matrix

Worley (1987) chose a 10-run Latin hypercube sampling (cf. Section 2.1). Morris et al. (1993) employed four types of 10-run designs, i.e., *Latin hypercube design*, *maximin design* (cf. Section 2.4.3), *maximin Latin hypercube design*, (cf. Section 2.4), and *modified maximin design*, while Ho and Xu (2000) chose a *uniform design* (cf. Chapter 3) that is given in Table 1.2. The y-values for all 32 runs are listed in the last column of Table 1.2.

3. Metamodel searching

In terms of statistical modeling, one wants to fit highly adaptive models by various modeling techniques. Section 1.6 gives a brief review of various modeling techniques. For a detailed treatment of modeling techniques, readers should refer to Chapter 5. In particular, Section 5.5 revisits this example using the Bayesian modeling method proposed by Sacks, Welch, Mitchell and Wynn (1989) and Morris et al. (1993). In this section, for the purpose of friendly

TABLE 1.2

Uniform Design and Related Output

No	r_w	r	T_u	T_l	H_u	H_l	L	K_w	y
1	0.0500	33366.67	63070	116.00	1110.00	768.57	1200	11732.14	26.18
2	0.0500	100.00	80580	80.73	1092.86	802.86	1600	10167.86	14.46
3	0.0567	100.00	98090	80.73	1058.57	717.14	1680	11106.43	22.75
4	0.0567	33366.67	98090	98.37	1110.00	734.29	1280	10480.71	30.98
5	0.0633	100.00	115600	80.73	1075.71	751.43	1600	11106.43	28.33
6	0.0633	16733.33	80580	80.73	1058.57	785.71	1680	12045.00	24.60
7	0.0700	33366.67	63070	98.37	1092.86	768.57	1200	11732.14	48.65
8	0.0700	16733.33	115600	116.00	990.00	700.00	1360	10793.57	35.36
9	0.0767	100.00	115600	80.73	1075.71	751.43	1520	10793.57	42.44
10	0.0767	16733.33	80580	80.73	1075.71	802.86	1120	9855.00	44.16
11	0.0833	50000.00	98090	63.10	1041.43	717.14	1600	10793.57	47.49
12	0.0833	50000.00	115600	63.10	1007.14	768.57	1440	11419.29	41.04
13	0.0900	16733.33	63070	116.00	1075.71	751.43	1120	11419.29	83.77
14	0.0900	33366.14	115600	116.00	1007.14	717.14	1360	11106.43	60.05
15	0.0967	50000.00	80580	63.10	1024.29	820.00	1360	9855.00	43.15
16	0.0967	16733.33	80580	98.37	1058.57	700.00	1120	10480.71	97.98
17	0.1033	50000.00	80580	63.10	1024.29	700.00	1520	10480.71	74.44
18	0.1033	16733.33	80580	98.37	1058.57	820.00	1120	10167.86	72.23
19	0.1100	50000.00	98090	63.10	1024.29	717.14	1520	10793.57	82.18
20	0.1100	100.00	63070	98.37	1041.43	802.86	1600	12045.00	68.06
21	0.1167	33366.67	63070	116.00	990.00	785.71	1280	12045.00	81.63
22	0.1167	100.00	98090	98.37	1092.86	802.86	1680	9855.00	72.54
23	0.1233	16733.33	115600	80.73	1092.86	734.29	1200	11419.29	161.35
24	0.1233	16733.33	63070	63.10	1041.43	785.71	1680	12045.00	86.73
25	0.1300	33366.67	80580	116.00	1110.00	768.57	1280	11732.14	164.78
26	0.1300	100.00	98090	98.37	1110.00	820.00	1280	10167.86	121.76
27	0.1367	50000.00	98090	63.10	1007.14	820.00	1440	10480.71	76.51
28	0.1367	33366.67	98090	116.00	1024.29	700.00	1200	10480.71	164.75
29	0.1433	50000.00	63070	116.00	990.00	785.71	1440	9855.00	89.54
30	0.1433	50000.00	115600	63.10	1007.14	734.29	1440	11732.14	141.09
31	0.1500	33366.67	63070	98.37	990.00	751.43	1360	11419.29	139.94
32	0.1500	100.00	115600	80.73	1041.43	734.29	1520	11106.43	157.59

introduction, we only consider polynomial regression models. A careful study by Ho and Xu (2000) suggested the following model:

$$
\begin{aligned}
\widehat{\log(y)} = {}& 4.1560 + 1.9903(\log(r_w) + 2.3544) - 0.0007292 * (L - 1400) \\
& - 0.003554 * (H_l - 760) + 0.0035068 * (H_u - 1050) \\
& + 0.000090868 * (K_w - 10950) \\
& + 0.000015325 * (H_u - 1050) * (H_l - 760) \\
& + 0.00000026487(L - 1400)^2 - 0.0000071759 * (H_l - 760)^2 \\
& - 0.0000068021 * (H_u - 1050)^2 \\
& - 0.00087286 * (\log(r) - 8.8914). \quad\quad (1.36)
\end{aligned}
$$

4. Verify the model

Since the response values are easily acquired from the flow through the borehole formula, we randomly generated new samples of $N = 2000$ to evaluate the model in (1.36). The MSE based on this model equals 0.2578156.

5. Interpret model and perform sensitivity analysis

In this example, we are also interested in ranking the importance of input variables to the output. While Chapter 6 gives a detailed discussion on this issue, here we employ the most straightforward approach using the sum of squares decomposition:

$$\text{SSTO} = \text{SSR} + \text{SSE},$$

where the sum of square total (SSTO) of output variation can be decomposed into the sum of square regression (SSR), the portion that can be an explainable component by the model, and the sum of square errors (SSE), the portion that is an unexplainable component by the model. We have

$$\text{SSTO} = \sum_{i=1}^{n}(y_i - \bar{y})^2,$$

where $\bar{y}$ is the average response;

$$\text{SSR} = \sum_{i=1}^{n}(\hat{y}_i - \bar{y})^2,$$

where $\hat{y}_i$ is the fitted value of y_i; and

$$\text{SSE} = \sum_{i=1}^{n}(y_i - \hat{y}_i)^2.$$

In conventional physical experiments, SSE represents lack of fit and/or random errors from the experiment. In computer experiments, there is no random error and SSE is a result of lack of fit either due to the limitations of the metamodel itself or the exclusion of some input variables from the model.

The increase of regression sum of squares can be used as a measure of the marginal effects of adding one or several covariates in the regression equation. Here, we decompose the regression sum of squares by sequentially adding new terms in the regression equation and calculating the corresponding increase in SSR. Let $\text{SSR}(x_i)$ be the regression sum of squares when only x_i is included in the regression mode. Thus, $\text{SSR}(x_i)$ stands for marginal reduction of total variation due to x_i. Readers should consult Section 6.2 for a detailed discussion. Here, we can use sequential SSR to rank the importance of variables. Table 1.3 presents the summary of the sequential SSR decomposition, where the last column shows the relative importance of the term with respect to SSTO, based on the SSR decomposition. In other words, the SSR decomposition shows rank order of importance of the input variables.

TABLE 1.3

Sum of squares decomposition

Terms	Seq.SSR	%Seq.SSR
r_w	11.2645	85.74
L	0.7451	5.67
H_l	0.4583	3.49
H_u	0.5170	3.94
K_w	0.1401	1.07
$H_u * H_l$	0.0064	0.05
L^2	0.0036	0.03
H_u^2	0.0007	0.01
H_l^2	0.0013	0.01
r	0.0001	0.00
SSTO	13.1373	100.00

1.9 Remarks on Computer Experiments

This section gives some general remarks on relationships between design and analysis of computer experiments (DACE) and other statistical designs. A computer experiment has its own model

$$y = f(x_1, \cdots, x_s) = f(\mathbf{x}), \quad \mathbf{x} = (x_1, \cdots, x_s)' \in T, \tag{1.37}$$

where the known true model f is known but is complicated or has no analytic formula; the input space T is large in most experiments. For a given input $\mathbf{x} \in T$, one can find the output without random error. The model (1.37) is deterministic, although calculating the output may involve round-off errors. Santner et al. (2003) distinguish three types of input variables:

Control variables: The variables are of interest in the experiment and can be set by an engineer or scientist to "control" the system. Control variables are presented in physical experiments as well as in computer experiments.

Environmental variables: The variables are not the main interest, but we have to consider their effects on the output. Temperature outside and highway quality when a car is running are examples of environmental variables. Environmental variables are also called *noise variables.*

Model variables: The variables are parameters to describe the uncertainty in the mathematical modeling that relates other inputs to output. In this book we prefer to use "model parameters," because we can estimate these parameters using the collected data.

Here control variables and environmental variables correspond to control factors and noise factors in the robust design and reliability-based design mentioned in Section 1.3.2. The number of input factors, s, can be very large in computer experiments. Santner, Williams and Notz (2003) consider two types of the inputs:

Homogeneous-input: All components of **x** are either control variables or environmental variables or model variables.

Mixed-input: **x** contains at least two of the three different input variables: control, environmental, and model.

However, when the intent of the experiment is to build a metamodel, we make no distinction between homogeneous and mixed input in setting up the design experiment or between the types of input variables.

There is a close relation between computer experiments and physical experiments with a model unknown. In Example 7, the experimenter does not know the true model representing the relationships between the mortality of the cells and the contents of six chemicals. For the nonparametric model

$$y = f(\mathbf{x}) + \varepsilon, \tag{1.38}$$

the model $f(\mathbf{x})$ is unknown and ε is the random error. The experimenter wish to estimate the model f by a physical experiment. Clearly, the two kinds of models, (1.37) and (1.38), have many common aspects: 1) the true model is complicated (known or unknown) and needs a simple metamodel to approximate the true model; 2) the experimental domain should be large so that we can build a metamodel that is valid for a larger design space; 3) space-filling designs are useful for both kinds of experiments; and 4) various modeling techniques can be applied to the modeling for the both cases.

The difference between the two models (1.37) and (1.38) is that the former does not have random error, but the latter involves random error. When we treat model (1.38) all the statistical concepts and tools can be employed. However, from a quick observation, one may argue that the statistical approach may no longer be useful for the model (1.37). On the other hand, due to common aspects of the two models (1.37) and (1.38), we may also argue that some useful statistical concepts/tools based on model (1.38) may still be useful for model (1.37). As an example, many ideas and concepts in sensitivity analysis come from statistics. The statistical approach is also useful to deal with complex experiments. In particular, when the number of input variables is large, the "sparsity principle" in factorial design mentioned in Section 1.2 can be applied. It is reasonable to believe that the number of relatively important input variables is small, and a statistical model is developed to include only these important variables; the relatively small effects of other input variables are not included in the model and are treated as random errors from unobserved variables. Under this consideration, a model (1.37) can be split into two parts: one relates to the set of important variables and the other relates to the set of insignificant variables. Thus, model (1.37) becomes model (1.38).

Computer-based simulation and analysis, such as probabilistic design, has been used extensively in engineering to predict performance of a system. Computer-based simulations have a model with the same format of the model (1.38), where the function f is known, and ε is random error simulated from a known distribution generated by a random number generator in the computer. Despite the steady and continuing growth of computing power and speed, the high computational costs in analysis of the computer simulation is still a crucial problem. Like computer experiments, computer-based simulation also involves a) a design to sample the domain of interest and b) finding a suitable metamodel.

Response surface methodology (RSM) has been successfully applied to many complex processes to seek "optimal" processing conditions. RSM employs sequential experimentation based on lower-order Taylor series approximation. RSM approach may be successfully applied to computer experiments when low-order polynomials (i.e., second order) are sufficient to capture the input-output relationship. However, in general, the approach may not be the best choice due to the following reasons: a) RSM typically considers only few factors while computer experiments may consider many input variables; b) the experimental domain in RSM is small while computer experiments need a large experimental space; c) the statistical model in RSM is well defined and simple while the model in most computer experiments is too complicated and has no analytic formula; d) RSM requires some prior knowledge to choose a good starting point in a reasonable neighborhood of the optimum while computer experiments are intended to provide a more global study on the model. Readers can refer to Myers and Montgomery (1995) for a comprehensive review on RSM.

1.10 Guidance for Reading This Book

This book is divided into three parts. Part I provides an introduction of design and modeling of computer experiments and some basic concepts that will be used throughout the book. Part II focuses on the design of computer experiments; here we introduce the most popular space-filling designs (Latin hypercube sampling and its modifications, and uniform design), including their definition, properties, construction, and related generating algorithms. Part III discusses modeling of data from computer experiments; here various modeling techniques as well as model interpretations (i.e., sensitivity analysis) are presented. Readers may elect to read each part of the book sequentially or readers interested in a specific topic may choose to go directly to a certain chapter. Examples are provided throughout the book to illustrate the methodology and to facilitate practical implementation.

Some useful concepts in statistics and matrix algebra are provided in the Appendix to make this book self-contained. The appendix reviews basic concepts in statistics, probability, linear regression analysis, and matrix algebra that will be required to understand the methodology presented in the book. The appendix may be useful as a reference even for a reader who has sufficient background and familiarity with the subject of the design of experiments.

Readers interested in applying the design of a computer experiment should read Part II carefully. Chapters 2 and 3 can be read independently. The former introduces the concept of *Latin hypercube sampling* and its various modifications, such as randomized orthogonal array, symmetric Latin hypercube sampling, and optimal Latin hypercube designs. Design optimality criteria for computer space-filling design are also introduced in the chapter. Chapter 3 discusses a different approach to space-filling design known as the *uniform design*. Various measures of uniformity are introduced in the context of the construction of space-filling designs. The most notable criteria are modified L_2-discrepancies. Algebraic approaches for constructing several classes of uniform design are presented. The techniques are very useful when one would like to construct uniform design without employing combinatorial optimization process. Readers interested in practical aspects of uniform design may want to focus on the construction of the design and skip the theoretical discussions.

Chapter 4 presents various stochastic optimization techniques for constructing optimal space-filling designs. These heuristic optimization algorithms can be employed to search for high-quality space-filling designs (e.g., optimal LHDs and uniform designs) under various optimality criteria discussed in Chapters 3 and 4. Several popular algorithms (i.e., Column-Pairwise, Simulated Annealing, Threshold Acceptance, and Stochastic Evolutionary) are presented. Readers may elect to read and implement one algorithm among the available alternatives.

Chapter 5 introduces various modeling techniques. The chapter starts with the fundamental concepts of prediction errors and model regularization. These concepts are crucial especially for dealing with the complexity of highly adaptive metamodels which are typically in the form of non-parametric regressions. A logical progression with increasing complexity of modeling techniques is presented including the well-known polynomial models, splines, Kriging, Bayesian approaches, neural networks (multilayer perceptron networks and radial basis function), and local polynomials. A unified view using basis function expansions is provided to relate the variety of modeling techniques. Again, readers who are interested in practical usage may choose to pay more attention to a particular modeling technique. Understanding the Kriging model may be necessary before reading about the Bayesian approach, radial basis function, and local polynomials as many references will be made to the Kriging model.

Because of the complexity of the structure of metamodels discussed in Chapter 5, special techniques for model interpretations are presented in Chapter 6. These techniques can be viewed as generalizations of the traditional Analysis of Variance (ANOVA) in linear models. It is highly recommended that readers study this chapter, especially those who are interested in understanding the sensitivity of input variables to the output. The chapter starts with the traditional sum of squares decomposition in linear models. The approach then is generalized to sequential sum of squares decomposition for general models. A Monte Carlo technique of Sobol' functional decompositions is introduced as a generalization of ANOVA decomposition. Additionally, analytic functional decomposition is also provided to take advantage of typical tensor product metamodel structures. An alternative computational technique known as the Fourier Amplitude Sensitivity Test (FAST) is also introduced.

The last chapter discusses computer experiments with functional responses. Here the response is in the form of a curve, where response data are collected over a range of time interval, space interval, or operation interval, for example, an experiment measuring engine radiated noise at a range of speeds. Functional responses are commonly encountered in today's computer experiments. The idea of analyzing functional response in the context of the design of experiments is a relatively new area. We introduce several possible approaches to deal with such cases. Real-life case studies will be used to introduce the concept.

The book emphasizes methodology and applications of design and modeling for computer experiments. Most of the theoretical proofs are omitted, but we give related references for interested readers.

The following flowchart figure gives some flexible strategy for reading this book. The reader may start to read Chapter 1 to see the motivation and strategies for computer experiments. Then the reader can go to Chapter 2 or Chapter 3 or both chapters. Readers can skip Chapter 4 if they do not need to construct LHSs and UDs. Chapters 5 and 6 are the most important chapters to read for modeling computer experiments. Chapter 7 is necessary only for readers interested in experiments in which the response is a curve, a function, or a trajectory over time The reader who needs some basic knowledge in matrix algebra, probability, statistics, regression analysis, and selection of variables in regression models will find the Appendix useful.

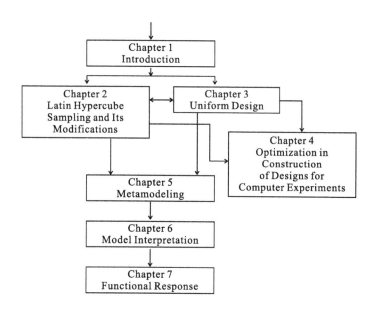

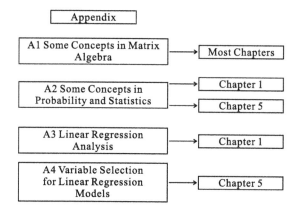

FIGURE 1.9
Flowchart for reading this book.

Part II Designs for Computer Experiments

In this part, we introduce the most popular *space-filling designs*, their definitions, properties, constructions, and related generating algorithms. In the text, n stands for the number of runs and s for the number of factors. For computer experiments, selecting an experimental design is a key issue in building an efficient and informative model. How best to construct a design depends on the underlying statistical model. The *overall mean model* introduced in Section 1.5 is the basis for the motivation of the *Latin hypercube sampling* and the *uniform design*. A good method of construction for space-filling designs should satisfy the following criteria: (a) it is optimal under the underlying statistical model; (b) it can serve various numbers of runs and input variables; (c) it can be easily generated with a reasonable computing time.

Two space-filling designs are called *equivalent* if one can be obtained by permuting rank of factors and/or runs of another. We do not distinguish between equivalent designs, so we may use the same notation for them. A space-filling design can be randomly generated, as in Latin hypercube sampling. In this case a design is a sample from a set of proposed designs. We shall use the same notation for this design class. For example, $LHS(n, s)$ can be a Latin hypercube sample, or the set of all such samples. A space-filling design can also be deterministic, like the uniform design. We use $U(n, q^s)$ for a symmetrical U-type (i.e., balanced) design of n runs and s factors, each having q levels, or the set of all such designs. Thus, a notation may stand for either a design or the set of the same type of designs without confusion.

Chapter 2 introduces Latin hypercube sampling and its modifications: randomized orthogonal array, symmetric Latin hypercube sampling, and optimal Latin hypercube designs. Optimal Latin hypercube designs under various criteria are given.

Chapter 3 concerns uniform design and its construction. Several measures of uniformity are introduced. Various methods of design construction, such as the good lattice point method, Latin square method, expending orthogonal array method, cutting method, and combinatorial design methods are considered.

Optimization plays an important role in the design and modeling of computer experiments. Chapter 4 gives a detailed introduction to known heuristic combinatorial optimization algorithms for generating optimal design, including local search, simulated annealing, threshold accepting, and stochastic evolutionary methods.

2

Latin Hypercube Sampling
and Its Modifications

This chapter introduces *Latin hypercube sampling* and its various modifications, such as randomized orthogonal array, symmetric Latin hypercube sampling, and optimal Latin hypercube designs. For simplicity, $C^s = [0,1]^s$ is chosen as the experimental domain.

2.1 Latin Hypercube Sampling

Consider model (1.8) and suppose one wants to estimate the overall mean of y, i.e., $E(y)$ on C^s in (1.17). A natural estimator would be the sample mean, $\bar{y}(\mathcal{P}) = \frac{1}{n}\sum_{i=1}^{n} f(\mathbf{x}_i)$, based on a set of experimental points $D_n = \{\mathbf{x}_1, \cdots, \mathbf{x}_n\}$ in C^s. From the statistical point of view one would like the sample mean to be unbiased and have minimal variance. For development of the theory, assume that

$$\int f(\mathbf{x})d\mathbf{x} < \infty \tag{2.1}$$

There are many ways to generate the design D_n, for instance,

(A) **Simply random sampling**: Experimental points $\mathbf{x}_1, \cdots, \mathbf{x}_n$ are independently identical samples from the uniform distribution $U(C^s)$. Obviously, the corresponding sample mean is unbiased with variance $\mathrm{Var}(f(\mathbf{x}))/n$, where $\mathbf{x} \sim U(C^s)$; the latter stands for the uniform distribution on C^s.

(B) **Latin hypercube sampling:** Divide the domain C^s of each $\mathbf{x}_k$ into n strata of equal marginal probability $1/n$, and sample once from each stratum. In fact, the LHS can be defined in terms of the Latin hypercube design.

Definition 1 A Latin hypercube design (LHD) with n runs and s input variables, denoted by $LHD(n, s)$, is an $n \times s$ matrix, in which each column is a random permutation of $\{1, 2, \cdots, n\}$.

An LHS can be generated by the following algorithm (LHSA).

Algorithm LHSA

Step 1. Independently take s permutations $\pi_j(1), \cdots, \pi_j(n)$ of the integers $1, \cdots, n$ for $j = 1, \cdots, s$, i.e., generate an $LHD(n, s)$;

Step 2. Take ns uniform variates (also called random numbers in Monte Carlo methods) $U_k^j \sim U(0, 1)$, $k = 1, \cdots, n$, $j = 1, \cdots, s$, which are mutually independent. Let $\mathbf{x}_k = (x_k^1, \cdots, x_k^s)$, where

$$x_k^j = \frac{\pi_j(k) - U_k^j}{n}, \ k = 1, \cdots, n, \ j = 1, \cdots, s.$$

Then $D_n = \{\mathbf{x}_1, \cdots, \mathbf{x}_n\}$ is a LHS and is denoted by $LHS(n, s)$.

Example 9 For generating an LHS for $n = 8, s = 2$, in the first step we generate two permutations of $\{1, 2, \cdots, 8\}$ as (2,5,1,7,4,8,3,6) and (5,8,3,6,1,4,7,2) to form an $LHD(8, 2)$ that is the matrix on the left below. Then we generate 16=8*2 random numbers to form an 8×2 matrix on the right below:

$$
\begin{bmatrix}
2 & 5 \\
5 & 8 \\
1 & 3 \\
7 & 6 \\
4 & 1 \\
8 & 4 \\
3 & 7 \\
6 & 2
\end{bmatrix},
\begin{bmatrix}
0.9501 & 0.8214 \\
0.2311 & 0.4447 \\
0.6068 & 0.6154 \\
0.4860 & 0.7919 \\
0.8913 & 0.9218 \\
0.7621 & 0.7382 \\
0.4565 & 0.1763 \\
0.0185 & 0.4057
\end{bmatrix}.
$$

Now an LHS is given by

$$
\frac{1}{8}\left(
\begin{bmatrix}
2 & 5 \\
5 & 8 \\
1 & 3 \\
7 & 6 \\
4 & 1 \\
8 & 4 \\
3 & 7 \\
6 & 2
\end{bmatrix}
-
\begin{bmatrix}
0.9501 & 0.8214 \\
0.2311 & 0.4447 \\
0.6068 & 0.6154 \\
0.4860 & 0.7919 \\
0.8913 & 0.9218 \\
0.7621 & 0.7382 \\
0.4565 & 0.1763 \\
0.0185 & 0.4057
\end{bmatrix}
\right)
=
\begin{bmatrix}
0.1312 & 0.5223 \\
0.5961 & 0.9444 \\
0.0491 & 0.2981 \\
0.8143 & 0.6510 \\
0.3886 & 0.0098 \\
0.9047 & 0.4077 \\
0.3179 & 0.8530 \\
0.7477 & 0.1993
\end{bmatrix}.
$$

Figure 2.1 gives plots of the design, where eight points are assigned in a grid of $64 = 8^2$ cells, satisfying that each row and each column has one and only one point, and each point is uniformly distributed in the corresponding cell. Note that $(\pi_1(k), \pi_2(k))$ determines in which cell $\mathbf{x}_k$ is located and (U_k^1, U_k^2) determines where $\mathbf{x}_k$ is located within that cell.

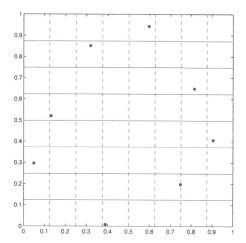

FIGURE 2.1
Two LHSs with six runs.

LHS was proposed by McKay, Beckman and Conover (1979) in what is widely regarded to be the first paper on design of computer experiments. LHS is an extension of stratified sampling which ensures that each of the input variables has all portions of its range represented (Sacks, Welch, Mitchell and Wynn (1989)). McKay, Beckman and Conover (1979) showed that the LHS has a smaller variance of the sample mean than the simply random sampling.

Theorem 1 *Denote by $\bar{y}_{SRS}$ and $\bar{y}_{LHS}$ the sample mean of n points generated by simple random sampling and LHS, respectively. If the function $f(x_1, \cdots, x_s)$ in model (1.8) is monotonic in each of its arguments, then $Var(\bar{y}_{SRS}) \geq Var(\bar{y}_{LHS})$.*

Stein (1987) and Owen (1992*a*) found an expression for the variance of $\bar{y}_{LHS}$ and showed that

$$\text{Var}(\bar{y}_{LHS}) = \frac{1}{n}\text{Var}(f(\mathbf{x})) - \frac{c}{n} + o(\frac{1}{n}),$$

where c is a positive constant. Sacks, Welch, Mitchell and Wynn (1989) commented on the advantages of the LHS: "Iman and Helton (1988) compare Latin hypercube sampling with Monte Carlo sampling for generating a response surface as an empirical surrogate for a computer model. The response surface was fitted by least squares to data from a fractional-factorial design. They found in a number of examples that the response surface could not adequately represent the complex output of the computer code but could be useful for ranking the importance of the input variables. Because Latin hypercube sampling excises the code over the entire range of each input variable, it can also be a systematic way of discovering scientifically surprising behavior as noted in Iman and Helton (1988)."

In fact, the second step for the use of random numbers U_k^j can be removed to reduce the complexity of the sampling. Many authors suggested that LHS has a lattice structure

$$x_k^j = \frac{\pi^j(k) - 0.5}{n},$$

that is, to put experimental point $\mathbf{x}_k$ at the center of the cell. The corresponding LHS is called *midpoint Latin hypercube sampling (MLHS)* or *centered Latin hypercube sampling*. A midpoint LHS with n runs and s input variables is denoted by $MLHS(n, s)$. The midpoint LHS has a close relationship with the so-called *U-type design* (Fang and Hickernell (1995)).

Definition 2 A U-type design with n runs and s factors, each having respective $q_1, \cdots, q_s$ levels, is an $n \times s$ matrix such that the q_j levels in the jth column appear equally often. This design is denoted by $U(n, q_1 \times \cdots \times q_s)$. When some q_j's are equal, we denote it by $U(n, q_1^{r_1} \times \cdots \times q_m^{r_m})$ where integers $r_1 + \cdots + r_m = s$. If all the q_j's are equal, denoted by $U(n, q^s)$, it is said to be symmetrical; otherwise it is asymmetrical.

U-type design is also called *balanced design* (Li and Wu (1997)) or *lattice design* (Bates, Riccomagno, Schwabe and Wynn (1995)). It is clear that each q_j should be a divisor of n. Very often the q entries take the values $\{1, \cdots, q\}$ or $\{(2i - 1)/2q, i = 1, \cdots, q\}$. Tables 2.1 and 2.2 are two U-type designs, $U(12, 12^4)$ and $U(9, 3^4)$, respectively. When we write $U \in U(n, q^s)$, it means that U is a design $U(n, q^s)$ and the notation $U(n, q^s)$ also denotes the set of all the U-type $U(n, q^s)$ designs. Many situations require that the elements of U fall in [0,1]. This requirement leads to the following concept.

Definition 3 Let $\mathbf{U} = (u_{ij})$ be a U-type design, $U(n, q^n)$ with entries $\{1, \cdots, q\}$. Let

$$x_{ij} = \frac{2u_{ij} - 1}{2q}, i = 1, \cdots, n, j = 1, \cdots, s. \tag{2.2}$$

Then the set $\mathbf{x}_i = (x_{i1}, \cdots, x_{i,s}), i = 1, \cdots, n$ is a design on C^s and called the **induced design** *of U, denoted by D_U.*

In fact, D_U is a U-type design $U(n, q^s)$ with entries

$$\frac{1}{2q}, \frac{3}{2q}, \cdots, \frac{2q - 1}{2q}.$$

There is a link between U-type designs and LHD.

TABLE 2.1
U-type Design in $U(12, 12^4)$

No	1	2	3	4
1	1	10	4	7
2	2	5	11	3
3	3	1	7	9
4	4	6	1	5
5	5	11	10	11
6	6	9	8	1
7	7	4	5	12
8	8	2	3	2
9	9	7	12	8
10	10	12	6	4
11	11	8	2	10
12	12	3	9	6

TABLE 2.2
$U(9, 3^4)$

No	1	2	3	4
1	1	1	1	1
2	1	2	2	2
3	1	3	3	3
4	2	1	2	3
5	2	2	3	1
6	2	3	1	2
7	3	1	3	2
8	3	2	1	3
9	3	3	2	1

LEMMA 2.1
The set of $LHD(n, s)$ is just the set of $U(n, n^s)$ with entries $\{1, 2, \cdots, n\}$ and the set of $MLHS$ is just the set of $U(n, n^s)$ with entries $\{\frac{1}{2n}, \frac{3}{2n}, \cdots, \frac{2q-1}{2n}\}$.

Scatter-box plots are useful for visualization of the projection of the distribution of all the design pairs of any two variables. If some scatter plots do not appear reasonably uniform, we should generate another LHS instead of the previous one. In the next section we shall discuss this issue further. Figures 2.2 and 2.3 gives scatter-box plots for an $LHS(30, 3)$ and its corresponding $LHD(30, 3)$. We can see that the two figures are very close to each other. The box plots are generated automatically by the software but this is not necessary as points in each column of the design are uniformly scattered.

2.2 Randomized Orthogonal Array

The LHS has many advantages, such as: (a) it is computationally cheap to generate; (b) it can deal with a large number of runs and input variables; (c) its sample mean has a smaller variance than the sample mean of a simply random sample. However, LHS does not reach the smallest possible variance for the sample mean. Many authors tried hard to improve LHS, i.e. by reducing of the variance of the sample mean. Owen (1992b, 1994b) and Tang (1993) independently considered randomized orthogonal arrays. Tang (1993) called this approach *orthogonal array-based Latin hypercube design* (OA-based LHD). The concept of orthogonal array was introduced in Section 1.2. The so-called *randomized orthogonal array* or *OA-based LHD* is an LHD that is

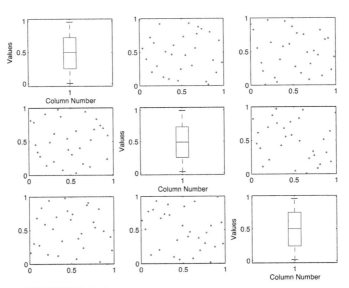

FIGURE 2.2
The scatter plots of $LHS(30, 3)$.

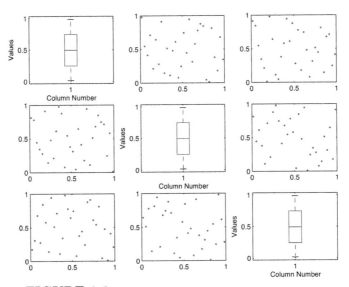

FIGURE 2.3
The scatter plots of $LHD(30, 3)$.

generated from an orthogonal array. Let us start with an illustrative example. Begin with the first column of $L_9(3^4)$ in Table 1.1. Randomly choose a permutation, for example, $\{2, 1, 3\}$ of $\{1, 2, 3\}$ to replace the three level 1's and a permutation, $\{6, 4, 5\}$ of $\{4, 5, 6\}$ to replace the three level 2's, and a permutation $\{9, 7, 8\}$ of $\{7, 8, 9\}$ to replace the three level 3's. Similarly, applying the above process to columns 2, 3, and 4 of Table 1.1 results in a new design, denoted by $OH(9, 9^4)$, with 9 runs and 4 factors, each having 9 levels in Table 2.3. The latter is an OA-based LHD. Figure 2.4 shows plots of the first two columns of $L_9(3^4)$ and $OH(9, 9^4)$. The experimental points of the former are denoted by '•' and of the latter by '×.' We can see that the original $L_9(3^4)$ has three levels for each factor, but the new $OH(9, 9^4)$ has nine levels for each factor. The latter spreads experiment points more uniformly on the domain. Now, we can write an algorithm to generate an OA-based LHD. Obviously, $OH(n, n^s)$ is an LHD.

TABLE 2.3
$OH_9(9^4)$ Design from an $L_9(3^4)$ Design

No	1	2	3	4
1	$1 \to 2$	$1 \to 2$	$1 \to 1$	$1 \to 2$
2	$1 \to 1$	$2 \to 6$	$2 \to 5$	$2 \to 4$
3	$1 \to 3$	$3 \to 8$	$3 \to 8$	$3 \to 8$
4	$2 \to 6$	$1 \to 3$	$2 \to 6$	$3 \to 9$
5	$2 \to 4$	$2 \to 5$	$3 \to 9$	$1 \to 1$
6	$2 \to 5$	$3 \to 7$	$1 \to 2$	$2 \to 5$
7	$3 \to 9$	$1 \to 1$	$3 \to 7$	$2 \to 6$
8	$3 \to 7$	$2 \to 4$	$1 \to 3$	$3 \to 7$
9	$3 \to 8$	$3 \to 9$	$2 \to 4$	$1 \to 3$

Algorithm ROA

Step 1. Choose an $OA(n, s, q, r)$ as needed and let **A** be this orthogonal array and let $\lambda = n/q^r$;

Step 2. For each column of **A**, replace the λq^{r-1} positions with entry k by a random permutation of $\{(k-1)\lambda q^{r-1} + 1, (k-1)\lambda q^{r-1} + 2, \cdots, (k-1)\lambda q^{r-1} + \lambda q^{r-1} = k\lambda q^{r-1}\}$, for $k = 1, \cdots, q$. The replacement process results in an LHD that is an OA-based LHD, $OH(n, n^s)$.

Owen (1992b) said "Tang (1991) independently and contemporaneously had the idea of using orthogonal arrays to construct designs for computer experiments. He constructs Latin hypercube samples with q^2 runs that become orthogonal arrays of strength 2 after grouping observations using q bins on each axis. Tang further shows that these designs achieve the same order of variance reduction as orthogonal arrays of strength 2, while their Latin hypercube property means that they use q^2 distinct values on each variable instead

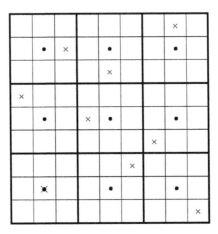

FIGURE 2.4
Experimental points of two factors by OA and OA-based LHD.

of the q used by the straight orthogonal array. This advantage becomes more important with the smaller computer experiments appropriate for the more expensive to run simulators. For larger experiments one might prefer the orthogonal arrays on account of their better balance."

Randomized orthogonal arrays can improve performance of LHD by achieving good combinatorial balance among any two or more columns of an orthogonal array. In fact, the balance properties of a (t, m, s)-net, proposed by Niederreiter (1992), are superior to those of an orthogonal array. It can be expected that randomized or scrambled (t, m, s)-nets have a better performance than that of the corresponding randomized orthogonal array. Owen (1997) found a variance for the sample mean over randomized (t, m, s)-nets. He concludes that for any square-integrable integrand f, the resulting variance surpasses any of the usual variance reduction techniques.

2.3   Symmetric and Orthogonal Column Latin Hypercubes

Since the LHS is only a form of stratified random sampling and is not directly related to any criterion, it may perform poorly in estimation and prediction of the response at untried sites (Ye, Li and Sudjianto (2003)). Figure 2.5 gives plots of the two $LHS(6,2)$ designs below:

$$D_{6-1} = \begin{bmatrix} 1 & 3 \\ 2 & 6 \\ 3 & 2 \\ 4 & 5 \\ 5 & 1 \\ 6 & 4 \end{bmatrix}, \text{ and } D_{6-2} \begin{bmatrix} 1 & 5 \\ 2 & 4 \\ 3 & 3 \\ 4 & 2 \\ 5 & 1 \\ 6 & 6 \end{bmatrix}.$$

The design on the left is perfect, while the design on the right has no experimental points in a certain area of the domain $C^2 = [0,1]^2$. As a result, the latter may pose difficulty in modeling. A one-dimensional projection would make both designs look balanced and adequate, but viewed in two dimensions, the second design is obviously deficient; one factor is almost confounded with the other. Therefore, it is logical to further enhance LHD not only to fill space in a one-dimensional projection but also in higher dimensions. Generating such design, however, requires a large combinatorial search of $(n!)^s$. Because of these large possibilities, several authors suggested narrowing the candidate design space. If we delete some poor candidates and narrow the design domain for sampling, the corresponding LHD can have better performance in practice. Another consideration is to find a special type of LHD, which has some good "built-in" properties in terms of sampling optimality that will be discussed in the next section. The so-called *symmetric Latin hypercube design* (SLHD) and *orthogonal column Latin designs* (OCLHD) are two such modified LHDs. It was reported by Park (1994) and Morris and Mitchell (1995) that many optimal LHDs they obtained have some symmetric property that leads to the following concept.

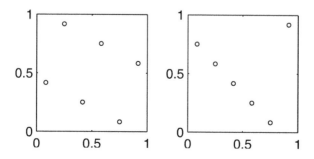

FIGURE 2.5
Two LHSs with six runs.

*Definition 4 A midpoint LHD is called a symmetric Latin hypercube design,
denoted by $SLHD(n,s)$, if it has the reflection property: for any row $\mathbf{x}_k$ of
the matrix, there exists another row such that it is the reflection of $\mathbf{x}_k$ through
the center $\frac{n+1}{2}$.*

By reflection, we mean that $(a_1, \cdots, a_s)$ is a row of an $MLHD(n,s)$, then
the vector $(n+1-a_1, \cdots, n+1-a_s)$ should be another row in the design
matrix. By permuting rows of a symmetric LHD we can always assume that
the kth row is the reflection of the $(n-k+1)th$ row and the summation of
these two rows forms a vector $(n+1, n+1, \cdots, n+1)$. Note that the design
D_{6-1} is an $SLHD(6,2)$, but the design D_{6-2} is not. The following is another
example for $SLHD(10,5)$:

$$
\begin{bmatrix}
1 & 6 & 6 & 5 & 9 \\
2 & 2 & 3 & 2 & 4 \\
3 & 1 & 9 & 7 & 5 \\
4 & 3 & 4 & 10 & 3 \\
5 & 7 & 1 & 8 & 10 \\
6 & 4 & 10 & 3 & 1 \\
7 & 8 & 7 & 1 & 8 \\
8 & 10 & 2 & 4 & 6 \\
9 & 9 & 8 & 9 & 7 \\
10 & 5 & 5 & 6 & 2
\end{bmatrix}.
$$

where the summation of row pairs 1 & 10, 2 & 9, 3 & 8, 4 & 7, or 5 & 6 is
$(11, 11, 11, 11, 11, 11)$.

Iman and Conover (1982) and Owen (1994a) proposed using LHD with
small off-diagonal correlations. Such designs for computer experiments would
benefit linear regression models, as the estimates of linear effects of the input
variables are only slightly correlated. Based on this consideration, Ye (1998)
proposed the orthogonal column LHD. Two column vectors in an LHD are
called *column-orthogonal* if their correlation is zero.

*Definition 5 An $n \times s$ matrix is called an orthogonal column LHD, denoted by
$OCLHD(n, n^s)$, if it is an LHD and its column-pairs are orthogonal.*

To easily check whether two columns are column-orthogonal or not, we
convert the levels of each column to be symmetric about zero. For example, we convert two column-orthogonal columns $\mathbf{a} = (4, 5, 3, 2, 1)'$ and
$\mathbf{b} = (1, 4, 3, 5, 2)'$ into $\mathbf{a}_* = (1, 2, 0, -1, -2)'$ and $\mathbf{b}_* = (-2, 1, 0, 2, 1)'$. It
is easy to find that $\mathbf{a}'_* \mathbf{b}_* = 0$, i.e., $\mathbf{a}_*$ and $\mathbf{b}_*$ are column-orthogonal, or equiv-
alently $\mathbf{a}$ and $\mathbf{b}$ are column-orthogonal. Ye (1998) suggested an algorithm
for construction of $OCLHD(n, n^s)$ with $n = 2^m + 1$, where m is a positive
integer. In this case n levels are $\{-2^{m-1}, \cdots, -1, 0, 1, \cdots, 2^{m-1}\}$. Table 2.4
shows an $OCLHS(9, 9^4)$. Note that the bottom half of the design is precisely

the mirror image of the top half. Denote the top half of the design by

$$\mathbf{T} = \begin{bmatrix} 1 & -2 & -4 & 3 \\ 2 & 1 & -3 & -4 \\ 3 & -4 & 2 & -1 \\ 4 & 3 & 1 & 2 \end{bmatrix}$$

and note that it can be expressed in terms of two same size matrices:

$$\mathbf{M} = \begin{bmatrix} 1 & 2 & 4 & 3 \\ 2 & 1 & 3 & 4 \\ 3 & 4 & 2 & 1 \\ 4 & 3 & 1 & 2 \end{bmatrix} \text{ and } \mathbf{H} = \begin{bmatrix} 1 & -1 & -1 & 1 \\ 1 & 1 & -1 & -1 \\ 1 & -1 & 1 & -1 \\ 1 & 1 & 1 & 1 \end{bmatrix},$$

where $\mathbf{M}$ is the matrix of which each entry is the absolute value of the corresponding entry in $\mathbf{T}$, and $\mathbf{H}$ is the matrix of which each entry is taken as 1 or -1 according to whether the corresponding entry in $\mathbf{T}$ is positive or negative. It is easy to see that $\mathbf{T}$ is the *element-wise product* (also known as the *Hadamard product*) of $\mathbf{M}$ and $\mathbf{H}$, i.e., $\mathbf{T} = \mathbf{M} \odot \mathbf{H}$ (cf. Section A.1).

TABLE 2.4
$OCLHD(9, 9^4)$

No	1	2	3	4
1	1	-2	-4	3
2	2	1	-3	-4
3	3	-4	2	-1
4	4	3	1	2
5	0	0	0	0
6	-4	-3	-1	-2
7	-3	4	-2	1
8	-2	-1	3	4
9	-1	2	4	-3

Notice that $\mathbf{M}$ is an $LHD(4, 4^3)$ that can be constructed by two matrices

$$\mathbf{I} = \begin{bmatrix} 1 & 0 \\ 0 & 1 \end{bmatrix} \text{ and } \mathbf{Q} = \begin{bmatrix} 0 & 1 \\ 1 & 0 \end{bmatrix}. \tag{2.3}$$

Let

$$\mathbf{m}_2 = \mathbf{I} \otimes \mathbf{Q} \cdot \mathbf{4} = \begin{bmatrix} 0\,1\,0\,0 \\ 1\,0\,0\,0 \\ 0\,0\,0\,1 \\ 0\,0\,1\,0 \end{bmatrix} \begin{bmatrix} 1 \\ 2 \\ 3 \\ 4 \end{bmatrix} = \begin{bmatrix} 2 \\ 1 \\ 4 \\ 3 \end{bmatrix},$$

$$\mathbf{m}_3 = \mathbf{Q} \otimes \mathbf{I} \cdot \mathbf{4} = \begin{bmatrix} 0\,0\,0\,1 \\ 0\,0\,1\,0 \\ 0\,1\,0\,0 \\ 1\,0\,0\,0 \end{bmatrix} \begin{bmatrix} 1 \\ 2 \\ 3 \\ 4 \end{bmatrix} = \begin{bmatrix} 4 \\ 3 \\ 2 \\ 1 \end{bmatrix},$$

where $\otimes$ denotes the Kronecker product (cf. Section A.1) and $\mathbf{4} = (1, 2, 3, 4)'$. The first column of $\mathbf{M}$ is $\mathbf{4}$ and the second and third columns of $\mathbf{M}$ are $\mathbf{m}_2$ and $\mathbf{m}_3$, respectively. The last column of $\mathbf{M}$ is the product of $[\mathbf{I} \otimes \mathbf{Q}][\mathbf{Q} \otimes \mathbf{I}] \cdot \mathbf{4}$.

Now, let us introduce the construction of a Hadamard matrix $\mathbf{H}$ of order 4 (cf. Section A.1). Let

$$\mathbf{b}_0 = \begin{bmatrix} -1 \\ 1 \end{bmatrix}, \text{ and } \mathbf{b}_1 = \begin{bmatrix} 1 \\ 1 \end{bmatrix}. \tag{2.4}$$

The first column of $\mathbf{H}$ is a 4-vector of 1's; the remaining three columns of $\mathbf{H}$ are

$$\mathbf{h}_1 = \mathbf{b}_1 \otimes \mathbf{b}_0 = (-1, 1, -1, 1)',$$
$$\mathbf{h}_2 = \mathbf{b}_0 \otimes \mathbf{b}_1 = (-1, -1, 1, 1)', \text{ and }$$
$$\mathbf{h}_3 = \mathbf{h}_1 \odot \mathbf{h}_2 = (1, -1, -1, 1)',$$

respectively, that is, $\mathbf{H} = (\mathbf{1}, \mathbf{h}_1, \mathbf{h}_2, \mathbf{h}_3) = (\mathbf{1}, \mathbf{h}_1, \mathbf{h}_2, \mathbf{h}_1 \odot \mathbf{h}_2)$.

From the above illustration example, we now introduce the following algorithms proposed by Ye (1998).

Algorithm OCLHD with $n = 2^m + 1$

Step 1. Let $n = 2^m + 1$, $\mathbf{n} = (1, 2, \cdots, 2^{m-1})'$ and

$$\mathbf{A}_k = \underbrace{\mathbf{I} \otimes \cdots \otimes \mathbf{I}}_{m-k-1} \otimes \underbrace{\mathbf{Q} \otimes \cdots \otimes \mathbf{Q}}_{k}, k = 1, \cdots, m-1.$$

Step 2. Columns of the matrix $\mathbf{M}$ are

$$\{\mathbf{n}, \mathbf{A}_k \mathbf{n}, \mathbf{A}_{m-1} \mathbf{A}_j \mathbf{n}, k = 1, \cdots, m-1, j = 1, \cdots, m-2\}.$$

Step 3. For $k = 1, \cdots, m-1$, define the vector $\mathbf{h}_k$ as

$$\mathbf{h}_k = \mathbf{q}_1 \otimes \mathbf{q}_2 \otimes \cdots \otimes \mathbf{q}_{m-1},$$

where $\mathbf{q}_k = \mathbf{b}_0$ and $\mathbf{q}_i = \mathbf{b}_1, i \neq k$, and $\mathbf{b}_0$ and $\mathbf{b}_1$ are defined in (2.4).

Step 4. Columns of the matrix **H** are

$$\{1, \mathbf{h}_i, \mathbf{h}_1 \odot \mathbf{h}_{j+1}, i = 1, \cdots, m - 1, j = 1, \cdots, m - 2\}.$$

Step 5. Calculate the matrix $\mathbf{T} = \mathbf{M} \odot \mathbf{H}$, the top half of the design, and make the mirror image of **T** the bottom half of the design. Finally, add a $(m + 1)$-vector of 0's between the top half and the bottom half. The design with $n = 2^m + 1$ runs and $m + 1$ factors, $OCLHD(n, n^{m+1})$, is generated.

Algorithm OCLHD with $n = 2^m$

Step 1. Generate an $OCLHd(n + 1, (n + 1)^{m+1})$ and delete the midpoint row of 0's.

Step 2. Make the remaining n rows with equidistant levels. This results in an $OCLHD(n, n^{m+1})$.

From Table 2.4 we can easily find an $OCLHD(8, 8^4)$ that is listed in Table 2.5 where levels $1, \cdots, 8$ replace the original levels $-4, -3, -2, -1, 1, 2, 3, 4$.

TABLE 2.5
$OCLHD(8, 8^4)$

No	1	2	3	4
1	5	3	1	7
2	6	5	2	1
3	7	1	6	4
4	8	7	5	6
5	1	2	4	3
6	2	8	3	5
7	3	4	7	8
8	4	6	8	2

Consider a polynomial regression model

$$\hat{y} = \beta_0 + \sum_{i=1}^{s} \beta_i x_i + \sum_{1 \le i \le j \le s} \beta_{ij} x_i x_j$$
$$+ \cdots + \sum_{1 \le i_1 \le i_2 \le \cdots \le i_q \le s} \beta_{i_1 \cdots i_q} x_{i_1} \cdots x_{i_q}. \qquad (2.5)$$

Define β_i to be the linear effect of x_i, β_{ii} to be the quadratic effect of x_i, and β_{ij} to be the bilinear interaction between x_i and x_j. Ye (1998) pointed out that the OCLHDs ensure the independence of the estimates of linear effects and the quadratic effects of each factor and bilinear interactions are uncorrelated with the estimates of the linear effects. However, for an OCLHD, the number of runs must be a power of two or a power of two plus one; this increases the number of runs dramatically as the number of columns increases. In contrast, the SLHDs and uniform designs introduced in Chapter 3 enjoy the advantage of a more flexible run size.

2.4 Optimal Latin Hypercube Designs

In the previous section, we used a strategy for improving the performance of LHDs by constructing an LHD from a narrow set of candidates with desirable properties. An alternative idea is to adopt some optimality criterion for construction of LHS, such as entropy (Shewry and Wynn (1987)), integrated mean squared error (IMSE)(Sacks, Schiller and Welch (1989)), and maximin or minimax distance (Johnson, Moore and Ylvisaker (1990)). An optimal LHD optimizes the criterion function over the design space. For a given number of runs and number of input variables, (n, s), the resulting *design space*, denoted by $\mathcal{D}$, can be the set of $U(n, n^s)$ or some subset.

2.4.1 IMSE criterion

In this section we introduce the integrated mean squares error (IMSE) criterion based on Sacks, Schiller and Welch (1989). Consider a Kriging model (see Sections 1.6 and 5.4)

$$\hat{y} = \sum_{i=1}^{m} \beta_i h_j(\mathbf{x}) + z(\mathbf{x}),$$

where $h_j(\mathbf{x})$s are known functions, the β_js are unknown coefficients to be estimated, and $z(\mathbf{x})$ is a stationary Gaussian random process (cf. Section A.2) with mean $E(z(\mathbf{x})) = 0$ and covariance

$$\text{Cov}(z(\mathbf{x}_i), z(\mathbf{x}_j)) = \sigma^2 R(\mathbf{x}_i - \mathbf{x}_j),$$

where σ^2 is the unknown variance of the random error and the correlation function R is given. The function R may have many choices, for example, a Gaussian correlation function is given by $R(d) = \exp\{-\theta d^2\}$, where θ is unknown. In this case, we have

$$\text{Cov}(z(\mathbf{x}_i), z(\mathbf{x}_j)) = \sigma^2 \exp[-\theta(\mathbf{x}_i - \mathbf{x}_j)'(\mathbf{x}_i - \mathbf{x}_j)]. \qquad (2.6)$$

Consider the linear predictor

$$\hat{y}(\mathbf{x}) = \mathbf{c}'(\mathbf{x})\mathbf{y}_D,$$

at an untried $\mathbf{x}$, where $\mathbf{y}_D = (y(\mathbf{x}_1), \cdots, y(\mathbf{x}_n))'$ is the response column vector collected according to the design $D_n = \{\mathbf{x}_1, \cdots, \mathbf{x}_n\}$. The column vector $\mathbf{y}_D$ can be regarded as an observation taken from $\mathbf{y} = (Y(\mathbf{x}_1), \cdots, Y(\mathbf{x}_n))'$. The best linear unbiased predictor (BLUE) is obtained by choosing the vector $\mathbf{c}(\mathbf{x})$ to minimize the mean square error

$$\mathrm{MSE}(\hat{y}(\mathbf{x})) = E[\mathbf{c}'(x)\mathbf{y}_D - Y(\mathbf{x})]^2 \tag{2.7}$$

subject to the unbiased constraint

$$E(\mathbf{c}'(\mathbf{x})\mathbf{y}_D) = E(Y(\mathbf{x})). \tag{2.8}$$

The BLUE of $\mathbf{y}(\mathbf{x})$ (cf. Section A.3.2) is given by

$$\hat{y}(\mathbf{x}) = \mathbf{h}'(\mathbf{x})\hat{\beta} + \mathbf{v}'_\mathbf{x}\mathbf{V}_D^{-1}(\mathbf{y}_D - \mathbf{H}_D\hat{\beta}), \tag{2.9}$$

where

$$\mathbf{h}(\mathbf{x}) = (h_1(\mathbf{x}), \cdots, h_m(\mathbf{x}))' : m \times 1,$$
$$\mathbf{H}_D = (h_j(\mathbf{x}_i)) : n \times m,$$
$$\mathbf{V}_D = (\mathrm{Cov}(Z(\mathbf{x}_i), Z(\mathbf{x}_j))); n \times n,$$
$$\mathbf{v}'_\mathbf{x} = (\mathrm{Cov}(Z(\mathbf{x}), Z(\mathbf{x}_1)), \cdots, \mathrm{Cov}(Z(\mathbf{x}), Z(\mathbf{x}_n)))' : n \times 1,$$

and

$$hpb = [\mathbf{H}'_D\mathbf{V}_D^{-1}\mathbf{H}_D]^{-1}\mathbf{H}_D\mathbf{V}_D^{-1}\mathbf{y}_D,$$

which is the generalized least squares estimate of β. The mean square error of $\hat{y}(\mathbf{x})$ is

$$\mathrm{MSE}(\hat{y}(\mathbf{x})) = \sigma^2 - (\mathbf{h}'(\mathbf{x}), \mathbf{v}'_\mathbf{x}) \begin{pmatrix} \mathbf{0} & \mathbf{H}'_D \\ \mathbf{H}_D & \mathbf{V}_D \end{pmatrix} \begin{pmatrix} \mathbf{h}(\mathbf{x}) \\ \mathbf{v}_\mathbf{x} \end{pmatrix}.$$

Let $\mathbf{r}(\mathbf{x}) = \mathbf{v}(\mathbf{x})/\sigma^2$ and $\mathbf{R}_D = \mathbf{V}_D/\sigma^2$; then the mean square error can be expressed as

$$\mathrm{MSE}(\hat{y}(\mathbf{x})) = \sigma^2 \left[1 - (\mathbf{h}'(\mathbf{x}), \mathbf{r}'_\mathbf{x}) \begin{pmatrix} \mathbf{0} & \mathbf{H}'_D \\ \mathbf{H}_D & \mathbf{R}_D \end{pmatrix} \begin{pmatrix} \mathbf{h}(\mathbf{x}) \\ \mathbf{r}_\mathbf{x} \end{pmatrix} \right]. \tag{2.10}$$

The maximum likelihood estimate of σ^2 is

$$\hat{\sigma}^2 = \frac{1}{n}(\mathbf{y}_D - \mathbf{h}'(\mathbf{x})\hat{\beta})'\mathbf{R}^{-1}(\mathbf{y}_D - \mathbf{h}'(\mathbf{x})\hat{\beta}).$$

The IMSE criterion chooses the design D_n to minimize

$$\mathrm{IMSE} = \int_T \mathrm{MSE}(\hat{y}(\mathbf{x})\phi(\mathbf{x})d\mathbf{x} \tag{2.11}$$

for a given weight function $\phi(\mathbf{x})$. From (2.10) the IMSE can be expressed as

$$\sigma^2\left[1 - \text{trace}\left[\left(\begin{array}{cc} \mathbf{0} & \mathbf{H}'_D \\ \mathbf{H}_D & \mathbf{R}_D \end{array}\right)^{-1}\int\left(\begin{array}{cc} \mathbf{h}(\mathbf{x})\mathbf{h}'(\mathbf{x}) & \mathbf{h}(\mathbf{x})\mathbf{r}'(\mathbf{x}) \\ \mathbf{r}(\mathbf{x})\mathbf{h}'(\mathbf{x}) & \mathbf{r}(\mathbf{x})\mathbf{r}'(\mathbf{x}) \end{array}\right)\phi(\mathbf{x})d\mathbf{x}\right]\right] \quad (2.12)$$

An optimal IMSE-optimal LHD minimizes IMSE over the set of $U(n, n^s)$ for a specific θ. To find the optimal design, we need a powerful optimization algorithm due to the large combinatorial design space. Various optimization techniques will be introduced in Chapter 4. Readers can find plots for many two-factor optimal IMSE LHDs in Koehler and Owen (1996). As the parameter θ is unknown, Sacks, Schiller and Welch (1989) gave a detailed discussion on how to choose a suitable θ. For example, they considered Example 4, in which all but two rate constants are hypothesized to have been established by previous work, so there are two input variables only. Miller and Frenklach (1983) took a nine-point design and approximated each of the five log concentrations with a quadratic model

$$y(x_1, x_2) = \beta_0 + \beta_1 x_1 + \beta_2 x_2 + \beta_{11} x_1^2 + \beta_{22} x_2^2 + \beta_{12} x_1 x_2 + z(x_1, x_2).$$

Another criterion related to MSE is *maximum mean squared error* (MMSE). This criterion chooses a design to minimize

$$\max \text{MSE}(\hat{y}(\mathbf{x})),$$

where the maximum is taken over all the candidate designs. This criterion is computationally more demanding. Section 5.4 will give a systematic study of the Kriging model and its estimation.

2.4.2 Entropy criterion

The entropy criterion proposed by Shannon in 1948 has been widely used in coding theory and statistics. Entropy measures the amount of information contained in the distribution of a set of data. Shewry and Wynn (1987) described this as the classical idea of 'the amount of information in an experiment.' Let $p(\mathbf{x})$ be a distribution in R^s; its entropy is defined by

$$\text{Ent}(p(\mathbf{x})) = -\int_{R^s} p(\mathbf{x})\log(p(\mathbf{x}))d\mathbf{x}. \quad (2.13)$$

It is easy to show $\text{Ent}(p(\mathbf{x})) \geq 0$. The lower the entropy, the more precise the knowledge is, or the higher the entropy, the more uncertainty there is.

The multivariate normal distribution of $N_p(\mu, \Sigma)$ (see Section A.2.2) has entropy

$$\frac{p}{2}(1 + \log(2\pi)) + \frac{1}{2}\log|\Sigma|. \quad (2.14)$$

When p is fixed, maximizing the entropy is equivalent to maximizing $\log|\Sigma|$ or $|\Sigma|$.

The **Bayesian approach** combines prior information and experimental data using an underlying model to produce a posterior distribution. Prediction will then be based on the posterior distribution. Let e be an experiment on a random system X with sampling density $p(\mathbf{x}|\theta)$ and parameter θ. Assume the prior distribution for θ is $\pi(\theta)$. If $\mathbf{x}$ is the data obtained on performing e, the posterior density for θ is

$$\pi(\theta|x) \propto \pi(\theta)p(x|\theta),$$

where "$\propto$" means "is proportional to." The amount of information on θ before and after the experiment is measured by their entropies

$$\mathrm{Ent}(\pi(\theta)) = -\int \pi(\theta)\log\pi(\theta)d\theta,$$

and

$$\mathrm{Ent}(\pi(\theta|x)) = -\int \pi(\theta|x)\log\pi(\theta|x)d\theta,$$

respectively. Thus, the information increase is

$$\mathrm{Ent}(\pi(\theta)) - \mathrm{Ent}(\pi(\theta|x)). \tag{2.15}$$

The expected value of (2.15) over the full joint distribution of X and θ is given by

$$g(e) = E_x E_\theta \log\left(\frac{\pi(\theta|\mathbf{x})}{\pi(\theta)}\right). \tag{2.16}$$

The *entropy criterion* maximizes $g(e)$.

If the experimental domain, denoted by T, is discrete (such as lattice points where each factor has a finite number of levels and the experimental domain is composed of all level-combinations), then the domain T is described by a random vector $\mathbf{y}_T = (Y_1, \cdots, Y_N)'$, where N is the cardinality of T and Y_i is attached to a point $x_i \in T$. Let D_n be a subset of n runs of T to be implemented and $\bar{D}_n$ be the complement of D_n, i.e., $D_n \cup \bar{D}_n = T$. Let $\mathbf{y}_{D_n}$ and $\mathbf{y}_{\bar{D}_n}$ (or $\mathbf{y}_D$ and $\mathbf{y}_{\bar{D}}$ for simplicity) be the vectors of outcomes from tried and untried points. Using the classical decomposition we have

$$\mathrm{Ent}(\mathbf{y}_T) = \mathrm{Ent}(\mathbf{y}_D) + E_{\mathbf{y}_D}[\mathrm{Ent}(\mathbf{y}_{\bar{D}}|\mathbf{y}_D)]. \tag{2.17}$$

The second item on the right-hand side of (2.17) is just the $-g(e)$. Since $\mathrm{Ent}(\mathbf{y}_T)$ is fixed, minimizing the second item is equivalent to maximizing $\mathrm{Ent}(\mathbf{y}_D)$, hence the term '*maximum entropy design.*'

For the Kriging model (1.27) with covariance functions (1.28) and (1.29), the entropy criterion suggests maximizing

$$\max_{D \in \mathcal{D}} \log|\mathbf{R}|, \tag{2.18}$$

where $\mathcal{D}$ is the design space and $\mathbf{R} = (r_{ij})$ is the correlation matrix of the design matrix $D = (\mathbf{x}_1, \cdots, \mathbf{x}_n)'$, where

$$r_{ij} = \exp\left\{-\sum_{k=1}^{s} \theta_k(x_{ki} - x_{kj})^2\right\}, \tag{2.19}$$

if the correlation function (1.29) is chosen. Shewry and Wynn (1987) demonstrated a linear model with stationary error (a simple Kriging model) and showed the corresponding maximum entropy designs. Readers can find some maximum entropy designs for two factors in Koehler and Owen (1996).

2.4.3 Minimax and maximin distance criteria and their extension

Let $d(\mathbf{u}, \mathbf{v})$ be a distance defined on $T \times T$ satisfying $d(\mathbf{u}, \mathbf{v}) \geq 0$, $d(\mathbf{u}, \mathbf{v}) = d(\mathbf{v}, \mathbf{u})$, and $d(\mathbf{u}, \mathbf{v}) \leq d(\mathbf{u}, \mathbf{w}) + d(\mathbf{w}, \mathbf{v}), \forall \mathbf{u}, \mathbf{v}, \mathbf{w} \in T$. Consider a design $D = \{\mathbf{x}_1, \cdots, \mathbf{x}_n\}$ on T.

A **minimax distance design** D^* minimizes the maximum distance between any $x \in T$ and D, $d(\mathbf{x}, D) = \max\{d(\mathbf{x}, \mathbf{x}_1), \cdots, d(\mathbf{x}, \mathbf{x}_n)\}$,, i.e.,

$$\min_{D} \max_{\mathbf{x} \in T} d(\mathbf{x}, D) = \max_{\mathbf{x} \in T} d(\mathbf{x}, D^*). \tag{2.20}$$

A **maximin distance design** (Mm) D_* maximizes the minimum inter-site distance $\min_{\mathbf{u}, \mathbf{v} \in D} d(\mathbf{u}, \mathbf{v})$, i.e.,

$$\max_{D} \min_{\mathbf{u}, \mathbf{v} \in D} d(\mathbf{u}, \mathbf{v}) = \min_{\mathbf{u}, \mathbf{v} \in D_*} d(\mathbf{u}, \mathbf{v}). \tag{2.21}$$

Minimax or maximin distance criteria, proposed by Johnson et al. (1990), measure how uniformly the experimental points are scattered through the domain. Designs based on these criteria ensure that no point in the domain is too far from a design point. Hence we can make reasonable predictions anywhere in the domain. Johnson et al. (1990) explored several connections between certain statistical and geometric properties of designs. They established equivalence of the maximin distance design criterion and an entropy criterion motivated by function prediction in a Bayesian setting. The latter criterion has been used by Currin et al. (1991) to design experiments for which the motivating application is to find an approximation of a complex deterministic computer model. They also gave many minimax and maximin distance designs, among which the simplest are as follows:

Example 10 (Johnson et al. (1990)) With L_1-distance the minimax distance design on [0,1] is

$$\frac{2i - 1}{2n}, i = 1, \cdots, n$$

*with the minimum distance $1/2n$, while the maximin distance design on $[0,1]$
is given by*

$$\frac{i}{n-1}, i = 1, \cdots, n$$

with the maximin distance $1/(n-1)$.

For a given design with n runs sort the distinct inter-site distances by
$d_1 < d_2 < \cdots < d_m$. Their associate indices are denoted by $(J_1, J_2, \cdots, J_m)$,
where J_i is the number of pairs of points in the design separated by distance d_i.
Morris and Mitchell (1995) extended the definition of the maximin distance
criterion to propose that a design D_n is called a maximin design if, among
the available designs, it

    (1a) maximizes d_1 among all designs meeting this criterion;
    (1b) minimizes J_1 among all designs meeting this criterion;
    (2a) maximizes d_2 among all designs meeting this criterion;
    (2b) minimizes J_2 among all designs meeting this criterion;

        ⋮

    (ma) maximizes d_m among all designs meeting this criterion;
    (mb) minimizes J_m.

Morris and Mitchell (1995) pointed out "Because statements (1a) and (1b)
alone specify the definition of Johnson et al. (1990) of a Mm design, our more
elaborate definition for Mm optimality essentially only breaks ties among multiple designs which would be Mm (and of minimum index) by their definition."
Here, Mm means maximin. Although this extended definition of Mm is intuitively appealing, practically it would be easier to use a scalar-value criterion.
This leads to the ϕ_p-**criterion**,

$$\phi_p(D_n) = \left[\sum_{i=1}^{m} J_i d_i^{-p}\right]^{1/p}, \tag{2.22}$$

where p is a positive integer, and J_i and d_i characterize the design D_n. In-
terestingly, Morris and Mitchell (1995) pointed out that for the case in which
$n = 4m + 1$ where m is a positive integer, the induced designs (cf. Definition
3) of orthogonal arrays such as Plackett-Burman designs (see Hedayat et al.
(1999)) are maximin designs for C^{4m} with respect either to L_1- (rectangular)
or L_2-distance. When the design domain is all U-type designs $U(n, n^s)$, the
maximin LHD is denoted by MmLH. Morris and Mitchell (1995) proposed
using a *simulated annealing* search algorithm to construct Mmlh's. Chapter
4 gives a detailed discussion on various powerful optimization algorithms.

2.4.4 Uniformity criterion

If you choose a uniform measure as the criterion and $U(n, q^s)$ as the design
space, the corresponding optimal designs are *uniform designs*. This concept

will be defined and discussed further in Chapter 3. As we will see, however, uniform design was motivated by a quite different philosophy from optimal LHD. Section 3.2 will introduce many existing measures of uniformity. Different measures of uniformity may lead to different uniform designs.

3

Uniform Experimental Design

This chapter gives a detailed introduction to uniform design, one kind of space-filling design. It involves various measures of uniformity, construction of symmetrical and asymmetrical uniform designs, and characteristics of the uniform design.

3.1 Introduction

The *uniform experimental design*, or *uniform design* for short, is a major kind of space-filling design that was proposed by Fang (1980) and Wang and Fang (1981) were involved in three large projects in system engineering in 1978. In those projects the output of the system had to be obtained by solving a system differential equation, but each run required a day of calculation since the computer speed was so slow at that time. Therefore, they needed to find a way of experiment so that as much information as possible could be found using relatively few experimental points. Unlike Latin hypercube sampling the uniform design is motivated by a *deterministic approach* to the overall mean model mentioned in Section 1.5. Let $D_n = \{\mathbf{x}_1, \cdots, \mathbf{x}_n\}$ be a set of n points in the experimental domain, which we assume is the s-dimensional unit cube C^s. The sample mean of y on D_n is denoted by $\bar{y}(D_n)$ and the overall mean of y is $E(y)$. With the *Koksma-Hlawka inequality*, an upper bound of the difference diff-mean $= |E(y) - \bar{y}(D_n)|$ is given by

$$\text{diff-mean} = |E(y) - \bar{y}(D_n)| \leq V(f)D(D_n), \tag{3.1}$$

where $D(D_n)$ is the *star discrepancy* of the design D_n, not depending on f, and $V(f)$ is the total variation of the function f in the sense of Hardy and Krause (see Niederreiter (1992)). The star discrepancy is a measure of uniformity and will be defined in the next section. The lower the star discrepancy, the better uniformity the set of points has. For instance, if all of the points were clustered at one corner of the sphere, so that uniformity would be violated and the sample mean would represent the population mean rather poorly, the star discrepancy would be very large. Obviously, we should find a design with the smallest star discrepancy, i.e., a *uniform design* (UD) with n runs and s input variables.

In fact, we may define more measures of uniformity that satisfy the Koksma-Hlawka inequality, and find the related uniform designs. In this chapter we shall introduce many useful measures of uniformity, such as the centered L_2-discrepancy, wrap-around L_2-discrepancy, and the categorical discrepancy.

For construction of a uniform design with n runs and s factors, it is an NP hard problem in the sense of computation complexity. Therefore, many construction methods, such as the good lattice point method, the Latin square method, the expending orthogonal array method, and the cutting method, have been proposed; these are introduced in Section 3.3. Section 3.4 describes characteristics of uniform design, such as admissibility, minimaxity, and robustness. Section 3.5 presents a new construction method for uniform designs via resolvable balanced incomplete block designs. The most attractive advantage of this method is that it does not require optimization computations for constructing UDs. These designs give the user more flexibility as in many experiments, the number of levels for the factors has to be small. In this case q is much smaller than the number of runs, n. The last section of this chapter introduces several methods for construction of asymmetrical uniform designs.

Optimization techniques have played an important role in constructing optimal Latin hypercube designs and uniform designs. They can be used for generating symmetrical and asymmetrical uniform designs. Chapter 4 will give a detailed introduction to various powerful optimization algorithms.

Several papers gave comprehensive reviews on uniform design. See, for example, Fang and Wang (1994), Fang and Hickernell (1995), Fang, Lin, Winker and Zhang (2000), Liang, Fang and Xu (2001), Fang (2002), and Fang and Lin (2003).

3.2   Measures of Uniformity

In this section we shall introduce various measures of uniformity. Let $D_n = \{\mathbf{x}_1, \cdots, \mathbf{x}_n\}$ be a set of n points in the s-dimensional unit cube C^s, where $\mathbf{x}_k = (x_{k1}, \cdots, x_{ks})$. Often D_n is expressed as an $n \times s$ matrix. A reasonable measure of uniformity should be *invariant under reordering the runs and relabeling the factors*. Two designs are called *equivalent* if one can be obtained from the other by relabeling the factors and/or reordering the runs.

3.2.1   The star L_p-discrepancy

Let $F(\mathbf{x})$ be the uniform distribution on C^s and $F_{D_n}(\mathbf{x})$ be the *empirical distribution* of the design D_n, i.e.,

$$F_{D_n}(\mathbf{x}) = \frac{1}{n} \sum_{k=1}^{n} I\{x_{k1} \leq x_1, \cdots, x_{ks} \leq x_s\}, \qquad (3.2)$$

where $\mathbf{x} = (x_1, \cdots, x_s)$ and $I\{A\} = 1$ if A occurs, or 0 otherwise. The *star L_p-discrepancy* is the L_p-norm of $||F_{D_n}(\mathbf{x}) - F(\mathbf{x})||_p$, where $p > 0$. The star L_p-discrepancy has been widely used in quasi-Monte Carlo methods (Hua and Wang (1981) and Niederreiter (1992)) with different statements of the definition. For each $\mathbf{x} = (x_1, \cdots, x_s)$ in C^s, let $[\mathbf{0}, \mathbf{x}) = [0, x_1) \times \cdots \times [0, x_s)$ be the rectangle determined by $\mathbf{0}$ and $\mathbf{x}$ and let $N(D_n, [\mathbf{0}, \mathbf{x}))$ be the number of points of D_n falling in $[\mathbf{0}, \mathbf{x})$. The ratio $\frac{N(D_n, [\mathbf{0}, \mathbf{x}))}{n}$ should be close to the volume of the rectangular $[\mathbf{0}, \mathbf{x})$, denoted by $\text{Vol}([\mathbf{0}, \mathbf{x}))$, if points in D_n are uniformly scattered on C^s (see Figure 3.1). The difference between the above ratio and volume

$$\left| \frac{N(D_n, [\mathbf{0}, \mathbf{x}))}{n} - Vol([\mathbf{0}, \mathbf{x})) \right|$$

is called the *discrepancy at point* $\mathbf{x}$. The L_p-norm average of all the discrepancies over C^s is just the star L_p-discrepancy and is given by

$$D_p(D_n) = \left\{ \int_{C^s} \left| \frac{N(D_n, [\mathbf{0}, \mathbf{x}))}{n} - \text{Vol}([\mathbf{0}, \mathbf{x})) \right|^p \right\}^{1/p}. \tag{3.3}$$

The following special cases have been used in the literature:

(i) **The star discrepancy:** Let $p \to \infty$ in (3.3) and we have

$$D(D_n) \equiv \lim_{p \to \infty} D_p(D_n) = \max_{\mathbf{x} \in C^s} \left| \frac{N(D_n, [\mathbf{0}, \mathbf{x}))}{n} - \text{Vol}([\mathbf{0}, \mathbf{x})) \right|$$

$$= \max_{\mathbf{x} \in C^s} |F_{D_n}(\mathbf{x}) - F(\mathbf{x})|. \tag{3.4}$$

The star discrepancy was first suggested by Weyl (1916) and is the *Kolmogorov-Smirnov statistic* in goodness-of-fit testing. The star discrepancy has played an important role in quasi-Monte Carlo methods and statistics, but it is not easy to compute. An algorithm for exact calculation of the star discrepancy in small dimensions is given by Bundschuh and Zhu (1993). Winker and Fang (1997) employed the threshold accepting method (cf. Section 4.2.4) to give an approximation of the star discrepancy.

(ii) **The star L_2-discrepancy:** Warnock (1972) gave an analytic formula for calculating the star L_2-discrepancy:

$$(D_2(D_n))^2 = \int_{C^s} \left| \frac{N(D_n, [\mathbf{0}, \mathbf{x}))}{n} - \text{Vol}([\mathbf{0}, \mathbf{x})) \right|^2$$

$$= 3^{-s} - \frac{2^{1-s}}{n} \sum_{k=1}^{n} \prod_{l=1}^{s} (1 - x_{kl}^2) + \frac{1}{n^2} \sum_{k=1}^{n} \sum_{j=1}^{n} \prod_{i=1}^{s} [1 - \max(x_{ki}, x_{ji})], \tag{3.5}$$

where $\mathbf{x}_k = (x_{k1}, \cdots, x_{ks})$. Obviously, the star L_2-discrepancy is much easier to calculate numerically than the star L_p-discrepancy. However, it has been shown that the star L_2-discrepancy ignores differences $|\frac{N(D_n, [\mathbf{0}, \mathbf{x}))}{n} - \text{Vol}([\mathbf{0}, \mathbf{x}))|$ on any low-dimensional subspace and is not suitable for design of computer experiments.

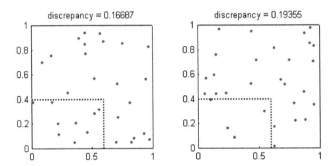

FIGURE 3.1
Discrepancy and data rotation

3.2.2   Modified L_2-discrepancy

Fang, Lin, Winker and Zhang (2000) found that both the star discrepancy and star L_2-discrepancy are unsuitable for searching UDs. The star discrepancy is not sensitive enough, while the star L_p-discrepancy ($p < \infty$) ignores differences $|\frac{N(D_n,[\mathbf{0},\mathbf{x}))}{n} - \text{Vol}([\mathbf{0},\mathbf{x}))|$ on any low-dimensional subspace. It is known that the lower dimensional structure of a design is more important, from the so-called *hierarchical ordering principle*: lower-order effects are more likely to be important than higher-order effects; main effects are more likely to be important than interactions; and effects of the same order are equally likely to be important (cf. Section 1.2). Moreover, the L_p-discrepancy is not invariant under rotating coordinates of the points as the origin $\mathbf{0}$ plays a special role. Observe the set of points in Figure 3.1(a) with the star discrepancy $D = 0.16687$. After rotation clockwise 90^0, the star discrepancy of the set in Figure 3.1(b) becomes $D = 0.19355$. Therefore, we have to put more requirements on the measure of uniformity. A measure of uniformity is reasonable if it satisfies the following requirements:

[C1] It is invariant under permuting factors and/or runs.

[C2] It is invariant under rotation of the coordinates.

[C3] It can measure not only uniformity of $\mathcal{P}$ over C^s, but also the projection uniformity of $\mathcal{P}$ over C^u, where u is a non-empty subset of $\{1, \cdots, s\}$.

[C4] There is some reasonable geometric meaning.

[C5] It is easy to compute.
[C6] It satisfies the Koksma-Hlawka-like inequality.
[C7] It is consistent with other criteria in experimental design.

Therefore, several modified L_p-discrepancies were proposed by Hickernell (1998a), among which the **centered L_2-discrepancy** (CD) has good properties. The CD and some other L_2-discrepancies are defined by

$$(D_{modified}(D_n))^2 = \sum_{u \neq \emptyset} \int_{C^u} \left| \frac{N(D_{n_u}, J_{\mathbf{x}_u})}{n} - \text{Vol}(J_{\mathbf{x}_u}) \right|^2 du, \qquad (3.6)$$

where u is a non-empty subset of the set of coordinate indices $S = \{1, \cdots, s\}$, $|u|$ denotes the cardinality of u, C^u is the $|u|$-dimensional unit cube involving the coordinates in u, D_{n_u} is the projection of D_n on C^u, $J_{\mathbf{x}}$ is a rectangle uniquely determined by $\mathbf{x}$ and is chosen with the geometric consideration for satisfying [C2], and $J_{\mathbf{x}_u}$ is the projection of $J_{\mathbf{x}}$ on C^u. Obviously, the new discrepancy measure considers discrepancy between the empirical distribution and the volume of $J_{\mathbf{x}_u}$ in any low dimensional rectangles $J_{\mathbf{x}_u}$.

3.2.3 The centered discrepancy

The centered L_2-discrepancy is considered due to the appealing property that it becomes invariant under reordering the runs, relabeling factors, or reflecting the points about any plane passing through the center of the unit cube and parallel to its faces. The latter is equivalent to the invariance under coordinate rotations. There is a summation for $u \neq \emptyset$ in (3.6); thus, the CD considers the uniformity not only of D_n over C^s, but also of all the projection uniformity of D_n over C^u. The choice of J_u ensures invariance of rotating coordinates. Figure 3.2 gives an illustration of $J_{\mathbf{x}}$ for two-dimensional case. The unit cube is split into four square cells (in general, 2^s cells). Corner points of each cell involve one corner point of the original unit cube, the central point $(\frac{1}{2}, \frac{1}{2})$ and others. Consider a point $\mathbf{x} \in C^s$ and its corresponding rectangle $J_{\mathbf{x}}$ depends on the cell in which $\mathbf{x}$ falls. Four cases in two-dimensional case are demonstrated in Figure 3.2.

Furthermore, CD has a formula for computation as follows:

$$(CD(D_n))^2 = \left(\frac{13}{12} \right)^s - \frac{2}{n} \sum_{k=1}^{n} \prod_{j=1}^{s} \left[1 + \frac{1}{2}|x_{kj} - 0.5| - \frac{1}{2}|x_{kj} - 0.5|^2 \right]$$

$$+ \frac{1}{n^2} \sum_{k=1}^{n} \sum_{j=1}^{n} \prod_{i=1}^{s} \left[1 + \frac{1}{2}|x_{ki} - 0.5| + \frac{1}{2}|x_{ji} - 0.5| - \frac{1}{2}|x_{ki} - x_{ji}| \right]. \qquad (3.7)$$

For comparing designs, the lower the CD-value is, the more uniform, and in that sense more desirable, the design is. The CD-value of the two designs D_{6-1} and D_{6-2} in Section 2.3 are 0.0081 and 0.0105, respectively. Obviously, then, the design D_{6-1} is better.

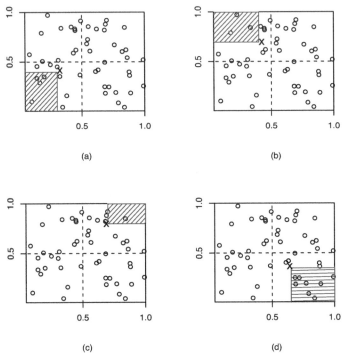

FIGURE 3.2
Illustration for centered discrepancy.

3.2.4 The wrap-around discrepancy

Both the star L_p-discrepancy and the centered L_2-discrepancy require the corner points of the rectangle $J_{\mathbf{x}}$ to involve one corner point of the unit cube. A natural idea allows that the rectangle J falls inside of the unit cube and does not involve any corner point of the unit cube (see Figure 3.3(a)). This consideration leads to the so-called *unanchored discrepancy* in the literature. In this case each rectangle is determined by two points $\mathbf{x}_1$ and $\mathbf{x}_2$ in C^s, denoted by $J\mathbf{x}_1, \mathbf{x}_2$. If we want the defined discrepancy to have the property [C2], we should allow wrapping the unit cube for each coordinate. As a consequence the rectangle $J\mathbf{x}_1, \mathbf{x}_2$ can be split into two parts. Figure 3.3 gives an illustration. The corresponding discrepancy is called the *wrap-around* discrepancy. The wrap-around L_2-discrepancy (WD), proposed also by Hickernell (1998*b*), is another modified L_2-discrepancy and has nice properties. Its analytical formula is given by

$$(WD(\mathcal{P}))^2 = \left(\frac{4}{3}\right)^s + \frac{1}{n^2} \sum_{k,j=1}^{n} \prod_{i=1}^{s} \left[\frac{3}{2} - |x_{ki} - x_{ji}|(1 - |x_{ki} - x_{ji}|)\right]. (3.8)$$

The WD-values of designs D_{6-1} and D_{6-2} are 4.8137 and 4.8421, respectively. Again, the design D_{6-1} is better in the sense of lower WD-value.

Hickernell (1998a) proved that the centered L_2-discrepancy and wrap-around L_2-discrepancy satisfy requirements [C1]–[C6]. Section 3.2.7 will show that these two discrepancies also have close relationships with many criteria, such as resolution and minimum aberration, in fractional factorial designs.

From now on, M denotes a measure of uniformity defined on C^s and $\mathcal{M}$ denotes the space of all signed measures M. The measure M can be the star discrepancy, L_2-discrepancy, CD, WD, or other measure of uniformity discussed in the literature. For a U-type design U, we always define $M(U) = M(D_U)$, where D_U is the induced design of U (cf. Definition 3 in Section 2.1).

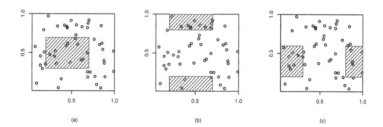

FIGURE 3.3
Illustration for centered discrepancy with wrap.

3.2.5 A unified definition of discrepancy

All the discrepancies mentioned so far can be defined by the concept of the *reproducing kernel* of a Hilbert space. This approach can be easily used to develop new theory and to propose new discrepancies. The categorical discrepancy discussed in Section 3.2.6 is defined in terms of the reproducing kernel of a Hilbert space. It may be difficult to understand for non-mathematicians/statisticians. It does not cause a serious problem if the reader decides to skip this section. The reader who skips this section only needs to be aware of formula (3.13).

Let $D_n = \{\mathbf{x}_1, \cdots, \mathbf{x}_n\}$ be a design in the canonical experimental domain C^s and $F_{D_n}(\mathbf{x})$ be its empirical distribution (cf. (3.2)). Let $F(\mathbf{x})$ be the uniform distribution on C^s. A norm of $||F_{D_n}(\mathbf{x}) - F(\mathbf{x})||_M$ under the measure M measures difference between the empirical distribution and the uniform distribution. The discrepancies mentioned in this section can be defined by a norm $||F_{D_n}(\mathbf{x}) - F(\mathbf{x})||_M$ with a different definition. When the L_∞-norm is chosen, we have the star discrepancy; when the L_p-norm is used, it leads to the L_p-discrepancy. In fact, the centered L_2- and wrap-around L_2-discrepancy can be expressed as a norm of the difference $F_{D_n}(\mathbf{x}) - F(\mathbf{x})$, but we need some tools in functional analysis. The concept of a reproducing kernel (cf. Saitoh (1988) and Wahba (1990)) is a very useful tool for this purpose.

Let T be the experimental domain in R^s and $K(\mathbf{x}, \mathbf{w})$ be a symmetric and positive definite function on $T \times T$, i.e.,

$$\begin{cases} K(\mathbf{x}, \mathbf{w}) = K(\mathbf{w}, \mathbf{x}), & \text{for any } \mathbf{x}, \mathbf{w} \in T, \\ \sum_{i,j=1}^{n} a_i a_j K(\mathbf{x}_i, \mathbf{x}_j) > 0, & \text{for any } a_i \in R, \mathbf{x}_i \in T, i = 1, \cdots, n, \end{cases} \tag{3.9}$$

and a_i cannot all equal zero. Let $\mathcal{W}$ be a Hilbert space of real-valued functions on T. An inner product $[\cdot, \cdot]_K$ is defined with the kernel K by

$$[G, H]_K = \int_{T \times T} K(\mathbf{x}, \mathbf{w}) dG(\mathbf{x}) dH(\mathbf{w}), \text{ for any } G, H \in \mathcal{M},$$

where $\mathcal{M}$ is the space of all signed measures G on T such that

$$\int_{T \times T} K(\mathbf{x}, \mathbf{w}) dG(\mathbf{x}) dG(\mathbf{w}) < \infty.$$

If the kernel K satisfies

$$f(\mathbf{x}) = [K(., \mathbf{x}), f]_K, \text{ for all } f \in \mathcal{W}, \text{ and } \mathbf{x} \in T,$$

then the kernel K is called a *reproducing kernel*. A norm of G under the kernel K is defined by

$$||G||_K = \int_{T \times T} K(\mathbf{x}, \mathbf{w}) dG(\mathbf{x}) dG(\mathbf{w}).$$

When T is the unit cube C^s, F is the uniform distribution on C^s, and G is the empirical distribution of D_n, the L_2-discrepancy with respect to a given K is defined by

$$\begin{aligned} D^2(D_n, K) &= ||F - F_{D_n}||_K^2 \\ &= \int_{C^s \times C^s} K(\mathbf{x}, \mathbf{w}) d(F - F_{D_n})(\mathbf{x}) d(F - F_{D_n})(\mathbf{w}) \\ &= \int_{C^s \times C^s} K(\mathbf{x}, \mathbf{w}) dF(\mathbf{x}) dF(\mathbf{w}) \\ &\quad - \frac{2}{n} \sum_{i=1}^{n} K(\mathbf{x}, \mathbf{x}_i) dF(\mathbf{x}) + \frac{1}{n^2} \sum_{i,j=1}^{n} K(\mathbf{x}_i, \mathbf{x}_j). \end{aligned} \tag{3.10}$$

Different kernels imply different measures of uniformity. For example, the CD and WD have the respective kernels

$$K_c(\mathbf{x}, \mathbf{w}) = \prod_{i=1}^{s} \left[1 + \frac{1}{2}|x_i - 0.5| + \frac{1}{2}|w_i - 0.5| - \frac{1}{2}|x_i - w_i| \right],$$

and

$$K_w(\mathbf{x}, \mathbf{w}) = \prod_{i=1}^{s} \left[\frac{2}{3} - |x_i - w_i| + |x_i - w_i|^2 \right],$$

where $\mathbf{x} = (x_1, \cdots, x_s)$ and $\mathbf{w} = (w_1, \cdots, w_s)$. By this unified approach Hickernell (1998a,b) proposed many new discrepancies, including the CD and WD we have mentioned. All the discrepancies defined in this way satisfy the Koksma-Hlawka type inequality (cf. (3.1)), i.e., the definition of the total variation of the function f depends on the measure M. The next section discusses another discrepancy.

3.2.6 Descrepancy for categorical factors

The discrepancy measures mentioned so far have assumed that the factors were continuous. However, discrepancy measures are also available for categorical factors. Here, the possible design points can be considered as points in a discrete lattice rather than in a continuous space. Lattice points have been widely considered in the construction of LHDs and UDs. In this case we assume that the experimental domain is all possible combinations of factor levels. More precisely, let $T = \{1, \ldots, q_1\} \times \cdots \times \{1, \ldots, q_s\}$ comprising all possible level combinations of the s factors and let F be the discrete uniform distribution on T. Choosing

$$K_j(x, w) = \begin{cases} a & \text{if } x = w, \\ b & \text{if } x \neq w, \end{cases} \text{ for } x, w \in \{1, \ldots, q_j\}, a > b > 0, \qquad (3.11)$$

and

$$K_d(\mathbf{x}, \mathbf{w}) = \prod_{j=1}^{s} K_j(x_j, w_j), \text{ for any } \mathbf{x}, \mathbf{w} \in T, \qquad (3.12)$$

then $K_d(\mathbf{x}, \mathbf{w})$ is a kernel function and satisfies conditions in (3.9). The corresponding discrepancy, called *categorical discrepancy* or *discrete discrepancy*, was suggested by Hickernell and Liu (2002). Fang, Lin and Liu (2003) gave a computational formula for categorical discrepancy and the lower bound of categorical discrepancy for any U-type design as follows:

Theorem 2 *For a U-type design, $D_n \in U(n; q_1 \times \ldots \times q_s)$, we have*

$$D^2(D_n, K_d) = \frac{a^s}{n} + \frac{b^s}{n^2} \sum_{k,l=1, l \neq k}^{n} \left(\frac{a}{b}\right)^{\lambda_{kl}} - \prod_{j=1}^{s} \left[\frac{a + (q_j - 1)b}{q_j}\right], \quad (3.13)$$

$$\geq \frac{a^s}{n} + \frac{n-1}{n} b^s \left(\frac{a}{b}\right)^{\lambda} - \prod_{j=1}^{s} \left[\frac{a + (q_j - 1)b}{q_j}\right], \quad (3.14)$$

where λ_{kl} is the number of coincidences between the the kth and lth columns of D_n, and the lower bound on the right-hand side of (3.14) can be achieved if and only if $\lambda = (\sum_{j=1}^{s} n/q_j - s)/(n-1)$ is a positive integer and all the $\lambda_{kl}s$ for $k \neq l$ are equal to λ. For the symmetrical designs $q_1 = \cdots = q_s = q$, we have $\lambda = s(n-q)/q(n-1)$ and the above lower bound becomes

$$\frac{a^s}{n} + \frac{n-1}{n} b^s \left(\frac{a}{b}\right)^{\lambda} - \left[\frac{a + (q-1)b}{q}\right]^s.$$

The categorical discrepancy and its lower bound in Theorem 2 play an important role in the connection between combinatorial designs and uniform designs. We will introduce this connection and its application to the construction of uniform designs in Sections 3.5 and 3.6. The categorical discrepancy is essentially useful for two- or three-level designs. A comprehensive discussion on the categorical discrepancy and relationships between the categorical discrepancy and centered/wrap-around discrepancy is given by Qin and Fang (2004).

Remark: The lower bound in (3.14) can be improved into

$$D^2(D_n, K_d) \geq \frac{a^s}{n} + \frac{n-1}{n} b^s \left[(\gamma + 1 - \lambda)\left(\frac{a}{b}\right)^{\gamma} + (\lambda - \gamma)\left(\frac{a}{b}\right)^{\gamma+1}\right]$$

$$- \prod_{j=1}^{s} \left[\frac{a + (q_j - 1)b}{q_j}\right], \quad (3.15)$$

where $\gamma = [\lambda]$ is the integer part of λ. The lower bound on the right-hand side of (3.15) can be achieved if and only if all $\lambda_{ij}s$ take the same value γ, or take only two values γ and $\gamma + 1$.

3.2.7 Applications of uniformity in experimental designs

Section 3.2.2 raises seven requirements for a good measure of uniformity. The requirement [C7] specifies that the measure of uniformity should be consistent with other criteria for experimental design. The existing criteria in factorial designs are related to statistical inference, while the measure of uniformity focuses on geometrical consideration of the experimental points. It appears at first that they are unrelated. However, in the past decade many authors found links between the measures of uniformity and some criteria in factorial designs and supersaturated designs. As a result, many applications of the concept of uniformity to factorial designs have been explored. This section gives a brief review. For the details the reader can refer to Fang (2001) and Fang and Lin (2003).

A. Uniformity and isomorphism: Two factorial designs with n runs, s factors each having q levels are called *isomorphic* if one can be obtained from the other by relabeling the factors, reordering the runs, or switching the levels of one or more factors. For identifying two such designs a complete search must be done to compare $n!(q!)^s s!$ designs and it is an NP hard problem. However, it is easy to check whether two isomorphic factorial designs have the same discrepancy, for example, the same CD-value. If two designs have different CD-values, they are not isomorphic. For a given $m < s$ and two designs d_1 and d_2 there are $\binom{s}{m}$ subdesigns whose CD-values form a distribution, denoted by $CD_{n,m,q}(d_i), i = 1, 2$. If d_1 and d_2 are isomorphic, we have $CD_{n,m,q}(d_1) = CD_{n,m,q}(d_2)$. Using the above facts Ma, Fang and Lin (2001) proposed a powerful algorithm based on uniformity that can easily detect non-isomorphic designs.

Hadamard matrices (see Section A.1) have played an important role in experimental design and coding theory. Two Hadamard matrices are called equivalent if one can be obtained from the other by some sequence of row and column permutations and negations. Fang and Ge (2004) proposed an algorithm based on uniformity for detecting inequivalent Hadamard matrices, and they discovered, for example, that there are at least 382 pairwise inequivalent Hadamard matrices of order 36.

B. Uniformity and orthogonality: Fang and Winker (1998) and Fang, Lin, Winker and Zhang (2000) found that many existing orthogonal designs (see Section 1.2 for the definition) are of uniform design under the centered L_2-discrepancy. They conjectured that any orthogonal design is a uniform design under a certain discrepancy. Later, Ma, Fang and Lin (2002) proved that the conjecture is true in many cases, but it is not always true. Tang (2005) provided some further discussion. Fang, Ma and Mukerjee (2001) also describe some relationships between orthogonality and uniformity.

C. Uniformity and aberration: The word length pattern for a fractional factorial design is one of the most important criteria for describing the estimation ability for the main effects and interactions of the factors. Based on the word length pattern two criteria, *resolution* and *minimum aberration*, have been widely used in the literature as well as in applications. Fang and Mukerjee (2000) found an analytic connection between two apparently unrelated areas, the uniformity and the aberration, in regular fractions of two-level factorials. This connection has opened a new research direction and inspired much further work. For further discussions the reader can refer to Ma and Fang (2001), Fang, Ma and Mukerjee (2001), and Fang and Ma (2002). Analogously to the word length pattern, we can define the so-called uniformity pattern. Fang and Qin (2004a) gave such a version for the two-level designs and Hickernell and Liu (2002) proposed an alternative version.

D. Uniformity and supersaturated design: Supersaturated designs introduced in Section 1.2. Many authors, including Lin (1993), Lin (1995), Lin (1999), Wu (1993), Li and Wu (1997), Cheng (1997), and Yamada and Lin (1999), proposed criteria for comparing and constructing supersaturated designs. Recently, uniformity has been recognized as a useful criterion and many ways have been suggested for constructing multi-level supersaturated designs under uniformity criteria. For detailed discussions the reader can refer to Fang, Lin and Ma (2000), Liu and Hickernell (2002), Fang, Lin and Liu (2003), and Fang, Ge, Liu and Qin (2004).

E. Majorization framework: Note that fractional designs, supersaturated designs, and uniform designs are all based on U-type designs. Zhang, Fang, Li and Sudjianto (2005) proposed a general theory for all of these designs. The theory is based on the so-called *majorization theory* that has been widely used in various theoretic approaches. They present a general majorization framework for assessing designs, which includes a stringent criterion of majorization via pairwise coincidences and flexible surrogates via convex functions. Classical orthogonality, aberration, and uniformity criteria are unified by choosing combinatorial and exponential kernels. This approach links the three kinds of designs, and would help a user who is already familiar with factorial and/or supersaturated designs feel more comfortable using uniform design.

3.3 Construction of Uniform Designs

Consider the canonical experimental domain C^s. Let $\mathcal{D}$ be the design space. In this chapter the design space is a subset of U-type designs, more precisely, a subset of $U(n, q^s)$. A design $D_n \in \mathcal{D}$ is called a *uniform design* under the measure of uniformity M, if it minimizes the M-value of the design over $\mathcal{D}$. In this section we choose the CD as the measure of uniformity for illustration. A uniform design of n runs and s factors each having q levels is denoted by $U_n(q^s)$. Section 3.3.1 gives $U_n(n^1)$ for the one-factor case, Sections 3.3.2–3.3.6 introduce several methods for constructing $U_n(n^s)$, and the construction of $U_n(q^s)$ will be discussed in Section 3.5. Uniform designs with mixed levels will be considered in Section 3.6.

3.3.1 One-factor uniform designs

Fang, Ma and Winker (2000) pointed out that the set of equidistant points is the unique UD on [0,1] for one-factor experiments.

Theorem 3 *For a one-factor experiment, the unique UD on [0,1] is*

$$\left\{\frac{1}{2n}, \frac{3}{2n}, \cdots, \frac{2n-1}{2n}\right\} \tag{3.16}$$

with $CD_2^2 = \frac{1}{12n^2}$.

In fact, the assertion of the above theorem holds if the measure of uniformity is the star discrepancy (see Example 1.1 of Fang and Wang (1994)), or many others. From Theorem 3 the lattice points in C^s are suggested, where $T = \left\{\frac{1}{2n}, \frac{3}{2n}, \cdots, \frac{2n-1}{2n}\right\}^s$, that is, $\mathcal{L}_{n,s}$ in the next section.

3.3.2 Symmetrical uniform designs

The following facts are noted in construction of multi-factor UDs:

- The UD is not unique when $s > 1$. We do not distinguish between equivalent designs (cf. Section 3.2).
- Finding a uniform design is a complex task and can be computationally intractable, even for moderate values of n and s. Fang, Ma and Winker (2000) give a detailed discussion on this point.

For the reduction of the computer computation complexity, we can narrow down the design space. Fang and Hickernell (1995) suggested using the U-type design set $U(n, n^s)$ as the design space. From Theorem 3 in this section, Fang, Ma and Winker (2000) gave some justification of Fang-Hickernell's suggestion. Now the canonical experimental domain becomes

$$\mathcal{L}_{n,s} = \left\{\frac{1}{2n}, \frac{3}{2n}, \cdots, \frac{2n-1}{2n}\right\}^s.$$

Here, $\mathcal{L}_{n,s}$ is the lattice point set where each marginal design is the uniform design (3.16). A design of n runs each belonging to $\mathcal{L}_{n,s}$ is called a uniform design if it has the smallest CD-value among all such designs.

In fact, the design space $\mathcal{L}_{n,s}$ is just the class of U-type designs $U(n, n^s)$. For any symmetrical design $U \in U(n, n^s)$, its induced design D_U is a U-type design with entries $\{(2i-1)/2n, i = 1, \cdots, n\}$. We define $CD(U) = CD(D_U)$ (cf. Section 3.2.4). Therefore, we want to find a design $D_n \in U(n, n^s)$ such that it has the smallest CD-value on $U(n, n^s)$.

From the above discussion the design space can be changed as follows:

Set of designs with n points on C^s	$\Rightarrow$	Set of designs with n runs on $\mathcal{L}_{n,s}$	$\Rightarrow$	Set of $U(n, n^s)$ with entries $\frac{2k-1}{2n}$ $k = 1, \cdots, n$	$\Leftrightarrow$	Set of $U(n, n^s)$ with entries $1, \cdots, n$

The first step reduces the design space $\mathcal{D}$ into a lattice point set, where each experimental point belongs to $\mathcal{L}_{n,s}$ (there are n^n points to form the experimental domain), the second step further reduces the design space into $U(n, n^s)$ with entries in (3.16), and the third step changes the design space into $U(n, n^s)$ with entries $1, \cdots, n$. More generally, the design space can be chosen as $U(n, q^s)$, where q is a divisor of n. In this case the marginal experimental points for each factor are $\frac{1}{2q}, \frac{3}{2q}, \cdots, \frac{2q-1}{2q}$. The design space $U(n, n^s)$ is its special case of $q = n$. For the rest of the book, the uniform design is understood as in the following definition.

Definition 6 A design in $U(n, q^s)$ is called a uniform design if it has the smallest CD-value on the set of $U(n, q^s)$. We write it as $U_n(q^s)$.

The remaining of the section focuses on construction of UDs with $q = n$. However, searching for a uniform design for moderate (n, s) is still computationally costly. To further reducing the computational complexity for searching UDs based on U-type designs, there are several methods such as the *good lattice method*, the *Latin square method*, the *expending orthogonal array method*, the *cutting method*, and the *optimization method* that can provide a good approximation to the uniform design. A good approximation to uniform design is also called a *nearly uniform design* (Fang and Hickernell (1995)). In the literature a good nearly uniform design is just simply called a uniform design.

3.3.3   Good lattice point method

The good lattice point (*glp*) method is an efficient quasi-Monte Carlo method, proposed by Korobov (1959*a,b*), and discussed by many authors such as Hua and Wang (1981), Shaw (1988), Joe and Sloan (1994), and Fang and Wang (1994).

Algorithm GLP

Step 1. Find the candidate set of positive integers for given n

$$\mathcal{H}_n = \{h : h < n, \text{the great common divisor of } n \text{ and } h \text{ is one}\}.$$

Step 2. For any s distinct elements of $\mathcal{H}_n$, $h_1, h_2, \cdots, h_s$, generate an $n \times s$ matrix $U = (u_{ij})$ where $u_{ij} = ih_j \pmod{n}$ and the multiplication operation modulo n is modified as $1 \leq u_{ij} \leq n$. Denote U by $U(n, \mathbf{h})$, where $\mathbf{h} = (h_1, \cdots, h_s)$ is called the *generating vector* of the U. Denote by $GLP_{n,s}$ the set of all such matrices $U(n, \mathbf{h})$.

Step 3. Find a design $\mathbf{U}(n, \mathbf{h}^*) \in GLP_{n,s}$ that minimizes the CD-value over the set $GLP_{n,s}$ with respect to the generating vector $\mathbf{h}$. The design $\mathbf{U}(n, \mathbf{h}^*)$ is a (nearly) uniform design $U_n(n^s)$.

Example 11 For $n = 21$ and $s = 2$, we have

$$\mathcal{H}_{21} = \{1, 2, 4, 5, 7, 8, 10, 11, 13, 15, 16, 17, 19, 20\}.$$

There are 12 integers in $\mathcal{H}_{21}$; as a consequence there are $66 = \binom{12}{2}$ candidate designs in $GLP_{21,2}$. If the CD is chosen as the measure of uniformity, the design matrix $\mathbf{U}(21, \mathbf{h}^*)$ with $\mathbf{h}^* = (1, 13)$ has the smallest D-value over $GLP_{21,2}$ and is a (nearly) uniform design $U_{21}(21^2)$ given as follows:

$$\begin{pmatrix} 1 & 2 & 3 & 4 & 5 & 6 & 7 & 8 & 9 & 10 & 11 & 12 & 13 & 14 & 15 & 16 & 17 & 18 & 19 & 20 & 21 \\ 13 & 5 & 18 & 10 & 2 & 15 & 7 & 20 & 12 & 4 & 17 & 9 & 1 & 14 & 6 & 19 & 11 & 3 & 16 & 8 & 21 \end{pmatrix}'.$$

The generating vector (1,3) is called the best generating vector.

A. The cardinality of $\mathcal{H}_n$: Obviously, the computation complexity of using the *glp* method depends on the cardinality of $\mathcal{H}_n$. From number theory (Hua and Wang (1981)) the cardinality of $\mathcal{H}_n$ is just the *Euler function* $\phi(n)$. Let $n = p_1^{r_1} \cdots p_t^{r_t}$ be the prime decomposition of n, where $p_1, \cdots, p_t$ are distinct primes and $r_1, \cdots, r_t$ are positive integers. Then

$$\phi(n) = n(1 - \frac{1}{p_1}) \cdots (1 - \frac{1}{p_t}).$$

For example, $\phi(21) = 21(1 - \frac{1}{3})(1 - \frac{1}{7}) = 12$ as $21 = 3 \cdot 7$ is the prime decomposition of 21. Further, $\phi(n) = n - 1$ if n is a prime, and $\phi(n) < n/2$ if n is even. The number of columns of U-type designs generated by the *glp* method is limited to $\phi(n)/2 + 1$ or $\phi(n+1)/2 + 1$ (p.208 of Fang and Wang (1994)). The maximum number of factors is less than $n/2 + 1$ if the UD is generated by the *glp* method.

B. The power generator: The *glp* method with the best generating vector can be applied only for the moderate $(\phi(n), s)$. For example, for the case of $n = 31$ and $s = 14$ there are $\binom{30}{14} \approx 3 \times 10^{21}$ generating vectors to compare as $\phi(31) = 30$. It is impossible to implement such comparisons to find the best generating vector. Therefore, the power generator is recommended. For given (n, s), let

$$\mathcal{P}_{n,s} = \{k : k, k^2, \cdots, k^{s-1}(\text{mod } n) \text{ are distinct}, 1 \le k \le n - 1\}.$$

The generating vector domain

$$\mathcal{G}_{n,s}^{power} = \{\mathbf{h}_k = (1, k, k^2, \cdots, k^{s-1})(\text{mod } n), k \in \mathcal{P}_{n,s}\}$$

is suggested. A vector in $\mathcal{G}_{n,s}^{power}$ is called a *power generator*. The set of designs in $GLP_{n,s}$ with a power generator is denoted by $GLP_{n,s}^{power}$. A design $\mathbf{U}(n, \mathbf{h}^*) \in GLP_{n,s}^{power}$ minimizing the CD-value over the set $GLP_{n,s}^{power}$ is a (nearly) uniform design $U_n(n^s)$.

It has been shown that uniform designs generated by the *glp* method with a power generator have a lower discrepancy (Hua and Wang (1981)). It is conjectured that the best possible order of the star discrepancy of D_n is $n^{-1}(\log(n))^{s-1}$, i.e,

$$D(D_n) \ge c(s)\frac{(\log n)^{s-1}}{n}, \quad \text{for any } D_n \text{ in } \mathcal{D},$$

where the constant $c(s)$ depends on s only. The conjecture is true for $s = 2$, and is an open problem for high dimensional cases. The best order of the star discrepancy of D_n generated by the *glp* method is

$$O(n^{-1}(\log n)^{s-1}),$$

and the best order of the star discrepancy of D_n generated by the *glp* method with a power generator is

$$O(n^{-1}(\log n)^{s-1}\log(\log n)),$$

which is slightly worse than the previous one. That is why the *glp* method has been appreciated by many authors.

Most of UDs used in the past are generated by the *glp* method with a power generating vector. Table 3.1 gives the generating vector and CD-value of UD's for $4 \le n \le 31$ and $s = 2, 3, 4, 5$ and was computed by Dr. C.X. Ma.

C. Modification of the *glp* method: When the cardinality of $\mathcal{H}_n$ is too small, the nearly uniform design obtained by the *glp* method may be far from the uniform design; see Example 12 below. In this case some modifications of the *glp* method are necessary.

One modification was suggested by Wang and Fang (1981) and Fang and Li (1994). Instead of $\mathcal{H}_n$, we work on $\mathcal{H}_{n+1}$ in Steps 1 and 2 of Algorithm GLP and delete the last row of $\mathbf{U}(n + 1, h_1, \cdots, h_s)$ to form an $n \times s$ matrix, say U. All such matrices U form a set, denoted by $GLP_{n,s}^*$. In Step 3, $GLP_{n,s}$ is replaced by $GLP_{n,s}^*$. Clearly, the power generator can be applied to this modification. Table 3.2, which is also calculated by Dr. C.X. Ma, gives the generating vector and CD-value of UDs by the use of this modified *glp* method.

Example 12 Consider constructing a UD with 6 runs and 2 factors. The number of elements in $\mathcal{H}_6 = \{1, 5\}$ is 2. The nearly uniform design $U_6(6^2)$ is not uniform as can be seen from its plot (see the right plot in Figure 2.5). Besides, it is impossible to obtain $U_6(6^s)$ for $s > 2$ based on $\mathcal{H}_6 = \{1, 5\}$. Note $\mathcal{H}_7 = \{1, 2, 3, 4, 5, 6\}$. All nearly uniform designs $U_6(6^s), s \le 6$ can be generated based on $\mathcal{H}_7$. The left plot in Figure 2.5 gives plots for the above designs.

When both n and s are large, Wang and Fang (2005) proposed a novel approach to the construction of uniform designs by the *glp* method, where the generating vector is found by a forward procedure. They use the wrap-around discrepancy as the measure of uniformity, which can be computed rapidly by a simple formula if the design is from the *glp* method. The cost of searching the generating vector with n points for all dimensions up to s requires approximately $O(n\phi(n)s^2)$ operations where $\phi(n)$ is the Euler function of n. Theoretical results show that the wrap-around discrepancy of such uniform design has the optimal order of convergence. The theoretical findings are supported by empirical evidence. Empirical studies also show that the forward

TABLE 3.1

Generating Vector and CD-Value of UDs by the *glp* Method

n	$s=2$	$s=3$	$s=4$	$s=5$
4	0.1350 (1 3)			
5	0.1125 (1 2)	0.1762 (1 2 3)	0.2490 (1 2 3 4)	
6	0.1023 (1 5)			
7	0.0812 (1 3)	0.1336 (1 2 3)	0.1993 (1 2 3 5)	0.2729 (1 2 3 4 5)
8	0.0713 (1 5)	0.1315 (1 3 5)	0.1904 (1 3 5 7)	
9	0.0650 (1 4)	0.1044 (1 4 7)	0.1751 (1 2 4 7)	0.2435 (1 2 4 5 7)
10	0.0614 (1 3)	0.1197 (1 3 7)	0.1749 (1 3 7 9)	
11	0.0528 (1 7)	0.0879 (1 5 7)	0.1364 (1 2 5 7)	0.1919 (1 2 3 5 7)
12	0.0506 (1 5)	0.1112 (1 5 7)	0.1628 (1 5 7 11)	
13	0.0472 (1 5)	0.0796 (1 4 6)	0.1191 (1 4 5 11)	0.1653 (1 3 4 5 11)
14	0.0432 (1 9)	0.0710 (1 9 11)	0.1418 (1 3 5 13)	0.2021 (1 3 5 9 11)
15	0.0400 (1 11)	0.0812 (1 4 7)	0.1186 (1 4 7 13)	0.1904 (1 2 4 7 13)
16	0.0381 (1 7)	0.0676 (1 5 9)	0.0997 (1 5 9 13)	0.1666 (1 3 5 9 13)
17	0.0362 (1 10)	0.0626 (1 4 10)	0.0958 (1 4 5 14)	0.1384 (1 4 10 14 15)
18	0.0341 (1 7)	0.0580 (1 7 13)	0.1247 (1 5 7 13)	0.1784 (1 5 7 11 13)
19	0.0321 (1 8)	0.0560 (1 6 8)	0.0857 (1 6 8 14)	0.1228 (1 6 8 14 15)
20	0.0313 (1 9)	0.0563 (1 9 13)	0.0835 (1 9 13 17)	0.1580 (1 3 7 11 19)
21	0.0292 (1 13)	0.0570 (1 4 5)	0.0917 (1 5 8 19)	0.1310 (1 4 10 13 16)
22	0.0284 (1 13)	0.0529 (1 5 13)	0.0808 (1 5 7 13)	0.1157 (1 3 5 7 13)
23	0.0272 (1 9)	0.0480 (1 7 18)	0.0745 (1 7 18 20)	0.1088 (1 4 7 17 18)
24	0.0258 (1 17)	0.0455 (1 11 17)	0.0679 (1 11 17 19)	0.1422 (1 5 7 13 23)
25	0.0248 (1 11)	0.0437 (1 6 16)	0.0672 (1 6 11 16)	0.0946 (1 6 11 16 21)
26	0.0238 (1 11)	0.0456 (1 11 17)	0.0700 (1 5 11 17)	0.1049 (1 3 5 11 17)
27	0.0240 (1 16)	0.0420 (1 8 10)	0.0670 (1 8 20 22)	0.1013 (1 8 20 22 23)
28	0.0234 (1 11)	0.0446 (1 9 11)	0.0703 (1 9 11 15)	0.0993 (1 9 11 15 23)
29	0.0214 (1 18)	0.0381 (1 9 17)	0.0606 (1 8 17 18)	0.0901 (1 7 16 20 24)
30	0.0206 (1 19)	0.0369 (1 17 19)	0.0559 (1 17 19 23)	0.1301 (1 7 11 13 29)
31	0.0206 (1 22)	0.0357 (1 18 24)	0.0583 (1 6 14 22)	0.0849 (1 6 13 20 27)

TABLE 3.2
The Generating Vector and CD-Value of UDs by the Modified *glp* Method

n	$s = 2$	$s = 3$	$s = 4$	$s = 5$
4	0.1275 (1 2)	0.2083(1 2 3)	0.2858(1 2 3 4)	
5	0.1392 (1 5)			
6	0.0902 (1 2)	0.1365(1 2 3)	0.2140(1 2 3 4)	0.2881(1 2 3 4 5)
7	0.0763 (1 3)	0.1539(1 3 5)	0.2194(1 3 5 7)	
8	0.0738 (1 2)	0.1150(1 2 4)	0.1914(1 2 4 5)	0.2616(1 2 4 5 7)
9	0.0615 (1 3)	0.1407(1 3 7)	0.2033(1 3 7 9)	
10	0.0576 (1 3)	0.0965(1 2 3)	0.1397(1 2 3 4)	0.1858(1 2 3 4 5)
11	0.0497 (1 5)	0.1305(1 5 7)	0.1892(1 5 7 11)	
12	0.0456 (1 5)	0.0782(1 3 4)	0.1211(1 2 3 5)	0.1656(1 2 3 4 5)
13	0.0485 (1 3)	0.0790(1 3 5)	0.1568(1 3 5 9)	0.2211(1 3 5 9 11)
14	0.0445 (1 4)	0.0889(1 2 4)	0.1270(1 2 4 7)	0.2029(1 2 4 7 8)
15	0.0387 (1 7)	0.0665(1 3 5)	0.0954(1 3 5 7)	0.1724(1 3 5 7 9)
16	0.0357 (1 5)	0.0641(1 3 5)	0.0937(1 3 4 5)	0.1406(1 2 3 5 8)
17	0.0339 (1 5)	0.0544(1 5 7)	0.1359(1 5 7 11)	0.1951(1 5 7 11 13)
18	0.0318 (1 7)	0.0519(1 7 8)	0.0916(1 3 4 5)	0.1295(1 2 5 6 8)
19	0.0333 (1 9)	0.0637(1 3 7)	0.0920(1 3 7 9)	0.1692(1 3 7 9 11)
20	0.0284 (1 8)	0.0538(1 4 5)	0.0973(1 2 5 8)	0.1382(1 2 4 5 8)
21	0.0292 (1 5)	0.0534(1 3 5)	0.0811(1 3 5 7)	0.1103(1 3 5 7 9)
22	0.0268 (1 7)	0.0492(1 4 10)	0.0764(1 4 5 7)	0.1075(1 3 4 5 7)
23	0.0263 (1 7)	0.0492(1 5 7)	0.0714(1 5 7 11)	0.1512(1 5 7 11 13)
24	0.0247 (1 7)	0.0442(1 4 11)	0.0660(1 4 6 9)	0.0901(1 4 6 9 11)
25	0.0245 (1 7)	0.0496(1 3 7)	0.0742(1 3 5 7)	0.1050(1 3 5 7 9)
26	0.0225 (1 8)	0.0385(1 8 10)	0.0725(1 4 5 7)	0.1088(1 2 5 7 8)
27	0.0251 (1 5)	0.0496(1 3 5)	0.0737(1 3 5 11)	0.1022(1 3 5 9 11)
28	0.0208 (1 12)	0.0357(1 8 12)	0.0559(1 8 9 12)	0.0892(1 4 5 7 13)
29	0.0209 (1 11)	0.0351(1 7 11)	0.0521(1 7 11 13)	0.1371(1 7 11 13 17)
30	0.0194 (1 12)	0.0353(1 7 9)	0.0592(1 4 13 14)	0.0855(1 4 5 6 14)
31	0.0201 (1 7)	0.0385(1 7 9)	0.0570(1 7 9 15)	0.0906(1 3 5 11 13)

procedure does surprisingly well, and outperforms the existing space-filling experimental designs. Moreover, the uniform design generated by the forward procedure has a favorable property: a uniform design U of "large size" is suited to all experiments having a number of factors no larger than the number of the columns of U. Furthermore, if a uniform design of larger size is needed, then we only need to find further columns, keeping all the existing ones unchanged.

3.3.4 Latin square method

Definition 7 An $n \times n$ matrix with n symbols as its elements is called a Latin square of order n if each symbol appears on each row as well as on each column once and only once.

Any Latin square of order n is a U-type design $U(n; n^n)$ and any s columns of a Latin square form a U-type design $U(n; n^s)$. The concept of a "Latin hypercube" is just the higher-dimensional generalization of a Latin square. Given n, a Latin square is always available. For example, a *left cyclic Latin square* of order n is an $n \times n$ Latin square such that

$$\mathbf{x}_{i+1} = L\mathbf{x}_i, i = 1, \cdots, n-1,$$

where $\mathbf{x}_i$ is the ith row of the square and L is the left shift operator defined by

$$L(a_1, a_2, \cdots, a_n) = (a_2, a_3, \cdots, a_n, a_1).$$

An example for generating a left cyclic Latin square of order 8 is given in Example 13. We start at a row vector (1 2 5 4 7 3 8 6), the first shift gives row (2 5 4 7 3 8 6 1), the second shift gives (5 4 7 3 8 6 1 2), and shifting this row again and again, eventually we have a Latin square. However, for a given n there are $n!$ left cyclic Latin squares. We want to find a left cyclic Latin square with the lowest discrepancy among all the $n!$ left cyclic Latin squares. The following algorithm was proposed by Fang, Shiu and Pan (1999), where $\mathcal{P}_n$ denotes the set of all $n!$ permutations of $(1, 2, \cdots, n)$.

Algorithm LUD

Step 1. Randomly choose an initial permutation, P^0, of $(1, 2, \cdots, n)$ and generate a left cyclic Latin square $\mathbf{L}_0$. Apply an optimization algorithm introduced in Chapter 4 and find a Latin square, $\mathbf{L} = (l_{ij}) = (\mathbf{l}_1, \cdots, \mathbf{l}_n)$ that has the smallest CD-value among all $n!$ left cyclic Latin squares of order n.

Step 2. Search s of n columns of the square, $\mathbf{l}_{i_1}, \cdots, \mathbf{l}_{i_s}$, to form a U-type design $U(n; n^s)$ such that this design has the smallest CD-value among all such $U(n; n^s)$ designs. This design is a (nearly) uniform design $U_n(n^s)$.

Both steps need a powerful optimization algorithm (cf. Chapter 4) in order to find a solution. It is easy to see that both the optimal Latin square in step 1 and the obtained UD in Step 2 may be a local minimum. One of the advantages of the Latin square method is that it can generate UDs for all $s \le n$.

Example 13 For $n = 8$ and $s = 3$ we find the cyclic Latin square

$$
\begin{bmatrix}
1 & 2 & 5 & 4 & 7 & 3 & 8 & 6 \\
2 & 5 & 4 & 7 & 3 & 8 & 6 & 1 \\
5 & 4 & 7 & 3 & 8 & 6 & 1 & 2 \\
4 & 7 & 3 & 8 & 6 & 1 & 2 & 5 \\
7 & 3 & 8 & 6 & 1 & 2 & 5 & 4 \\
3 & 8 & 6 & 1 & 2 & 5 & 4 & 7 \\
8 & 6 & 1 & 2 & 5 & 4 & 7 & 3 \\
6 & 1 & 2 & 5 & 4 & 7 & 3 & 8
\end{bmatrix}
$$

with the smallest CD-value 0.4358. For $U_8(8^s)$, columns of the design are chosen from columns of the above Latin square:

  $s = 2$: columns 1 and 4 with CD-value 0.0696
  $s = 3$: columns 1, 2, and 6 with CD-value 0.1123
  $s = 4$: columns 1, 2, 5, and 6 with CD-value 0.1601
  $s = 5$: columns 1, 2, 3, 6, and 7 with CD-value 0.2207

3.3.5 Expanding orthogonal array method

There are some relationships between orthogonal arrays (cf. Section 1.2) and uniform designs, such as

- Many orthogonal arrays are uniform designs; this was discovered by Fang, Lin, Winker and Zhang (2000).
- Any orthogonal design $L_n(q^s)$ can be extended to a number of $U(n, n^s)$ designs (cf. Table 2.3), among which the design with the smallest discrepancy is a nearly uniform design.

In fact, Algorithm ROA in Section 2.2 can produce all possible $U(n, n^s)$s. By some optimization algorithm in Chapter 4, we can find one with the smallest discrepancy.

The above algorithm was proposed by Fang (1995). The reader can refer to Fang and Hickernell (1995) for a discussion on the strengths and weaknesses of the above three methods in Sections 3.3.3–3.3.5. The expanding orthogonal array method requires powerful optimization algorithms, some of which will be introduced in Chapter 4.

3.3.6 The cutting method

It is known that the good lattice point (*glp*) method has been appreciated by many authors due to its economic computation and good performance. In particular, the *glp method* with a *power generator* has the lowest quasi-Monte Carlo computation complexity among comparable methods and also has good performance in the sense of uniformity (cf. Section 3.3.3).

Suppose that we want to have a uniform design $U_n(n^s)$, where n is not a prime and the uniform design constructed by the traditional *glp* method may have a poor uniformity. Ma and Fang (2004) proposed the *cutting method* that cuts a larger uniform design; the latter is easy to generate, especially generated by the *glp* method with a power generator.

Let U_p be a uniform design $U_p(p^s)$, where $n < p$ or $n << p$ and p or $p + 1$ is a prime, and let D_p be its induced design. Let $\mathcal{V}$ be a proper subset of C^s such that there are exactly n points of D_p falling in $\mathcal{V}$. Denote by D_n these n points. Then the points in D_n are approximately uniformly scattered on $\mathcal{V}$, from the theory of *quasi-Monte Carlo methods*. This is the key idea of the cutting method. Specifically, we choose a rectangle as $\mathcal{V}$ such that there are exactly n points of D_p falling in this rectangle. These n points will form a (nearly) uniform design $U_n(n^s)$ by some linear transformations. Let us consider an illustration example.

Example 14 Construction of a $U_{10}(10^2)$ design. A uniform design $U_{30}(30^2)$ was obtained by Fang, Ma and Winker (2000) as follows:

$$U_{30} = \begin{pmatrix} 24 & 23 & 1 & 9 & 5 & 11 & 19 & 6 & 21 & 3 & 12 & 15 & 20 & 18 & 17 \\ 25 & 6 & 12 & 18 & 16 & 7 & 4 & 9 & 11 & 3 & 14 & 20 & 30 & 15 & 24 \end{pmatrix}$$

$$\begin{pmatrix} 26 & 7 & 4 & 28 & 27 & 25 & 13 & 14 & 29 & 22 & 8 & 2 & 16 & 30 & 10 \\ 2 & 29 & 21 & 13 & 28 & 17 & 27 & 1 & 8 & 19 & 5 & 26 & 10 & 22 & 23 \end{pmatrix}'.$$

Its reduced design on C^2 is plotted in Figure 3.4(a). If we wish to have a $U_{10}(10^2)$ from this U_{30}, we can choose ten successive points according to the first coordinate (Figure 3.4(b)) or according to the second coordinate (Figure 3.4(c)). For each coordinate we can wrap the square such that position 0 and position 1 are at the same position. By this wrapping consideration, ten successive points can be separated in two rectangles: some points are near to 0 while others are near to 1 (Figure 3.4(d)). There are 60=30*2 such subsets of ten points. The ten points in each cutting are uniformly scattered in the related (wrapped) rectangle. By a linear transformation the ten points in each cutting can be transformed into a unit square and the resulting points form a U-type design of ten runs on the unit square. So we have 60 sets of ten points, or 60 designs with 10 runs. Suppose that the centered L_2-discrepancy (CD) is chosen as measure of uniformity. Finally, we choose one design with the smallest CD-value among these 60 designs. This design is a nearly uniform design $U_{10}(10^2)$ with CD=0.0543 and is given by

$$CU_{10} = \begin{bmatrix} 1 & 2 & 3 & 4 & 5 & 6 & 7 & 8 & 9 & 10 \\ 5 & 9 & 1 & 7 & 3 & 8 & 4 & 10 & 2 & 6 \end{bmatrix}.$$

Its induced design is plotted in Figure 3.5(a). Fang, Ma and Winker (2000) found a uniform design $U_{10}(10^2)$ with CD=0.0543 by the threshold accepting algorithm as follows:

$$U_{10} = \begin{bmatrix} 1 & 2 & 3 & 4 & 5 & 6 & 7 & 8 & 9 & 10 \\ 5 & 9 & 1 & 7 & 3 & 10 & 4 & 6 & 2 & 8 \end{bmatrix}.$$

Its induced design is plotted in Figure 3.5(b). Note that CU_{10} obtained by the cutting method is a uniform design $U_{10}(10^2)$, as it has the same CD-value as U_{10}. If the *glp* method is applied directly, another nearly uniform design is found as follows

$$U_{glp,10} = \begin{bmatrix} 1 & 2 & 3 & 4 & 5 & 6 & 7 & 8 & 9 & 10 \\ 3 & 6 & 9 & 2 & 5 & 8 & 1 & 4 & 7 & 10 \end{bmatrix},$$

with $CD = 0.0614$, which is larger than the CD-value of CU_{10}. This experiment shows the advantages of the cutting method compared with the traditional *glp* method.

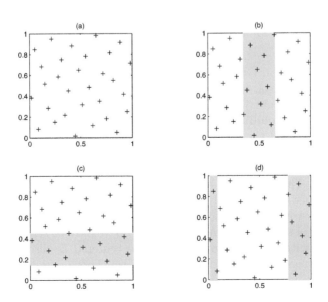

FIGURE 3.4
Illustration of the cutting method.

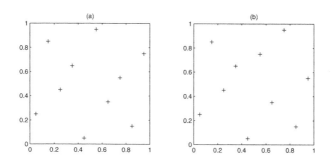

FIGURE 3.5
Plots of the induced designs of two $U_{10}(10^2)'s$.

Example 14 shows that the cutting method has good performance in construction of uniform designs. The algorithm for the cutting method is given below.

Algorithm CUT

Step 1. For given (n, s), find a $U_p(p^s)$, where $p >> n$, and calculate its induced design $D_p = \{\mathbf{c}_1, \cdots, \mathbf{c}_p\}$, denoted by $\mathbf{C} = (\mathbf{c}_{ij})$. The design $U_p(p^s)$ or D_p is called the *initial design*.

Step 2. For $l = 1, \cdots, s$ reorder the rows of $\mathbf{C}$ by sorting the jth column of $\mathbf{C}$ such that the elements in this column are sorted from small to large and denote the reordered matrix by $\mathbf{C}^{(l)} = (c_{kj}^{(l)})$.

Step 3. For $m = 1, \cdots, p$, let $\mathbf{C}^{(l,m)} = (c_{kj}^{(l,m)})$, where

$$
c_{kj}^{(l,m)} = \begin{cases} c_{k+m-n-1\ j}^{(l)}, & m > n, k = 1, \cdots, n, \\ c_{k\ j}^{(l)}, & m \leq n, k = 1, \cdots, m-1, \quad j = 1, \cdots, s. \\ c_{k+p-n\ j}^{(l)}, & m \leq n, k = m, \cdots, n, \end{cases}
$$

Step 4. Relabel elements of the jth column of $\mathbf{C}^{(l,m)}$ by $1, 2, \cdots, n$ according to the magnitude of these elements. The resulting matrix becomes a U-type design $U(n, n^s)$ and is denoted by $U^{(l,m)}$. There are ps U-type designs with such a structure.

Step 5. For a given measure of uniformity, M, compare the ps designs $U^{(l,m)}$ obtained in the previous step and choose the one with the smallest M-value. That one is a nearly uniform design $U_n(n^s)$.

There are some advantages to the cutting method: (a) the uniformity of the resulting designs is often better than designs directly generated by the *glp* method; (b) from one initial design D_p, we can find many $U_n(n^s), n < p, s = 2, 3, \cdots$; the computation complexity is lower. Ma and Fang (2004) gave an illustration example. They chose $p = 31$, $s = 2, 3, 4$, and $n = 7, 8, \cdots, 30$ based on a nearly uniform design $U_{31}(31^1 6)$. The results show that all the nearly uniform designs by the cutting method have a smaller CD-value than their corresponding designs generated by the *glp* method.

3.3.7 Construction of uniform designs by optimization

It is clear that searching a UD for a given (n, s, q) is an optimization problem. As the domain $U(n; q^s)$ is a finite set, all the classical Newton-Gauss type optimization methods do not work. In the past twenty years, many powerful **stochastic optimization algorithms** have been proposed. Popular among them are the *simulated annealing*, the *threshold accepting*, the *genetic algorithm* and *stochastic evolutionary* algorithms. Winker and Fang (1998) used the TA algorithm to find UDs under the star discrepancy. Fang, Ma and Winker (2000) applied the TA algorithm for finding UDs under the centered L_2- and wrap-around-discrepancies, respectively. Jin, Chen and Sudjianto (2005) proposed the enhanced stochastic evolutionary (ESE) algorithm for constructing optimal design of computer experiments including uniform designs. Chapter 4 gives a detailed introduction to some powerful optimization methods and their applications to construction of space-filling designs. The user can find many useful uniform designs on the web site http://www.math.hkbu.edu.hk/UniformDesign/ or http://www.stat.psu.edu /∽rli/UniformDesign/. More methods for generating UDs will be introduced in Sections 3.5 and 3.6.

3.4 Characteristics of the Uniform Design: Admissibility, Minimaxity, and Robustness

Each type of statistical design has its own characteristics under a certain statistical model. For example, the *optimal design* introduced in Section 1.2 is optimal under certain criteria (like D-optimality, A-optimality) for the linear regression model (Kiefer (1959) and Atkinson and Donev (1992)). Cheng (1980) established the universal optimality of a fractional factorial plan given by an *orthogonal array* of strength two (cf. Section 1.2 for the definition of an orthogonal array). Chan, Fang and Mukerjee (2001) characterized an orthogonal array of strength two in terms of D-optimality under a regression model where the factor levels are allowed to vary continuously in the design space. The *uniform design* was motivated by the overall mean model (cf. Section 1.5). However, good estimation of the overall mean is not enough for most computer experiments, when the experimenter wants to have a high quality metamodel. Therefore, several authors such as Wiens (1991), Hickernell (1999), Xie and Fang (2000), Yue (2001), and Hickernell and Liu (2002) gave many optimalities of the uniform design under different models.

Let $D_n = \{\mathbf{x}_1, \cdots, \mathbf{x}_n\}$ be a design. Its empirical distribution $F_{D_n}(\mathbf{x})$ can also be considered a design. Therefore, any distribution on the experimental domain can be considered a design. A measure can be defined by a distribution in measure theory. This idea was proposed by Kiefer (1959) and has been a powerful force in the development of the theory of optimal design and of uniform design.

Uniform design has been applied in computer experiments and in factorial plans with a model unknown. The former has the model

$$y = f(\mathbf{x})$$

and the model for the latter is

$$y = f(\mathbf{x}) + \varepsilon, \tag{3.17}$$

where $f(\mathbf{x})$ is unknown, but in a certain class of functions, denoted by $\mathcal{F}$, and ε is the random error. One wants to find a metamodel

$$y = g(\mathbf{x})$$

for the above two cases. Let $\mathbf{v}(\mathbf{x}) = (v_1(\mathbf{x}), \cdots, v_k(\mathbf{x}))$ be any $k(\leq n)$ linearly independent functions on $L_2(C^s)$, the set of L_2-integrable functions on C^s. If, as an approximation, the linear model

$$y = \sum_{i=1}^{k} \beta_i v_i(\mathbf{x}) + \epsilon \tag{3.18}$$

is misused as the true model, Xie and Fang (2000) proposed a framework in terms of the decision theory for model (3.18) as an approximation to (3.17) and proved that the uniform measure on C^s is *admissible* and *minimax* under this framework. If we put more conditions on the class $\mathcal{F}$, the uniform measure on C^s is optimal in a certain sense. Wiens (1991) obtained two optimality properties of the uniform measure. Hickernell (1999) combined goodness-of-fit testing, discrepancy, and robust design, and considered worst-case and average case models; we call them the *maximum mean-square-error model* and the *average mean-square-error models*, respectively. He proved that the uniform design is optimal and robust for both of these models. Yue and Hickernell (1999) considered an approximate model and decomposed the sum of squares into two components: variance and model bias. They investigated the importance of both variance and bias components and showed that the variance-minimizing designs can yield substantial bias, whereas bias-minimizing designs are rather efficient. Moreover, bias-minimizing designs tend to spread the points evenly over the domain. This gives another justification for the uniform design. Hickernell and Liu (2002) showed that categorical discrepancy (cf. Section 3.2.6) is an important criterion in study of orthogonal designs and found a number of interesting results.

Hickernell and Liu (2002) found that "When fitting a linear regression model to data, aliasing can adversely affect the estimates of the model coefficients and the decision of whether or not a term is significant. Optimal experimental designs give efficient estimators assuming that the true form of the model is known, while robust experimental designs guard against inaccurate estimates caused by model misspecification. Although it is rare for a single design to be both maximally efficient and robust, it is shown here that uniform designs limit the effects of aliasing to yield reasonable efficiency and robustness together." They also showed that the uniform designs limit aliasing. Yue (2001) applied the Bayesian approach to model (3.17) and found that low-discrepancy sets have good performance in such nonparametric models.

The above studies show the advantages of the uniform design (measure). It is interesting to note that both LHS and UD are motivated by the overall mean model at the beginning of the study. Although the overall mean model is far from the final goal of finding a good metamodel, both LHS and UD have excellent performance in computer experiments. One explanation for this phenomenon is that both approaches can work for a large class of functions $f(\cdot)$, not just for a specific $f(\cdot)$, i.e., both LHS and UD are robust against the model specification. Uniform design has been widely used in various fields in the past decades. Uniform design can be applied to

- computer experiments

- factorial plans with a model unknown

- experiments with mixtures

For a discussion of the latter, readers should refer to Fang and Wang (1994), Wang and Fang (1996), and Fang and Yang (1999). Users report that the uniform design is

- a friendly design in the sense of flexibility of the numbers of runs and levels of the factors

- a robust design which is reliable against changes in the model

- easy to understand and implement

- convenient in that designs with small size have been tabulated for general use

3.5 Construction of Uniform Designs via Resolvable Balanced Incomplete Block Designs

Due to the resource needs of the experiment or limitations of the experimental environment many factors in experiments can only allow a small number of levels. Therefore, designs $U_n(q^s)$, $q << n$, or $U_n(q_1^{r_1} \times, \cdots \times q_m^{r_m})$, with some $q_j << n$, are extremely useful. Fang, Lin, Winker and Zhang (2000) found that many existing orthogonal designs are uniform designs as the orthogonal designs satisfy some combinatorial balances. When orthogonality cannot be satisfied, some weaker combinatorial conditions can be set, that lead to many *combinatorial designs*, among which the *resolvable balanced incomplete block design* has been systematically developed. Lu and Meng (2000) and Fang, Lin and Liu (2003) independently found a relationship between resolvable balanced incomplete designs and uniform designs under the categorical discrepancy defined in Section 3.2.6. Along with this relationship many uniform designs can be obtained from the rich literature in combinatorial designs without any optimization computation. Lu and Sun (2001), Fang, Ge and Liu (2002), Fang, Ge, Liu and Qin (2002, 2004), Fang, Ge and Liu (2003), and Fang, Lu, Tang and Yin (2003) found many new UDs by this method. This section introduces the construction method via combinatorial designs. At the beginning we review some basic concepts in block designs.

3.5.1 Resolvable balanced incomplete block designs

Definition 8 A balanced incomplete block design *(BIBD) with parameters* (n, s, m, t, λ), *denoted by BIBD(n, s, m, t, λ), is an arrangement of n treatments into s blocks of size t, where $t < n$, such that each treatment appears in m blocks, and every pair of treatments appears in exactly λ blocks.*

The five parameters must satisfy the following relations:

$$nm = st \quad \text{and} \quad \lambda(n - 1) = m(t - 1). \tag{3.19}$$

Hence there are only three independent parameters in the definition. A BIBD(n, s, m, t, λ) is said to be *resolvable*, denoted by RBIBD(n, s, m, t, λ), if its blocks can be partitioned into m sets of blocks, called *parallel classes*, such that every treatment appears in each parallel class precisely once. The RBIBD is an important class of block designs. Table 3.3 gives a design RBIBD(10, 45, 9, 2, 1), where there are 10 treatments that are arranged in 45 blocks of size 2 such that each treatment appears in 9 blocks and every pair of treatments appears in exactly one block. There are 9 parallel classes $\mathbf{P}_1, \ldots, \mathbf{P}_9$ listed in 9 columns of Table 3.3. Clearly, this design has a *good balance among blocks, treatments, and pairs.*

3.5.2 RBIBD construction method

Now, let us establish the relationship between the above RBIBD(10, 45, 9, 2, 1) and $U(10, 5^9)$ in Table 3.4. We have 9 parallel classes $\mathbf{P}_1, \ldots, \mathbf{P}_9$ in the design RBIBD(10, 45, 9, 2, 1). Each parallel class corresponds to a column of $U(10, 5^9)$, and the five rows $b_1^j, b_2^j, b_3^j, b_4^j, b_5^j$ of Table 3.3 correspond to the five levels 1, 2, 3, 4, 5 of $U(10, 5^9)$. The pair $\{1, 10\}$ falls in column $\mathbf{P}_1$ and row b_1^j row, so we put level '1' at positions '1' and '10' of column **1** of $U(10, 5^9)$ in Table 3.4. Similarly, the pair $\{8, 9\}$ is located at $\mathbf{P}_1$ column and b_2^j row, so we put level '2' at positions 8 and 9 of column **1** in Table 3.4. In this way, 9 columns are then constructed from these 9 parallel classes, which form a U-type design $U(10; 5^9)$, as shown in Table 3.4.

TABLE 3.3
An RBIBD$(10, 45, 9, 2, 1)$

	$\mathbf{P}_1$	$\mathbf{P}_2$	$\mathbf{P}_3$	$\mathbf{P}_4$	$\mathbf{P}_5$	$\mathbf{P}_6$	$\mathbf{P}_7$	$\mathbf{P}_8$	$\mathbf{P}_9$
b_1^j	{1,10}	{2,10}	{4,9}	{3,7}	{2,8}	{5,7}	{5,6}	{1,7}	{1,6}
b_2^j	{8,9}	{5,8}	{3,10}	{4,10}	{6,9}	{2,4}	{3,4}	{2,5}	{2,7}
b_3^j	{4,5}	{3,6}	{7,8}	{1,2}	{5,10}	{1,9}	{1,8}	{4,6}	{4,8}
b_4^j	{6,7}	{7,9}	{2,6}	{5,9}	{1,3}	{3,8}	{7,10}	{3,9}	{3,5}
b_5^j	{2,3}	{1,4}	{1,5}	{6,8}	{4,7}	{6,10}	{2,9}	{8,10}	{9,10}

On the contrary, given an RBIBD(n, s, m, t, λ), where the n treatments are denoted by $1, \ldots, n$, and the m parallel classes are denoted by $\mathbf{P}_1, \ldots, \mathbf{P}_m$, each of which consists of $q = s/m$ disjoint blocks. Clearly, $t = n/q$. So this RBIBD can be expressed as RBIBD$(n, mq, m, n/q, \lambda)$. With this RBIBD we can construct a U-type design $U(n, q^m)$. The construction method, is called the RBIBD *method*, can be carried out as follows:

Algorithm RBIBD-UD

Step 1. Give a natural order $1, \ldots, q$ to the q blocks in each parallel class $\mathbf{P}_j, j = 1, \ldots, m$.

Step 2. For each $\mathbf{P}_j$, construct a q-level column $\mathbf{x}^j = (x_{kj})$ as follows:
Set $x_{kj} = u$, if treatment k is contained in the u-th block of $\mathbf{P}_j$, $u = 1, 2, \ldots, q$.

Step 3. The m q-level columns constructed from $\mathbf{P}_j$ $(j = 1, \ldots, m)$ form a $U(n; q^m)$ that is the output.

3.5.3 New uniform designs

There are many existing RBIBDs in the literature. If there exists an RBIBD $(n, mq, m, n/q, \lambda)$, a q-level design $U(n; q^m)$ can be constructed by the RBIBD method. The categorical discrepancy was defined in Section 3.2.6, in which a

TABLE 3.4

$U(10; 5^9)$

Row	1	2	3	4	5	6	7	8	9
1	1	5	5	3	4	3	3	1	1
2	5	1	4	3	1	2	5	2	2
3	5	3	2	1	4	4	2	4	4
4	3	5	1	2	5	2	2	3	3
5	3	2	5	4	3	1	1	2	4
6	4	3	4	5	2	5	1	3	1
7	4	4	3	1	5	1	4	1	2
8	2	2	3	5	1	4	3	5	3
9	2	4	1	4	2	3	5	4	5
10	1	1	2	2	3	5	4	5	5

lower bound of the categorical discrepancy of $U(n; q_1 \times \cdots \times q_m)$ is also given. Obviously, a U-type design whose categorical discrepancy attains the lower bound given in Theorem 2 is a uniform design. Therefore, Fang, Ge and Liu (2003) obtained the following theorem:

Theorem 4 *Given an RBIBD$(n, mq, m, n/q, \lambda)$, then the $U(n; q^m)$ constructed by the RBIBD method is a uniform design in the sense of the categorical discrepancy. And if $\lambda = 1$, then any of the possible level-combinations between any two columns appears at most once.*

Based on this theorem, we can use Algorithm RBIBD-UD to construct many new symmetric uniform designs. From the rich literature on RBIBDs, a number of uniform designs can be obtained as follows:

Theorem 5 *The following $U_n(q^s)$ can be constructed by the Algorithm RBIB-D-UD, where the categorical discrepancy is employed as the measure of uniformity:*

(a) If n is even, then a uniform $U_n((\frac{n}{2})^{k(n-1)})$ exists, where k is a positive integer.

(b) If $n \equiv 3 \pmod 6$, then a uniform $U_n((\frac{n}{3})^{\frac{n-1}{2}})$ exists.

(c) If $n \equiv 0 \pmod 3$ and $n \neq 6$, then a uniform $U_n((\frac{n}{3})^{n-1})$ exists.

(d) If $n \equiv 4 \pmod{12}$, then a uniform $U_n((\frac{n}{4})^{\frac{n-1}{3}})$ exists.

(e) If $n \equiv 0 \pmod 4$, then a uniform $U_n((\frac{n}{4})^{n-1})$ exists.

(f) If $n \equiv 0 \pmod 6$, then a uniform $U_n((\frac{n}{6})^{n-1})$ exists, except possibly for $n \in \{174, 240\}$.

(g) If $n \equiv 0 \pmod 6$, then a uniform $U_n((\frac{n}{6})^{2(n-1)})$ exists.

(h) If $n \equiv 5 \pmod{20}$, then a uniform $U_n((\frac{n}{5})^{\frac{n-1}{4}})$ exists, except possibly for $n \in \{45, 225, 345, 465, 645\}$.

(i) *If $n \equiv 5 \pmod{10}$ and $n \neq 15$, then a uniform $U_n((\frac{n}{5})^{\frac{n-1}{2}})$ exists, except possibly for $n \in \{45, 115, 135, 195, 215, 225, 235, 295, 315, 335, 345, 395\}$.*

(j) *If $n \equiv 0 \pmod{5}$ and $n \neq 10, 15$, then a uniform $U_n((\frac{n}{5})^{n-1})$ exists, except possibly for $n \in \{70, 90, 135, 160, 190, 195\}$.*

For more information on the RBIBDs in Case (a) the reader can refer to Rees and Stinson (1992), and for RBIBDs in Case (b)–(g) the reader can refer to Colbourn and Dinitz (1996); some new results also can be found on the web site http://www.emba.uvm.edu/~dinitz/newresults.html. For the RBIBDs in Cases (h)–(j), the reader can refer to Abel, Ge, Greig and Zhu (2001). Table 3.5 constructed by the case (b) of the theorem, gives $U_{15}(5^7)$. For more tables, the reader can visit the web site:

http://www.math.hkbu.edu.hk/UniformDesign.

TABLE 3.5
$U_{15}(5^7)$

Row	1	2	3	4	5	6	7
1	1	1	1	1	1	1	5
2	1	5	2	2	2	4	1
3	2	1	4	3	2	5	2
4	5	2	1	4	2	2	3
5	2	2	5	1	4	3	1
6	3	3	4	5	1	2	1
7	5	3	3	1	3	4	2
8	4	1	2	5	3	3	3
9	4	2	3	2	1	5	4
10	1	4	5	3	3	2	4
11	3	5	3	3	4	1	3
12	3	4	1	2	5	3	2
13	2	3	2	4	5	1	4
14	4	4	4	4	4	4	5
15	5	5	5	5	5	5	5

All the existing uniform designs $U_n(q^s)$ are constructed based on U-type designs, which require the number of experimental runs n to be a multiple of the number of factor levels q. When n is not a multiple of q, Fang, Lu, Tang and Yin (2003) developed a general construction method for UDs via resolvable packings and coverings. Several series of new uniform designs are then obtained. The reader can find the definition and related theory and method in their papers.

3.6 Construction of Asymmetrical Uniform Designs

In the previous section, we have introduced several methods for construction of symmetrical uniform designs. In this section, we shall concentrate on the construction of asymmetrical uniform designs. A design $U_n(q_1^{r_1} \times \cdots \times q_m^{r_m})$ is called an asymmetrical uniform design if $q_1, \cdots, q_m$ are different. Asymmetrical uniform designs are required when the experimental environment has some limitation, or the experimenter wants to include more levels for important factors and less levels for less important factors. Suppose we need a design $U_n(q_1 \times \cdots \times q_s)$, but it cannot be found in the literature. We can apply an optimization algorithm introduced in Chapter 4 and obtain a required design. If using an optimization method is too difficult (no computational code is available, for example), we have to find alternative ways. In this section we introduce the pseudo-level technique, collapsing, and combinatorial methods.

3.6.1 Pseudo-level technique

Let us start with an illustrative example. Suppose one needs a uniform design $U_{12}(6 \times 4 \times 3 \times 2)$ that cannot be found in the literature. Fortunately, we can find $U_{12}(12^4)$, which is listed in Table 2.1. Suppose that we want to change four columns of this design into a mixed level $6 \times 4 \times 3 \times 2$ design. Then we merge the original levels in column **1** by $(1,2) \Rightarrow 1, (3,4) \Rightarrow 2, \cdots,$ $(11, 12) \Rightarrow 6$. As a result, column **1** becomes a six-level column. Similarly, columns **2** through **4** become 4, 3, and 2 levels, respectively, and the new design is given in Table 3.6. In fact, we can assign the original four columns to have 3, 4, 6, 2 levels, respectively, or other choices. There are 4!=24 choices. We should choose one that has the minimum CD-value. The above method is called the *pseudo-level technique*. If the original design is UD, the new one obtained by the pseudo-level technique usually has good uniformity. From this example the reader can easily obtained a design with mixed levels from a symmetrical UD.

3.6.2 Collapsing method

The *collapsing method*, suggested by Fang and Qin (2002), merges two uniform designs into one with a larger size. Let $U = (u_{ij})$ be a U-type design, $U(n, q_1 \times \cdots \times q_s)$ and $V = (v_{kl})$ be another design, $U(m, m^t)$. Construct a new U-type design $D_{U,V}$ by collapsing U and V as follows

$$D_{U,V} = (\mathbf{1}_m \otimes U \vdots V \otimes \mathbf{1}_n),$$

TABLE 3.6
$U_{12}(6 \times 4 \times 3 \times 2)$

No	1	2	3	4
1	1	4	1	2
2	1	2	3	1
3	2	1	2	2
4	2	2	1	1
5	3	4	3	2
6	3	3	2	1
7	4	2	2	2
8	4	1	1	1
9	5	3	3	2
10	5	4	2	1
11	6	3	1	2
12	6	1	3	1

where $\otimes$ denotes the Kronecker product (cf. Section A.1), and $\mathbf{1}_m$ is the $m \times 1$ vector of ones. Obviously, the design $D_{U,V}$ is a U-type design $U(nm, q_1 \times \cdots \times q_s \times m^t)$.

For example, let

$$U = \begin{pmatrix} 1\,1\,1 \\ 2\,1\,2 \\ 3\,2\,2 \\ 4\,2\,1 \end{pmatrix} \quad \text{and} \quad V = \begin{pmatrix} 1\,3 \\ 2\,2 \\ 3\,1 \end{pmatrix}$$

be two U-type designs; then

$$D_{U,V} = \begin{pmatrix} 1 & 1 & 1 & 1 & 3 \\ 2 & 1 & 2 & 1 & 3 \\ 3 & 2 & 2 & 1 & 3 \\ 4 & 2 & 1 & 1 & 3 \\ 1 & 1 & 1 & 2 & 2 \\ 2 & 1 & 2 & 2 & 2 \\ 3 & 2 & 2 & 2 & 2 \\ 4 & 2 & 1 & 2 & 2 \\ 1 & 1 & 1 & 3 & 1 \\ 2 & 1 & 2 & 3 & 1 \\ 3 & 2 & 2 & 3 & 1 \\ 4 & 2 & 1 & 3 & 1 \end{pmatrix}$$

is a $U(12, 4 \times 2^2 \times 3^2)$. The above construction method is the *collapsing method* proposed by Fang and Qin (2002).

Denote by $U^{(s,t)}(mn, q_1 \times \cdots \times q_s \times m^t)$ the set of designs $D_{U,V}$. We wish to find a design with the minimum CD-value in the design space $U^{(s,t)}(mn, q_1 \times$

$\cdots \times q_s \times m^t)$. A natural idea would be to choose U and V to be uniform designs. Fang and Qin (2002), pointed out that:

- Under the wrap-around L_2-discrepancy (cf. Section 3.2.4) the design $D_{U,V}$ is a uniform design over $U^{(s,t)}(mn, q_1 \times \cdots \times q_s \times m^t)$ if and only if U is a uniform design $U_n(q_1 \times \cdots \times q_s)$ and V is a uniform design $U_m(m^t)$.
- Under the centered L_2-discrepancy, when $q_1 = \cdots = q_s = 2$ and $m = 2$, the design $D_{U,V}$ is a uniform design over $U^{(s,t)}(2n, 2^{s+t})$ if and only if U is a uniform design $U_n(2^s)$ and V is a uniform design $U_2(2^t)$.
- When the CD is chosen as a measure of uniformity and $t = 1$, we have
  (1) When m is odd, the design $D_{U,V}$ is a uniform design over $U^{(s,1)}(mn, q_1 \times \cdots \times q_s \times m)$ if and only if U is a uniform design $U_n(q_1 \times \cdots \times q_s)$.
  (2) When m is even and $q_1 = \cdots = q_s = 2$, the design $D_{U,V}$ is a uniform design on $U^{(s,1)}(mn, 2^s \times m)$ if and only if U is a uniform design $U_n(2^s)$.

More results are open for further study. Designs obtained by the collapsing method are nearly UDs, as the design space $U^{(s,t)}(mn, q_1 \times \cdots \times q_s \times m^t)$ is a proper subset of the domain $U(mn, q_1 \times \cdots \times q_s \times m^t)$. The WD- and CD-value of the new design $D_{U,V}$ can be calculated in terms of $WD(U), WD(V), CD(U)$, and $CD(V)$ as follows:

$$(WD(D_{U,V}))^2 = \left[(WD(U))^2 + \left(\frac{4}{3}\right)^s\right]\left[(WD(V))^2 + \left(\frac{4}{3}\right)^t\right] - \left(\frac{4}{3}\right)^{s+t}$$

and

$$(CD(D_{U,V}))^2 = \left[(CD(U))^2 - \left(\frac{13}{12}\right)^s + \alpha_U\right]\left[(CD(V))^2 - \left(\frac{13}{12}\right)^t + \beta_V\right]$$
$$+ \left(\frac{13}{12}\right)^{s+t} - \frac{1}{2}\alpha_U\beta_V,$$

where

$$\alpha_U = \frac{2}{n}\sum_{k=1}^{n}\prod_{j=1}^{s}\left(1 + \frac{1}{2}\left|x_{kj} - \frac{1}{2}\right| - \frac{1}{2}\left|x_{kj} - \frac{1}{2}\right|^2\right),$$

$$\beta_V = \frac{2}{m}\sum_{a=1}^{m}\prod_{l=1}^{t}\left(1 + \frac{1}{2}\left|y_{al} - \frac{1}{2}\right| - \frac{1}{2}\left|y_{al} - \frac{1}{2}\right|^2\right).$$

Note that a design $U(mn, q_1 \times \cdots \times q_s \times m^t)$ obtained by the collapsing method need not be a uniform design in the design space $U(mn, q_1 \times \cdots \times q_s \times m^t)$. The user can apply some optimization method to find a design with a lower discrepancy.

3.6.3   Combinatorial method

We introduced the relationships between resolvable balanced incomplete block designs and uniform designs and Algorithm RBIBD-UD for construction of symmetrical UDs in Section 3.5. However, this method can also be applied to construction of asymmetrical UDs. Note that all blocks in RBIBDs are of the same size; as a consequence, the UDs constructed by RBIBDs are symmetrical. We can construct asymmetrical UDs if we can find balanced block designs that have blocks with different sizes. Therefore, a new concept of *pairwise balanced designs* is helpful.

A. Uniformly resolvable block design: From now on, denote by $|\mathcal{A}|$ the cardinality of a set $\mathcal{A}$. Let $\mathcal{K}$ be a subset containing the different values of the block sizes, and $\mathcal{R}$ be a multiset satisfying $|\mathcal{R}| = |\mathcal{K}|$. Suppose that, for each $k \in \mathcal{K}$, there is a corresponding positive $r_k \in \mathcal{R}$ such that there are exactly r_k parallel classes of block size k. A *pairwise balanced design* of order n with block size from $\mathcal{K}$, denoted by $PBD(n, \mathcal{K}, \lambda)$ where n is the number of runs and λ is a positive integer, is a pair $(\mathcal{V}, \mathcal{B})$ which satisfies the following properties:

If $B \in \mathcal{B}$, then $|B| \in \mathcal{K}$, and every pair of distinct treatments of $\mathcal{V}$ appears in exactly λ blocks of $\mathcal{B}$.

A parallel class in a design is a set of blocks that partition the point set. A parallel class is *uniform* if every block in the parallel class is of the same size. A *uniformly resolvable block design*, denoted by $URBD(n, \mathcal{K}, \lambda, \mathcal{R})$, is a resolvable $PBD(n, \mathcal{K}, \lambda)$ such that all of the parallel classes are uniform. Let us cobsider an example.

Example 15 $URBD(6, \{3, 2\}, 1, \{1, 3\})$. Consider the case of $n = 6$ and $\mathcal{V} = \{1, 2, 3, 4, 5, 6\}$. Six treatments are arranged into $b = 11$ blocks of size k_j from $\mathcal{K} = \{3, 2\}$, such that each treatment appears in exactly four parallel classes, $\mathcal{P}_1, \mathcal{P}_2, \mathcal{P}_3$, and $\mathcal{P}_4$, that are given as follows (where in each parallel class, a $\{\cdots\}$ represents a block):

$$\mathcal{P}_1 = \{\{1, 2, 3\}, \{4, 5, 6\}\};$$
$$\mathcal{P}_2 = \{\{1, 4\}, \{2, 5\}, \{3, 6\}\};$$
$$\mathcal{P}_3 = \{\{3, 5\}, \{1, 6\}, \{2, 4\}\};$$
$$\mathcal{P}_4 = \{\{2, 6\}, \{3, 4\}, \{1, 5\}\}.$$

Note that every pair of treatments occurs in exactly $\lambda = 1$ block, so this block design is a $URBD(6, \{3, 2\}, 1, \{1, 3\})$.

B. Construction method: We now establish relationships between uniform designs and uniformly resolvable block designs.

To illustrate this, suppose that there exists a $URBD(n, \mathcal{K}, \lambda, \mathcal{R})$. For convenience, let $\mathcal{V} = \{1, 2, \cdots, n\}$, $\mathcal{K} = \{n/q_1, \cdots, n/q_m\}$, and $\mathcal{B} = \cup_{j=1}^{m} \mathcal{P}_j$, where each $\mathcal{P}_j$ with q_j blocks represents a parallel class of blocks. It can be shown that $\lambda = (\sum_{j=1}^{m} n/q_j - m)/(n-1)$. Then a uniform design $U_n(q_1 \times \cdots \times q_m)$ in the sense of the categorical discrepancy can be constructed with the following procedure proposed by Fang, Ge and Liu (2003).

Algorithm URBD-UD

Step 1. Assign a natural order $1, 2, \cdots, q_j$ to the q_j blocks in each parallel class $\mathcal{P}_j$ ($j = 1, 2, \cdots, m$).

Step 2. For each $\mathcal{P}_j$, construct a q_j-level column $\mathbf{x}^j = (x_{ij})$ as follows: Set $x_{ij} = u$, if treatment i is contained in the u-th block of $\mathcal{P}_j$ of $\mathcal{B}$ ($j = 1, 2, \cdots, m$).

Step 3. Combine these m columns constructed from $\mathcal{P}_j$ of $\mathcal{B}$ ($j = 1, 2, \cdots, m$) to form a U-type design $U(n, q_1 \times \cdots \times q_m)$.

Example 15 (Continued)

A UD $U_6(2 \times 3^3)$ can be obtained via $URBD(6, \{3, 2\}, 1, \{1, 3\})$, given in this example. Four columns of $U_6(2 \times 3^3)$ are constructed by the four parallel classes of $URBD(6, \{3, 2\}, 1, \{1, 3\})$ as follows: In $\mathcal{P}_1$, there are two blocks $\{1, 2, 3\}$ and $\{4, 5, 6\}$. We put "1" in the cells located in rows 1, 2, and 3 of the first column of the design and "2" in the cells located in rows 4, 5, and 6 of that column, thus we obtain a 2-level column of the design. Similarly, from $\mathcal{P}_2$ we put "1" in the cell located in rows 1 and 4 of the second column of the design and "2" and "3" in the cells located in rows 2, 5 and 3, 6 of that column, respectively. The second column of the design is thus generated. In this way, 4 columns are then constructed from these 4 parallel classes, which form a $U_6(2 \times 3^3)$ listed in Table 3.7.

TABLE 3.7

$U_6(2 \times 3^3)$

No	1	2	3	4
1	1	1	2	3
2	1	2	3	1
3	1	3	1	2
4	2	1	3	2
5	2	2	1	3
6	2	3	2	1

Let λ_{ij} be the number of blocks in which the pair of treatments i and j appears. The fact that $\lambda_{ij} = \psi$ for all $i \neq j$ means that any pair of distinct treatments in $\mathcal{V}$ appears in exactly λ blocks. Fang, Ge, Liu and

Qin (2004) pointed out that there exists a $U_n((\frac{n}{k_1})^{r_1} \cdots (\frac{n}{k_l})^{r_l})$ with $\lambda_{ij} = \lambda$ for all $1 \leq i \neq j \leq n$ if and only if there exists a $URD(n, \mathcal{K}, \lambda, \mathcal{R})$, where $\mathcal{K} = \{k_1, \ldots, k_l\}$, $\mathcal{R} = \{r_1, \ldots, r_l\}$, and $\lambda = (\sum_{j=1}^{l} r_j(k_j - 1))/(n - 1)$ is a positive integer.

The following example gives another UD with mixed levels. More asymmetrical UDs can be found in Fang, Ge and Liu (2003), Qin (2002), and Chatterjee, Fang and Qin (2004).

Example 16 A $URBD(12, \{2, 3\}, 1, \{5, 3\})$ can be found in combinatorial design theory as follows:

$$\mathcal{P}_1 = \{\{1, 2\}, \{3, 4\}, \{5, 6\}, \{7, 8\}, \{9, 10\}, \{11, 12\}\};$$
$$\mathcal{P}_2 = \{\{1, 3\}, \{2, 4\}, \{5, 7\}, \{6, 8\}, \{9, 11\}, \{10, 12\}\};$$
$$\mathcal{P}_3 = \{\{1, 4\}, \{2, 3\}, \{5, 8\}, \{6, 7\}, \{9, 12\}, \{10, 11\}\};$$
$$\mathcal{P}_4 = \{\{1, 7\}, \{2, 8\}, \{3, 12\}, \{4, 9\}, \{5, 10\}, \{6, 11\}\};$$
$$\mathcal{P}_5 = \{\{3, 5\}, \{4, 6\}, \{1, 10\}, \{2, 11\}, \{7, 12\}, \{8, 9\}\};$$
$$\mathcal{P}_6 = \{\{1, 5, 9\}, \{2, 6, 10\}, \{3, 7, 11\}, \{4, 8, 12\}\};$$
$$\mathcal{P}_7 = \{\{1, 6, 12\}, \{2, 7, 9\}, \{3, 8, 10\}, \{4, 5, 11\}\};$$
$$\mathcal{P}_8 = \{\{1, 8, 11\}, \{2, 5, 12\}, \{3, 6, 9\}, \{4, 7, 10\}\}.$$

From this URD and Algorithm URBD-UD, we can construct a $U_{12}(6^5 \times 4^3)$ shown in Table 3.6.

TABLE 3.8

$U_{12}(6^5 \times 4^3)$

No	1	2	3	4	5	6	7	8
1	1	1	1	1	3	1	1	1
2	1	2	2	2	4	2	2	2
3	2	1	2	3	1	3	3	3
4	2	2	1	4	2	4	4	4
5	3	3	3	5	1	1	4	2
6	3	4	4	6	2	2	1	3
7	4	3	4	1	5	3	2	4
8	4	4	3	2	6	4	3	1
9	5	5	5	4	6	1	2	3
10	5	6	6	5	3	2	3	4
11	6	5	6	6	4	3	4	1
12	6	6	5	3	5	4	1	2

3.6.4 Miscellanea

In Sections 3.5 and 3.6.3 we discussed construction of UDs via balanced block designs. There are many ways to generated resolvable balance incomplete block designs and uniformly resolvable block designs. The following are some approaches:

A. Room squares: Fang, Ge and Liu (2002) considered using *room squares*. The concept of the room square is a very useful tool in discrete mathematics.

B. Resolvable group divisible designs: Fang, Ge and Liu (2003) used the so-called *resolvable group divisible designs* to construct UDs. The reader can find the definition and related discussion in their paper.

TABLE 3.9

$U_{12}(3 \times 4^4)$

No	1	2	3	4	5
1	1	1	1	1	1
2	1	2	2	2	2
3	1	3	3	3	3
4	1	4	4	4	4
5	2	4	1	3	2
6	2	2	3	1	4
7	2	3	2	4	1
8	2	1	4	2	3
9	3	1	2	3	4
10	3	4	3	2	1
11	3	2	1	4	3
12	3	3	4	1	2

C. Orthogonal Latin squares/arrays: The concept of Latin squares is defined in Definition 7. Two Latin squares of order m are called orthogonal to each other if their m^2 pairs, which occur at the same location of the two Latin squares, are all different. A set of Latin squares of order m, any pair of which are orthogonal, is called a set of *pairwise orthogonal Latin squares*. Let $N(m)$ denote the maximum number of pairwise orthogonal Latin squares of order m. Fang, Ge and Liu (2003) proved that there exists a $U_{km}(k \times m^m)$ if $N(m) > k - 2$. It is known that there are three pairwise orthogonal Latin squares of order 4. So we can generate a $U_{12}(3 \times 4^4)$ that is listed in Table 3.9.

In fact, this approach is equivalent to the method proposed in Fang, Lin and Liu (2003). The latter starts an orthogonal array $L_n(q^s)$ and then generates a UD $U_{n-q}((q-1) \times q^{s-1})$, by deleting a row block with the same level in the first column of $L_n(q^s)$. Fang, Lin and Liu (2003) also proposed deleting more row blocks from $L_n(q^s)$ and obtained many (nearly) UDs such as $U_6(2 \times 3^3)$,

$U_8(2 \times 4^4)$, $U_{12}(3 \times 4^4)$, $U_{5k}(k \times 5^5)$, $k = 2, 3, 4$, $U_{7k}(k \times 7^7)$, $2 \le k \le 6$, $U_{8k}(k \times 8^8)$, $2 \le k \le 7$, and $U_{9k}(k \times 9^9)$, $2 \le k \le 8$.

4

Optimization in Construction of Designs for Computer Experiments

Various optimization techniques have been employed for finding optimal LHDs, uniform designs, and other space-filling designs. In the previous chapter we introduced many useful methods for construction of UDs, each focusing on a specific structure of designs (e.g., $U_n(n^s)$, $U_n(q^s)$, and UDs with mixed levels). Alternatively, one may choose to employ a more general optimization approach that can be used to deal with any kind of UDs. This generality, though at the expense of computational complexity, is an attractive advantage of the optimization approach. In the past decades many powerful *optimization algorithms* have been developed. Park (1994) developed an algorithm for constructing optimal LHDs based on IMSE criterion and entropy criterion, Morris and Mitchell (1995) adapted a version of *simulated annealing algorithm* for construction of optimal LHDs based on ϕ_p criterion, and Ye (1998) employed the *column-pairwise algorithm* for constructing optimal symmetrical LHDs based on the ϕ_p criterion and entropy criterion. The *threshold accepting heuristic* has been used for construction of uniform designs under the star discrepancy (Winker and Fang (1998)), centered L_2-discrepancy (Fang, Ma and Winker (2000), Fang, Lu and Winker (2003), and Fang, Maringer, Tang and Winker (2005)), and wrap-around L_2-discrepancy (Fang and Ma (2001b), Fang, Lu and Winker (2003), and Fang, Tang and Yin (2005)). Jin, Chen and Sudjianto (2005) demonstrated the use of an evolutionary algorithm to efficiently generate optimal LHD, and uniform designs under various criteria. In this chapter we shall introduce some useful optimization algorithms and their applications to the construction of space-filling designs.

4.1 Optimization Problem in Construction of Designs

Let $\mathcal{D}$ be the space of designs, each of which has n runs and s factors, with each factor having q levels. Note that the cardinality of $\mathcal{D}$ is finite, although it is very large in most cases. Let $h(D)$ be the objective function of $D \in \mathcal{D}$. The optimization problem in the construction of space-filling designs is to find a design $D^* \in \mathcal{D}$ such that

$$h(D^*) = \min_{D \in \mathcal{D}} h(D). \tag{4.1}$$

The objective function h can be the IMSE (2.11), the entropy (2.18), the ϕ_p criterion (2.22), the star discrepancy (A.2.2), the CD (3.6), or the WD (3.8). In cases where the objective function $h(\cdot)$ is to be maximized, we can use $-h(\cdot)$ instead and minimize the new objective function.

4.1.1 Algorithmic construction

The *stochastic optimization algorithms*, like column-pairwise exchange (Li and Wu (1997) and Ye (1998)), simulated annealing (Morris and Mitchell (1995)), threshold accepting heuristic (Winker and Fang (1998), Fang, Ma and Winker (2000), and Fang, Lu and Winker (2003)) and stochastic evolutionary (Jin et al. (2005)), have been widely used in searching for space-filling designs. Although these stochastic optimization methods have quite different names and appear in a different iteration process, we can use the following unified way of constructing them.

- Start with a randomly chosen initial design D^0, and let $D^c = D^0$;
- Construct the neighborhood of D^c, a set of candidate designs, according to the given definition of the neighborhood. Sample a new design D^{new} from the neighborhood;
- Compute the objective function value of the new design D^{new} and decide whether to replace the current design D^c with the new one D^{new}. Go back to the previous step until the stopping rule is reached.

4.1.2 Neighborhood

In the iteration process of an optimization algorithm, the current design will move to a nearby design, i.e., one which is in the same "neighborhood." There are many considerations for the definition of neighborhood in a given context. A necessary condition for defining a neighborhood is that the designs in the neighborhood of D^c should be a subset of $\mathcal{D}$; furthermore, we need to define a way in which members of this subset have similar features of $\mathcal{D}^c$. Let $\mathcal{D}$ be the set of $U(n, q^s)$. The smallest neighborhood is to exchange two levels in a row or a column of $D^c \in \mathcal{D}$. The column-exchange approach has been a more popular choice in the literature because it maintains the structure of the design such as the resulting design maintains the U-type to D^c. On the other hand, the design obtained by a row-exchange may not necessarily be the U-type. More precisely, to define the column-exchange approach, let $D^c \in U(n, q^s)$. Randomly choose one column of D^c and two entries in this column, then exchange these two entries to create a new design. All such designs form a neighborhood of D^c.

Example 17 The design D^c in (4.2) is a $U(5, 5^4)$. Randomly choose a column, say column 2, then randomly choose two entries in column 2, say entries (i.e., rows) 4 and 5. Exchange these two entries and obtain a new design on the right-hand side. The new design is still a $U(5, 5^4)$ and belongs to the neighborhood of D^c.

$$D^c = \begin{bmatrix} 1\,2\,4\,5 \\ 3\,4\,5\,3 \\ 2\,1\,3\,4 \\ 4\,5\,1\,2 \\ 5\,3\,2\,1 \end{bmatrix} \implies \begin{bmatrix} 1\,2\,4\,5 \\ 3\,5\,5\,3 \\ 2\,1\,3\,4 \\ 4\,4\,1\,2 \\ 5\,3\,2\,1 \end{bmatrix}. \tag{4.2}$$

It is easy to see that there are $4 \cdot \binom{5}{2} = 4 \cdot 10 = 40$ designs in the neighborhood of D^c.

Let us define another neighborhood of D^c. Randomly choose two columns of D^c, two entries from each column, and exchange the two entries. For example, randomly choose two columns, say 1 and 4, of D^c, then randomly choose two entries in columns 1 and 4, and exchange them as follows:

$$D^c = \begin{bmatrix} 1\,2\,4\,5 \\ 3\,4\,5\,3 \\ 2\,1\,3\,4 \\ 4\,5\,1\,2 \\ 5\,3\,2\,1 \end{bmatrix} \implies \begin{bmatrix} 5\,2\,4\,5 \\ 3\,4\,3\,3 \\ 2\,1\,5\,4 \\ 4\,5\,1\,2 \\ 1\,3\,2\,1 \end{bmatrix}.$$

Now, there are $\binom{4}{2} \cdot [\binom{5}{2}]^2 = 6 \cdot 10^2 = 600$ designs in the neighborhood.

Winker and Fang (1997) gave a detailed discussion on choice of the neighborhood. Many authors recommend choosing a smaller neighborhood so that there may be a higher possibility of reaching the global minimum.

4.1.3   Replacement rule

Let D^c be the current design and D^{new} be the new design chosen from the neighborhood of D^c. We need to decide whether the new design should replace the current one or not. Let $\nabla h = h(D^{new}) - h(D^c)$. The local search (LS) algorithm, like the column pairwise, compares $h(D^c)$ and $h(D^{new})$:

$$\text{LS Rule:} \begin{cases} 1, & \text{if } \nabla h \leq 0; \\ 0, & \text{if } \nabla h > 0. \end{cases} \tag{4.3}$$

Here, '1' means 'replace' and '0' means 'do not replace.' When the function $h(\cdot)$ has many local minima, this rule will terminate the optimization process at a local minimum point and might be far from the global minimum. Therefore, the optimization process needs to be repeated using various starting points. Alternatively, we can incorporate some heuristic rules for jumping out from a local minimum point. A threshold $Th (\geq 0)$ is used for this purpose in the *threshold accepting* rule:

$$\text{TA Rule:} \begin{cases} 1, & \text{if } \nabla h \leq Th; \\ 0, & \text{if } \nabla h > Th. \end{cases}$$

When $Th \geq$ h-range $\equiv \max_{D \in \mathcal{D}} h(D) - \min_{D \in \mathcal{D}} h(D)$, the optimal process is similar to a random walk; the case of $Th = 0$ reduces to the 'LS Rule.' Therefore, Th should be in the (0, h-range). The method used to choose a good value for Th is called the threshold accepting heuristic.

Choose the predefined sequence of threshold h-range$> T_1 > T_2 > \cdots > T_m = 0$, and change T_i to T_{i+1} in the iteration process according to some predefined rule. Often one chooses $T_i = (m - i + 1)\frac{Th}{m}$, where $Th = \frac{\text{h-range}}{10}$ and m is a positive integer, for example, $m = 8$. So the 'TA rule' becomes:

$$\text{TA Rule:} \begin{cases} 1, & \text{if } \nabla h \leq T_i; \\ 0, & \text{if } \nabla h > T_i. \end{cases} \tag{4.4}$$

The simulated annealing (SA) method uses a slightly different replacement rule; the D^{new} will replace the current design D^c if

$$\exp\left(-\frac{\nabla h}{Th}\right) \geq u,$$

where u is a random number (see Section 2.1.2), denoted by $u \sim U(0, 1)$, and Th is a parameter called 'temperature,' by analogy to the physical process of the annealing of solids. Like the TA heuristic, the parameter Th in the SA will be monotonically reduced. Set $Th = Th_0$ at the beginning of the optimal process, and let $Th = \alpha Th_0$, where α is a constant called the cooling factor. A slightly worse design is more likely to replace the current design than a significantly worse design. The performance of the SA largely depends on the selection of Th_0 and α. The replacement rule for the SA is given by

$$\text{SA Rule:} \begin{cases} 1, & \text{if } \exp(-\nabla h/Th) \geq u; \\ 0, & \text{if } \exp(-\nabla h/Th) < u. \end{cases} \tag{4.5}$$

For the stochastic evolutionary (SE) algorithm, the replacement rule combines LS and SA. When $\nabla h < 0$, replace $h(D^c)$ by $h(D^{new})$; otherwise $h(D^{new})$ will be accepted to replace $h(D^c)$ if it satisfies the following condition

$$\nabla h \leq Th \cdot u,$$

where u is a random number and Th is a control parameter. Like the SA and TA, the threshold will change in the optimization process. How to change the Th will be discussed in Section 4.2.5.

Each optimization algorithm has its own stopping rule that is controlled by ∇h and a few controlling parameters. We shall discuss the details in the next section.

4.1.4 Iteration formulae

To reduce computational complexity this section gives iteration formulae for different criteria, such as the entropy, ϕ_p, and centered L_2-discrepancy. In this section the neighborhood is chosen as the smallest one discussed in Example 17, where the kth column of D^c is chosen and two elements x_{ik} and x_{jk} will be exchanged. This strategy allows highly efficient calculation of the objective function of the new design. Since the search for optimal design involves a large scale combinatorial search, the ability to quickly calculate the objective function is crucial for optimizing a design within a reasonable time frame. Let us introduce iteration formulae for the following criteria:

Entropy criterion: We maximize $\log|\mathbf{R}|$, where $\mathbf{R} = (r_{ij})$ is the correlation matrix of the design D^c (cf. (2.18) and (2.19)), where r_{ij} is given by

$$r_{ij} = \exp\left\{-\sum_{k=1}^{s} \theta_k |x_{ki} - x_{kj}|^p\right\}, \quad 1 \le i, j \le n, 1 \le p \le 2.$$

As $\mathbf{R}$ is a positive definite matrix, it has the Cholesky decomposition

$$\mathbf{R} = \mathbf{V}'\mathbf{V},$$

where $\mathbf{V} = (v_{ij})$ is an upper triangle matrix, i.e., $v_{ij} = 0$ if $i < j$. The Cholesky factorization algorithm is:

REPEAT FROM $i = 1$ TO n
$v_{ii} = \sqrt{1 - \sum_{k=1}^{i-1} v_{ki}^2}$
   REPEAT FROM $j = i + 1$ TO n
$v_{ij} = (r_{ij} - \sum_{k=1}^{i-1} v_{ki}v_{kj})/v_{ii}$
   END LOOP
END LOOP

After an exchange of x_{ik} and x_{jk}, only the elements in row i and j and columns i and j of $\mathbf{R}$ are changed. Denote the new correlation matrix by $\mathbf{R}^* = (r_{ij}^*)$. For any $1 \le t \le n$ and $t \ne i, j$, let

$$s(i, j, k, t) = \exp\left\{\theta_k(|x_{jk} - x_{tk}|^p - |x_{ik} - x_{tk}|^p)\right\}.$$

It can be easily verified that

$$r_{it}^* = r_{ti}^* = r_{it}/s(i, j, k, t),$$
$$r_{ijt}^* = r_{ti}^* = r_{jt}s(i, j, k, t).$$

For calculating the determinant of $\mathbf{R}^*$, the Cholesky decomposition of $\mathbf{R}^*$ will be useful. We can use our knowledge about the above iteration process to reduce the computational load. Let $n_1 = min(i, j)$, and let $\mathbf{R}$ be re-expressed as

$$\mathbf{R} = \begin{bmatrix} \mathbf{R}_1 & \mathbf{R}_2 \\ \mathbf{R}_2' & \mathbf{R}_3 \end{bmatrix},$$

where $\mathbf{R}_1 : n_1 \times n_1, \mathbf{R}_2 : n_1 \times (n - n_1)$, and $\mathbf{R}_3 : (n - n_1) \times (n - n_1)$. Let $\mathbf{R}_1 = \mathbf{V}_1'\mathbf{V}_1$ be the Cholesky decomposition. Now the Cholesky factorization $\mathbf{V}$ of $\mathbf{R}$ can be computed based on $\mathbf{V}_1$:

$$\mathbf{V} = \begin{bmatrix} \mathbf{V}_1 & \mathbf{V}_2 \\ \mathbf{O} & \mathbf{V}_3 \end{bmatrix},$$

where $\mathbf{V}_3$ is also an upper triangle matrix. Note that the elements of $\mathbf{V}$ with index $1 \leq i \leq j \leq n_1$ are kept unchanged. The rest of the elements of $\mathbf{V}$ can be calculated by the following modified Cholesky factorization algorithm:

REPEAT FROM $i = 1$ TO n_1
   REPEAT FROM $j = n_1$ TO n
   $v_{ij} = (r_{ij} - \sum_{k=1}^{i-1} v_{ki}v_{kj})/v_{ii}$
   END LOOP
END LOOP
REPEAT FROM $i = n_1$ TO n
   $v_{ii} = \sqrt{1 - \sum_{k=1}^{i-1} v_{ki}^2}$
   REPEAT FROM $j = i + 1$ TO n
   $v_{ij} = (r_{ij} - \sum_{k=1}^{i-1} v_{ki}v_{kj})/v_{ii}$
   END LOOP
END LOOP

The computational complexity of Cholesky factorization (or decomposition) is $O(n^3)$. In addition, the calculation of the elements of $\mathbf{R}$ costs $O(sn^2)$, and therefore the computational complexity for totally re-evaluating the entropy will be $O(sn^2) + O(n^3)$.

While the determinant of the new $\mathbf{R}$ matrix cannot be directly evaluated based on the determinant of the old $\mathbf{R}$ matrix, some improvement in efficiency is achievable by modifying the Cholesky algorithm. The computational complexity of the modified Cholesky factorization algorithm will depend on both n and n_1. For example, if $n_1 = n - 1$, the computational complexity will be $O(n^2)$. On the other hand, if $n_1 = 1$, the computational complexity will still be $O(n^3)$. On average, the computational complexity will be smaller than $O(n^3)$ but larger than $O(n^2)$. The total computational complexity of the new method will be between $O(n) + O(n^2)$ and $O(n) + O(n^3)$, which is not dramatically better than $O(n^3) + O(sn^2)$.

ϕ_p **criterion:** The ϕ_p criterion was introduced in Section 2.4.3 and defined in (2.22). The calculation of ϕ_p includes three parts, i.e., the evaluation of all the inter-site distances, the sorting of those inter-site distances to obtain a distance list and index list, and the evaluation of ϕ_p. The evaluation of all the inter-site distances will take $O(mn^2)$ operations, the sorting will take $O(n^2 \log_2(n))$ operations (cf. Press et al. (1992)), and the evaluation of ϕ_p will take $O(s^2 \log_2(p))$ operations (since p is an integer, p-powers can be computed by repeated multiplications). In total, the computational complexity will be $O(sn^2) + O(n^2 \log_2(n)) + O(s^2 \log_2(p))$. Thus, calculating ϕ_p will be very time-consuming. Therefore, a more efficient computation technique is needed. Such a more efficient computation can be constructed by avoiding the sorting of inter-site distances. To this end, express ϕ_p as follows:

$$\phi_p = \left[\sum_{1 \leq i < j \leq n} d_{ij}^{-p} \right]^{1/p},$$

where $d_{ij} = (\sum_{k=1}^n (x_{ik} - x_{jk})^q)^{1/q}$ is the inter-site distance between $\mathbf{x}_i$ and $\mathbf{x}_j$ and $\mathbf{D} = (d_{ij})$. After an exchange between x_{ik} and x_{jk}, in the $\mathbf{D}$ matrix only elements from rows i and j and columns i and j are changed. Denote the new distance matrix as $\mathbf{D}^* = (d_{ij}^*)$. For any $1 \leq j \leq n$ and $j \neq i, j$, let

$$s(i, j, k, t) = |x_{jk} - x_{tk}|^q - |x_{ik} - x_{tk}|^q,$$

then

$$d_{it}^* = d_{ti}^* = [d_{it}^p + s(i, j, k, t)]^{1/q}$$

and

$$d_{jt}^* = d_{tj}^* = [d_{jt}^p - s(i, j, k, t)]^{1/q}.$$

The new ϕ_p-value becomes

$$\phi_p^* = \left[\phi_p^p + \sum_{1 \leq k \leq n, k \neq i, j} \left[(d_{ik}^*)^{-p} - d_{ik}^{-p} \right] + \sum_{1 \leq k \leq n, k \neq i, j} \left[(d_{jk}^*)^{-p} - d_{jk}^{-p} \right] \right]^{1/p}.$$

Now, the total computational complexity of the new calculation technique is $O(n) + O(n \log_2(p))$. This results in significant reduction of computation time compared to the original calculation of ϕ_p.

The centered L_2-discrepancy: This was defined in (3.7). The computational complexity of calculating the CL_2 discrepancy is $O(sn^2)$. We can use an idea similar to that of ϕ_p to improve the computational efficiency. Let $\mathbf{Z} = (z_{ij})$ be the centered design matrix, where $z_{ik} = x_{ik} - 0.5$. Let $\mathbf{C} = (c_{ij})$ be a symmetric matrix, whose elements are:

$$c_{ij} = \begin{cases} \frac{1}{n^2} \prod_{k=1}^s \frac{1}{2}(2 + |z_{ik}| + |z_{jk}| - |z_{ik} - z_{jk}|), & \text{if } i \neq j; \\ \frac{1}{n^2} \prod_{k=1}^s (1 + |z_{ik}|) - \frac{2}{n} \prod_{k=1}^s (1 + \frac{1}{2}|z_{ik}| - \frac{1}{2}z_{ik}^2), & \text{otherwise.} \end{cases}$$

It can be verified that

$$[CD(D_n)]^2 = \left(\frac{13}{12}\right)^2 + \sum_{i=1}^{n}\sum_{j=1}^{n} c_{ij},$$

where D_n is the current design. For any $1 \le t \le n$ and $t \ne i, j$, let

$$\gamma(i, j, k, t) = \frac{2 + |z_{jk}| + |z_{tk}| - |z_{jk} - z_{tk}|}{2 + |z_{ik}| + |z_{tk}| - |z_{ik} - z_{tk}|}.$$

After an exchange of x_{ik} and x_{jk}, the square CD-value of the new design D_n^* is given by

$$[CD(D_n^*)]^2 = [CD(D_n)]^2 + c_{ii}^* - c_{ii} + c_{jj}^* - c_{jj} + 2\sum_{t=1, t \ne i, j}^{n}(c_{it}^* - c_{it} + c_{jt}^* - c_{jt}),$$

where

$$c_{it}^* = \gamma(i, j, k, t)c_{it}, \quad \text{and} \quad c_{jt}^* = c_{jt}/\gamma(i, j, k, t).$$

The computational complexity of this new approach is $O(n)$. The total computational complexity is also $O(n)$, which is much less than the original $O(sn^2)$.

A comparison of the computational complexity of the original technique (i.e., totally recalculate the value using all elements in matrices) and those of the new methods are summarized in Table 4.1. From the table, we find that for the ϕ_p criterion and the CL_2 criterion, with the new approach, the efficiency can be significantly improved. The new computational complexity is close to $O(n)$ in both cases. However, for the entropy criterion, because of the involvement of matrix determinant calculation, the efficiency is not dramatically improved.

TABLE 4.1
Computational Complexity of Criterion Evaluation

Method	ϕ_p	CL_2	Entropy
Original	$O(mn^2) + O(n^2\log_2(n))$ $+O(s^2\log_2(p))$	$O(mn^2)$	$O(n^3) + O(mn^2)$
Proposed	$O(n) + O(n\log_2(p))$	$O(n)$	$O(n^2) + O(n) \sim O(n^3) + O(n)$

Table 4.2 provides examples of the performance comparison of computing times of the optimization criteria for various size of LHDs. The time ratio, T_o/T_n, where T_o and T_n are the times needed to calculate the criterion using the original methods (i.e., calculation using all matrix elements) and the new more efficient methods, respectively, show the improvement of computational efficiency. The empirical results in Table 4.2 match well with the analytical

examinations in Table 4.1. The larger the size of the design, the more the computational savings are. For example, for 100×10 LHDs, the calculation of the CL_2 criterion only requires about $1/82$ of the computation effort of the original method using the whole matrix. Compared to the other two criteria, the entropy criterion is much less efficient. It is also observed that the computing time needed for the ϕ_p criterion is up to 3.0 times as many as that of the CL_2 criterion. This result reinforces the attractiveness of uniform design using the CL_2 criterion from the design generation point of view. Interested readers should consult Jin et al. (2005).

TABLE 4.2

Computing Time (in seconds) of Criterion Values for $500,000$ LHDs (T_o = the time needed by the original calculation methods to evaluate $500,000$ LHDs; T_n = the time needed to calculate $500,000$ LHDs using the new more efficient methods; T_o/T_n is the ratio of T_o and T_n)

LHDs	$\phi_p(p=50, t=1)$			CL_2			Entropy($\theta=5, t=2$)		
	T_o	T_n	T_o/T_n	T_o	T_n	T_o/T_n	T_o	T_n	T_o/T_n
12×4	12.2	5.5	2.2	10.7	2.4	4.5	16.6	14.2	1.2
25×4	53.0	10.1	5.2	41.5	3.4	12.1	75.3	39.8	1.9
50×5	239.0	19.8	12.1	197.0	6.5	30.3	347.0	167.0	2.1
100×10	1378.0	45.2	30.5	1305.0	15.9	82.1	2116.0	1012.0	2.1

4.2   Optimization Algorithms

This section gives summaries of optimization algorithms for the four optimization methods, LS, SA, TA, and SE, introduced in the previous section.

4.2.1   Algorithms

For all four optimization methods, LS, SA, TA, and SE, their algorithms have the same structure:

Algorithm for stochastic optimization:

Step 1. Start from a randomly chosen initial design D^0 from $\mathcal{D}$, set $D^c = D^0$ and the threshold parameter Th to its start value T_c;

Step 2. Find the neighborhood of D^c, denoted by $N(D^c)$. Choose a design, denoted by D^{new}, from $N(D^c)$. Calculate $\nabla h = h(D^{new}) - h(D^c)$ for the given criterion h.

Step 3. Replace D^c with the new design D^{new} if
- $\nabla h \leq 0$;
- otherwise, replace D^c by D^{new} with a probability of $p(T_c, D^c, D^{new})$;

Step 4. Repeat *Step 2* and *Step 3* until either fulfilling the rule for changing the threshold value or fulfilling the stopping rule. Finally, deliver D^c.

The replacement probability for the above four algorithms is given by:

Local search

$$p(T_c, D^c, D^{new}) = \begin{cases} 1, & \text{if } \nabla h \leq T_c; \\ 0, & \text{otherwise}, \end{cases}$$

where $T_c = 0$.

Threshold Accepting

$$p(T_c, D^c, D^{new}) = \begin{cases} 1, & \text{if } \nabla h \leq T_c; \\ 0, & \text{otherwise}, \end{cases}$$

where T_c is chosen from the pre-determined series $T_1 > T_2 > \cdots > T_m$.

Simulated annealing

$$p(T_c, D^c, D^{new}) = \exp\{\nabla h / T_c\}, \quad T_c = \alpha T h \quad (0 < \alpha < 1).$$

Stochastic evolutionary

$$p(T_c, D^c, D^{new}) = \begin{cases} 1, & \text{if } \nabla h < 0; \\ 1 - \nabla h / T_c, & \text{else, if } \nabla h < T_c; \\ 0 & \text{otherwise}. \end{cases}$$

Here, T_c changes according to the current situation. For details the reader can refer to Section 4.2.5. It is clear that each stochastic optimization algorithm has its own controlling rule. Let us introduce them one by one.

4.2.2   Local search algorithm

The local search (LS) algorithm takes $Th = 0$ in the entire process. It is the simplest algorithm among the four algorithms introduced in this section.

Step 1. Start with a randomly chosen initial design D^0 from $\mathcal{D}$ and let $D^c = D^0$.

Step 2. Find the neighborhood of D^c, denoted by $N(D^c)$. Choose a design, denoted by D^{new}, from $N(D^c)$. Calculate $\nabla h = h(D^{new}) - h(D^c)$.

Step 3. Replace D^c with the new design D^{new} if $\nabla h \leq 0$; otherwise, stop the process.

Step 4. Deliver D^c.

The LS algorithm could quickly find a locally optimal design. The column pairwise algorithm is a local search algorithm. Repeating the local search algorithm from different initial designs can increase the chances of reaching the global minimum.

4.2.3   Simulated annealing algorithm

Simulated annealing (SA) algorithm was proposed by Kirkpatrick, Gelett and Vecchi (1983) for a cost function that may possess many local minima. It works by emulating the physical process whereby a solid is slowly cooled so that when eventually its structure is 'frozen,' the minimum energy configuration is obtained. Although the convergence rate of the SA is slow, it has been successfully applied to many famous problems, such as the traveling problem, the graph partitioning problem, the graph coloring problem, and the number partitioning problem.

Step 1. Start with a randomly chosen initial design D^0 from $\mathcal{D}$ and let $D^c = D^0$. Initialize $Th = T_0, \alpha$, and i_{max};

Step 2. Find the neighborhood of D^c, denoted by $N(D^c)$. Choose a design, denoted by D^{new}, from $N(D^c)$. Calculate $\nabla h = h(D^{new}) - h(D^c)$. Set $i = 1$ and $Flag = 0$;

Step 3. Replace D^c with the new design D^{new} and Flag=1 if $\nabla h < 0$; otherwise, if

$$\exp\{-\nabla h/Th \ge u\}, u \sim U(0, 1),$$

else $i = i + 1$.

Step 4. If $i > i_{max}$, let $Th := \alpha Th$. Repeat *Step 2* and *Step 3* until Flag=0. Deliver D^c.

As mentioned before, α is a cooling parameter. A given increment in criterion value is more likely to be accepted early in the search when temperature has a relatively high value than later in the search as the temperature is cooling. To achieve good results a careful cooling parameter tuning is required. A too large T_0 or α may lead to slow convergence and tremendous computational cost while a too small T_0 or α may lead to an inferior design. It is recommended to test a small-size design first in order to get experience in the choice of T_0 and α.

4.2.4   Threshold accepting algorithm

The threshold accepting (TA) algorithm is a modification of the SA, and was proposed by Dueck and Scheuer (1991). It has been shown to be simpler and more efficient than the SA in many applications. There are several versions of the TA algorithm; for details the reader can refer to Winker and Fang (1997), Winker and Fang (1998), Fang, Lin, Winker and Zhang (2000), Fang,

Lu and Winker (2003), and Winker (2001). The latter gives a comprehensive introduction to the TA and its applications. The following version was used in Fang, Lin, Winker and Zhang (2000).

Step 1. Start from a randomly chosen initial design D^0 from $\mathcal{D}$, let $D^c = D^0$, and give a sequence of the threshold parameter Th, $T_1 > T_2 \cdots > T_m$, and starting value $T_c = T_1$;

Step 2. Find the neighborhood of D^c, denoted by $N(D^c)$. Choose a design, denoted by D^{new}, from $N(D^c)$. Calculate $\nabla h = h(D^{new}) - h(D^c)$.

Step 3. Replace D^c with the new design D^{new} if $\nabla h \leq T_c$, else leave D^c unchanged;

Step 4. Repeat *Step 2* and *Step 3* a fixed number of times;

Step 5. If the given threshold sequence is not yet exhausted, take the next threshold value and repeat *Step 2* and *Step 4*; otherwise, deliver D^c.

Winker and Fang (1997) gave a detailed discussion on the choice of the neighborhoods and the two controlling parameters.

4.2.5   Stochastic evolutionary algorithm

The *stochastic evolutionary* (SE) algorithm, proposed by Saab and Rao (1991), is a stochastic optimization algorithm that decides whether to accept a new design based on a threshold-based acceptance criterion. The threshold Th is initially set to a small value Th_0. Its value is incremented based on certain 'warming' schedules only if it seems that the process is stuck at a local minimum; whenever a better solution is found in the process of warming up, Th is set back to Th_0. Saab and Rao (1991) show that the SE can converge much faster than the SA. However, it is often difficult to decide the value of Th_0 and the warming schedule for a particular problem. Therefore, Jin, Chen and Sudjianto (2005) developed an *enhanced stochastic evolutionary* (ESE) algorithm that uses a sophisticated combination of warming schedule and cooling schedule to control Th so that the algorithm can adjust itself to suit different experimental design problems. The ESE algorithm consists of an inner loop and an outer loop. The inner loop picks up new designs from the neighborhood of the current design and decides whether to accept them based on an acceptance criterion. The outer loop controls the entire optimization process by adjusting the threshold Th in the acceptance criterion.

The inner loop process is similar to that of the TA algorithm. In the outer loop, at the beginning, Th is set to a small value, e.g., $0.005 \times$criterion value of the initial design. The threshold Th is maintained on a small value in the improving stage and is allowed to be increased in the exploration stage. Let n_{acpt} be the number of accepted designs in the inner loop and n_{imp} be the better designs found in the inner loop. Th will be increased or decreased based on the following situations:

(a) In the improving stage, Th will be maintained at a small value so that only better designs or slightly worse designs will be accepted. Specifically, Th will be decreased if n_{acpt} is larger than a small percentage (e.g., 0.1) of the number of total designs J and n_{imp} is less than n_{acpt}; Th will be increased if n_{acpt} is less than the same percentage of the number of total designs J. The following equations are used in the algorithm to decrease or increase Th, respectively:

$$Th := \alpha_1 Th, \text{ or } Th := Th/\alpha_1,$$

where $0 < \alpha_1 < 1$, for example, set $\alpha_1 = 0.8$.

(b) In the exploration stage, Th will fluctuate within a range based on the value of n_{acpt}. If n_{acpt} is less than a small percentage (e.g., 0.1) of J, Th will be rapidly increased until n_{acpt} is larger than a large percentage (e.g., 0.8) of J. If n_{acpt} is larger than the large percentage, Th will be slowly decreased until n_{acpt} is less than the small percentage. This process will be repeated until an improved design is found. The following equations are used to decrease or increase Th, respectively:

$$Th := \alpha_2 Th, \text{ or } Th := Th/\alpha_3,$$

where $0 < \alpha_3 < \alpha_2 < 1$. For example, set $\alpha_2 = 0.9$ and $\alpha_3 = 0.7$. Th is increased rapidly (so that an increasing number of worse designs could be accepted) to help the search process move away from a locally optimal design. Th is decreased slowly to search for better designs after moving away from the local optimal design. The ESE algorithm is more complicated than the above three algorithms. Figure 4.1 presents the flow chart of the ESE algorithm.

4.3 Lower bounds of the discrepancy and related algorithm

Suppose that a discrepancy (for example, the categorical discrepancy, centered discrepancy, or wrap-around discrepancy) is employed in construction of uniform designs. For a given design space $U(n, q^s)$, if one can find a strict lower-bound of the discrepancy on the design space, this lower bound can be used as a benchmark for construction of uniform designs and can speed up the searching process by the use of some optimization algorithm. As soon as the discrepancy of the current design in the optimization process reaches the lower bound, the process will be terminated since no further improvement is possible. Therefore, many authors have invested much effort in finding some strict lower bounds for different discrepancies. The word 'strict' means the lower bound can be reached in some cases.

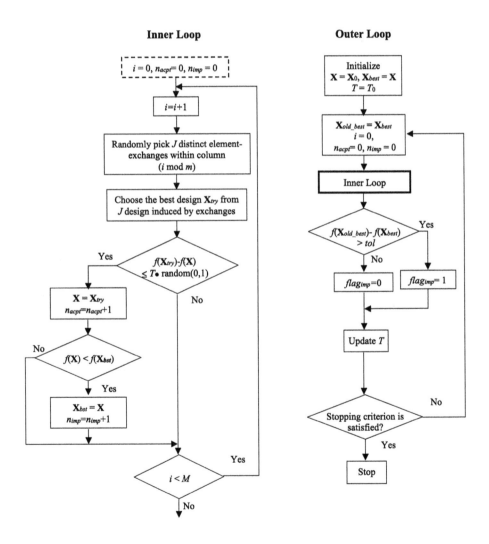

FIGURE 4.1
Flowchart of the enhanced stochastic evolutionary algorithm.

4.3.1 Lower bounds of the categorical discrepancy

Theorem 2 in Section 3.2.6 gives lower bounds for the categorical discrepancy that have played an important role in construction of uniform designs in Sections 3.5 and 3.6.

4.3.2 Lower bounds of the wrap-around L_2-discrepancy

Fang, Lu and Winker (2003) obtained strict lower bounds for centered and wrap-around L_2-discrepancies for two- and three-level U-type designs. They gave two kinds of lower bounds: one is related to the distribution of all the level-combinations among columns of the design, denoted by $LB^c_{WD_2}(n, q^s)$, and another is based on Hamming distances of any two rows of the design, denoted by $LB^r_{WD_2}(n, q^s)$. The *Hamming distance* between rows $\mathbf{x}_k = (x_{k1}, \cdots, x_{ks})$ and $\mathbf{x}_l = (x_{l1}, \cdots, x_{ls})$, is the number of positions where they differ, i.e. $D_H(\mathbf{x}_k, \mathbf{x}_l) = \#\{j : x_{kj} \neq x_{lj}, j = 1, \cdots, s\}$. For two-level U-type designs, their results are summarized in the following theorem.

Theorem 6 *Two lower bounds of the wrap-around L_2-discrepancy over the design space $U(n, 2^s)$ are given by*

$$LB^c_{WD_2}(n, 2^s) = \left(\frac{11}{8}\right)^s - \left(\frac{4}{3}\right)^s$$

$$+ \frac{1}{n^2}\left(\frac{5}{4}\right)^s \sum_{m=1}^{s}\left(\frac{1}{5}\right)^m \binom{s}{m} s_{n,m,2}\left(1 - \frac{s_{n,m,2}}{2^m}\right) \quad and \quad (4.6)$$

$$LB^r_{WD_2}(n, 2^s) = -\left(\frac{4}{3}\right)^k + \frac{1}{n}\left(\frac{3}{2}\right)^k + \frac{n-1}{n}\left(\frac{5}{4}\right)^k\left(\frac{6}{5}\right)^\lambda, \quad (4.7)$$

respectively, where $s_{n,m,2}$ is the remainder at division of n by 2^m ($n \mod 2^m$), $\lambda = s(n-2)/[2(n-10)]$, and $[x]$ is the integer part of x.

Fang, Lu and Winker (2003) pointed out that the lower bound $LB^c_{WD_2}(n, 2^s)$ is more accurate than the bound $LB^r_{WD_2}(n, 2^s)$ when $n \leq s$, while the ordering is reversed when $n > s$. The maximum of these two lower bounds can be used as a benchmark in searching process. For three-level U-type designs, two lower bounds are given as follows:

Theorem 7 *Two lower bounds of the wrap-around L_2-discrepancy over the design space $U(n, 3^s)$ are given by*

$$LB^c_{WD_2}(n, 3^s) = -\left(\frac{4}{3}\right)^s + \left(\frac{73}{54}\right)^s +$$

$$\frac{1}{n^2}\left(\frac{23}{18}\right)^s \sum_{m=1}^{k}\left(\frac{4}{23}\right)^m \binom{s}{m} s_{n,m,3}\left(1 - \frac{s_{n,m,3}}{3^m}\right) \quad and \quad (4.8)$$

$$LB^r_{WD_2}(n, 3^s) = -\left(\frac{4}{3}\right)^s + \frac{1}{n}\left(\frac{3}{2}\right)^s + \frac{n-1}{n}\left(\frac{23}{18}\right)^s\left(\frac{27}{23}\right)^\lambda, \quad (4.9)$$

respectively, where $s_{n,m,3}$ is the residual of n (mod 3^m) and $\lambda = \frac{k(n-3)}{3(n-1)}$.

Similar to the two-level case, the two lower bounds have different accuracy in different situations. Usually, when $k \leq (n-1)/2$, $LB_{WD_2}^c(n, 3^s)$ is more accurate than $LB_{WD_2}^r(n, 3^s)$; otherwise, the accuracy is reversed. The maximum of these two lower bounds can be used as a benchmark in the search process.

Recently, Fang, Tang and Yin (2005) proposed a lower bound on the design space $U(n, q^s)$ for each positive integer q. From the analytical expression of equation (3.8), it is easy to see that the wrap-around L_2-discrepancy is only a function of products of $\alpha_{ij}^k \equiv |x_{ik} - x_{jk}|(1 - |x_{ik} - x_{jk}|)$ $(i, j = 1, \cdots, n, i \neq j$ and $k = 1, \cdots, s)$. However, for a U-type design, its α-values can only be limited into a specific set. More precisely, for a U-type design $U(n, q^s)$, when q is even, α-values can only take $q/2+1$ possible values, i.e., 0, $2(2q-2)/(4q^2)$, $4(2q-4)/(4q^2)$, ..., $q^2/(4q^2)$; when q is odd, these products can only take $(q+1)/2$ possible values, i.e., 0, $2(2q-2)/(4q^2)$, $4(2q-4)/(4q^2)$, ..., $(q-1)(q+1)/(4q^2)$. The following table gives the distribution of α-values over the set $\{\alpha_{ij}^k : 1 \leq i < j \leq n, 1 \leq k \leq s\}$, for both even and odd q. Note that given (n, q, ms), this distribution is the same for each design in $U(n, q^s)$. We shall see that this fact is very useful in finding the lower bounds and create a new algorithm.

q even		q odd	
α − values	number	α − values	number
0	$\dfrac{sn(n-q)}{2q}$	0	$\dfrac{sn(n-q)}{2q}$
$\dfrac{2(2q-2)}{4q^2}$	$\dfrac{sn^2}{q}$	$\dfrac{2(2q-2)}{4q^2}$	$\dfrac{sn^2}{q}$
$\dfrac{4(2q-4)}{4q^2}$	$\dfrac{sn^2}{q}$	$\dfrac{4(2q-4)}{4q^2}$	$\dfrac{sn^2}{q}$
$\cdots$	$\cdots$	$\cdots$	$\cdots$
$\cdots$	$\cdots$	$\cdots$	$\cdots$
$\dfrac{(q-2)(q+2)}{4q^2}$	$\dfrac{sn^2}{q}$	$\dfrac{(q-3)(q+3)}{4q^2}$	$\dfrac{sn^2}{q}$
$\dfrac{q^2}{4q^2}$	$\dfrac{sn^2}{2q}$	$\dfrac{(q-1)(q+1)}{4q^2}$	$\dfrac{sn^2}{q}$

For any two different rows of the design $U(n, q^s)$, $\mathbf{x}_k = (x_{k1}, \cdots, x_{ks})$, and $\mathbf{x}_l = (x_{l1}, \cdots, x_{ls})$, denote by F_{kl}^α the distribution of their $\{\alpha_{kl}^j, j = 1, \cdots, s\}$. The F_{kl}^α's can characterize whether a U-type design is a uniform design or not.

Theorem 8 *A lower bound of the wrap-around L_2-discrepancy on $U(n, q^s)$ with even q and odd q is given by*

$$LB_{even} = \Delta + \frac{n-1}{n}\left(\frac{3}{2}\right)^{\frac{s(n-q)}{q(n-1)}}\left(\frac{5}{4}\right)^{\frac{sn}{q(n-1)}}\left(\frac{3}{2} - \frac{2(2q-2)}{4q^2}\right)^{\frac{2sn}{q(n-1)}}\cdots$$

$$\left(\frac{3}{2} - \frac{(q-2)(q+2)}{4q^2}\right)^{\frac{2sn}{q(n-1)}} \quad and$$

$$LB_{odd} = \Delta + \frac{n-1}{n}\left(\frac{3}{2}\right)^{\frac{s(n-q)}{q(n-1)}}\left(\frac{3}{2} - \frac{2(2q-2)}{4q^2}\right)^{\frac{2sn}{q(n-1)}}\cdots$$

$$\left(\frac{3}{2} - \frac{(q-1)(q+1)}{4q^2}\right)^{\frac{2sn}{q(n-1)}},$$

respectively, where $\Delta = -\left(\frac{4}{3}\right)^s + \frac{1}{n}\left(\frac{3}{2}\right)^s$. *A U-type design* $U(n, q^s)$ *is a uniform design under the wrap-around* L_2*-discrepancy, if all its* F_{kl}^α *distributions,* $k \neq l$, *are the same. In this case, the WD-value of this design achieves the above lower bound.*

When $q = 2, 3$, the above lower bounds are equivalent to $LB_{WD_2}^r(n, 2^s)$ and $LB_{WD_2}^r(n, 3^s)$ obtained in Theorems 6 and 7, respectively.

Theorem 8 provides not only the lower bounds that can be used for a benchmark, but also the importance of balance of $\{F_{kl}^\alpha\}$. Checking that all F_{kl}^α distributions are the same imposes a heavy computational load. Therefore, we define

$$\delta_{kl} = \sum_{j=1}^{m} \ln(\frac{3}{2} - \alpha_{kl}^j),$$

for any two rows k and l. Obviously, for any $1 \leq k \neq l, p \neq q \leq n$ the fact that $F_{kl}^\alpha = F_{pq}^\alpha$ implies $\delta_{kl} = \delta_{pq}$, but the inverse may not be true. Aiming to adjust those δ_{kl}s as equally as possible, Fang, Tang and Yin (2005) proposed a more powerful algorithm, which is called the *balance-pursuit heuristic*. It will be introduced in Section 4.3.4.

4.3.3 Lower bounds of the centered L_2-discrepancy

The first strict lower bound of the centered L_2-discrepancy was obtained by Fang and Mukerjee (2000).

Theorem 9 *Consider the design space* $U(n, 2^s)$ *with* $n = 2^{s-k}$ *for some positive integer* k *(*$k \leq s$*). A lower bound of the centered* L_2*-discrepancy on the design space is given by*

$$LB_{CD_2} = \left(\frac{13}{12}\right)^s - 2\left(\frac{35}{32}\right)^s + \sum_{r=0}^{s-k}\binom{s}{r}\frac{1}{8^r} + \frac{1}{n}\sum_{r=s-k+1}^{s}\binom{s}{r}\frac{1}{4^r}.$$

Late, Fang, Lu and Winker (2003) extended the above result to all two-level U-type designs as follows:

Theorem 10 *Consider the design space $U(n, 2^s)$. A lower bound of the centered L_2-discrepancy on the design space is given by*

$$LB_{CD_2} = \left(\frac{13}{12}\right)^s - 2\left(\frac{35}{32}\right)^s + \frac{1}{n}\left(\frac{5}{4}\right)^s + \frac{n-1}{n}\left(\frac{5}{4}\right)^\lambda, \qquad (4.10)$$

and $\lambda = s(n-2)/[2(n-1)]$.

Recently, Fang, Maringer, Tang and Winker (2005) used a different approach to find some lower bounds for the design space $U(n, 3^s)$ and $U(n, 4^s)$. The reader can find the details in their paper.

4.3.4   Balance-pursuit heuristic algorithm

The threshold accepting heuristic has been successfully applied for finding uniform designs by Winker and Fang (1998), Fang and Ma (2001b), Fang, Ma and Winker (2002). However, the lower bounds discussed in this section can be used as a benchmark. Furthermore, the above theory shows that the lower bound can be reached if the Hamming distances between any two rows are the same if the categorical discrepancy is being used, or if $\{F_{kl}^\alpha\}$ or $\{\delta_{kl}\}$ are the same if the wrap-around L_2-discrepancy is being used. That gives us a possibility to propose a more powerful algorithm, which we call the *balance-pursuit heuristic* (BPH, for simplicity), proposed by Fang, Tang and Yin (2005). Compared with the existing threshold accepting heuristic, for example Fang, Lu and Winker (2003), the BPH algorithm has more chances to generate better designs in the sense of lower discrepancy in each iteration, since it gives an approximate direction to the better status, which can save considerable time in the computational searching.

One of the advantages of the BPH algorithm is that it does not require a threshold accepting series, which plays an important role in the threshold accepting heuristic. As stated in Winker and Fang (1998), the aim of using a temporary worsening up to a given threshold value is to avoid getting stuck in a local minimum. But how to determine a proper threshold accepting series is itself a difficult problem, since it will depend on the structure and property of the design.

The BPH algorithm also uses a random warming-up procedure, but it uses a different way to jump out from a local minimum. Similar to the threshold accepting heuristic, the BPH algorithm starts with a randomly generated U-type design D^0. Then it goes into a large number, say τ, of iterations. In each iteration the algorithm tries to replace the current solution D^c with a new one. The new design D^{new} is generated in a given neighborhood of the current solution D^c. In fact, a neighborhood is a small perturbation of D^c. The difference between the discrepancies of D^{new} and D^c is calculated and compared in each iteration. If the result is not worse, or the design needs to be warmed-up, then we replace D^c with D^{new} and continue the iteration. Three key points are emphasized as follows:

(A) **Neighborhood**: Most authors choose a neighborhood of the current solution D^c so that each design in the neighborhood is still of the U-type. This requirement can be easily fulfilled by selecting one column of D^c and exchanging two elements in the selected column. To enhance convergence speed the BPH algorithm suggests two possible pre-selection methods for determining the neighborhood choice, instead of using random selection elements within a column for exchanging as done in the literature. According to Theorem 8, we should reduce differences among the current δ_{kl}s. So our two pre-selection methods both aim to distribute the distances δ_{kl}s as evenly as possible. Two methods, *maximal and minimal distances of row pairs* and *single row with maximal and minimal sum of distances*, are suggested by Fang, Tang and Yin (2005).

(B) **Iteration**: Instead of calculating two discrepancies of $WD_2(D^{new})$ and $WD_2(D^c)$, the BPH algorithm focuses on the difference between $WD_2(D^{new})$ and $WD_2(D^c)$, similarly to the way we introduced in Section 4.2.1. Based on the formula (3.8), we know that the wrap-around L_2-discrepancy can be expressed in terms of the sum of $e^{\delta_{ij}}$s. And for a single exchange of two elements in the selected column, there are altogether $2(n-2)$ distances (δ_{ij}s) updated. Suppose the k-th elements in rows $\mathbf{x}_i$ and $\mathbf{x}_j$ are exchanged, then for any row $\mathbf{x}_t$ other than $\mathbf{x}_i$ or $\mathbf{x}_j$, the distances of row pair $(\mathbf{x}_i, \mathbf{x}_t)$ and row pair $(\mathbf{x}_j, \mathbf{x}_t)$ will be changed. Denote the new distances between row pair $(\mathbf{x}_i, \mathbf{x}_t)$ and row pair $(\mathbf{x}_j, \mathbf{x}_t)$ as $\tilde{\delta}_{ti}$ and $\tilde{\delta}_{tj}$, then

$$\tilde{\delta}_{it} = \delta_{it} + \ln(3/2 - \alpha_{jt}^k) - \ln(3/2 - \alpha_{it}^k);$$

$$\tilde{\delta}_{jt} = \delta_{jt} + \ln(3/2 - \alpha_{it}^k) - \ln(3/2 - \alpha_{jt}^k).$$

Here α_{it}^k and α_{jt}^k are α-values as defined in Section 4.3.2, and the objective function change will be

$$\nabla = \sum_{t \neq i,j} \left(e^{\tilde{\delta}_{it}} - e^{\delta_{it}} + e^{\tilde{\delta}_{jt}} - e^{\delta_{jt}} \right).$$

(C) **Lower bound**: The last key point is the lower bound. As soon as the lower bound is reached, the process will be terminated.

The following gives a pseudo-code of the BPH algorithm proposed by Fang, Tang and Yin (2005):

The BPH Algorithm for searching uniform designs under WD_2

1 Initialize τ

2 Generate starting design $D^c \in \mathcal{U}(n, q^m)$ **and let** $D^{min} := D^c$

3 for $i = 1$ **to** τ **do**

4   Generate $D^{new} \in \mathcal{N}(D^c)$ **by randomly using two pre-selection methods**

5   if $WD_2(D^{new})$ **achieves the lower bound then**

6   return(D^{new}**)**

7   end if

8   if $WD_2(D^{new}) \leq WD_2(D^c)$ **then**

9 $D^c := D^{new}$

10   if $WD_2(D^{new}) < WD_2(D^{min})$ **then**

11 $D^{min} := D^{new}$

12   end if

13   else if rand(1000)< 3 then

14 $D^c := D^{new}$

15   end if

16 end for

17 return(D^{min}**)**

Part III Modeling for Computer Experiments

In this part, we introduce diverse modeling techniques for computer experiments. Chapter 5 begins with fundamental concepts of modeling computer experiments, and describes the commonly used metemodeling approaches, including polynomial regression, spline regression, Kriging method, neural network approach, and local polynomial regression. Various methods for sensitivity analysis are presented in Chapter 6. These methods are very useful tools for interpreting the resulting metamodels built by using methods introduced in Chapter 5. The last chapter treats computer experiments with functional responses. Analysis of functional response in the context of design of experiment is a relatively new topic. We introduce several possible approaches to deal with such kinds of data. Real-life case studies will be used to demonstrate the proposed methodology.

5

Metamodeling

In the previous chapters various types of designs for computer experiments
were introduced. Once the data have been collected from an experiment, we
wish to find a metamodel which describes empirical relationships between the
inputs and outputs. Notice that the outputs of computer experiments are
deterministic (i.e., no random errors); therefore, we describe the relationship
between the input variables and the output variable by the model (1.1) or

$$\text{output variable} = f(\text{input variables}), \qquad (5.1)$$

where f is an unspecified smooth function to be approximated. Note that compared to nonparametric regression models in the statistical literature, (5.1)
does not have a random error term on the right-hand side. In this chapter, we aim to extend regression techniques existing in statistical literature to
model computer experiments. Let us begin with some fundamental concepts
in statistical modeling.

5.1 Basic Concepts

Modeling computer experiment can be viewed as regression on data without
random errors. This view suggests that many fundamental ideas and concepts
in statistical modeling may be applied and extended to modeling computer
experiments. Some concepts and methods of modeling were briefly reviewed
in Section 1.6. For some useful techniques in regression, the reader can refer
to Appendix A.3. In this section, we will refresh some useful ideas and concepts on modeling; thereafter, we will provide more detailed discussion on the
methodologies.

5.1.1 Mean square error and prediction error

Let y denote the output variable and $\mathbf{x} = (x_1, \cdots, x_s)$ consist of the input
variables. The primary goal of metamodeling is to predict the true model
$f(\mathbf{x})$ at an untried site $\mathbf{x}$ by $g(\mathbf{x})$ in the metamodel (1.9) built on a computer
experiment sample $\{(\mathbf{x}_i, y_i), i = 1, \cdots, n\}$. Thus, it is a natural question

how to assess the performance of $g(\mathbf{x})$. Intuitively, we want the residual or approximate error, defined as $f(\mathbf{x}) - g(\mathbf{x})$, as small as possible over the whole experimental region T. Define *mean square error* (MSE) as

$$\text{MSE}(g) = \int_T \{f(\mathbf{x}) - g(\mathbf{x})\}^2 \, d\mathbf{x}.$$

Because there is no random error in the outputs of computer experiments, when $\mathbf{x}$ is uniformly scattered over the experiment region T, the MSE equals the *prediction error* (PE) in the statistical literature. More generally, we may define *weighted mean square error* (WMSE)

$$\text{WMSE}(g) = \int_T \{f(\mathbf{x}) - g(\mathbf{x})\}^2 w(\mathbf{x}) \, d\mathbf{x},$$

where $w(\mathbf{x}) \geq 0$ is a weighted function with $\int_T w(\mathbf{x}) \, d\mathbf{x} = 1$. The weighted function $w(\mathbf{x})$ allows one to easily incorporate prior information about the distribution of $\mathbf{x}$ over the experimental domain. When such prior information is not available, it is common to assume that $\mathbf{x}$ is uniformly distributed over T. Then WMSE becomes equal to MSE.

From the above definition of the prediction error, we need the value of $f(\mathbf{x})$ over unsampled sites to calculate the prediction error. When computer experiments are neither time-consuming nor computationally intensive, we may collect a large number of new samples, and then calculate the prediction error. We now give a simple illustrative example.

Example 18 Let $\{(x_k, y_k), k = 1, \cdots, 11\}$ be a computer experiment sample from the true model

$$f(x) = 2x \cos(4\pi x)$$

at $x_k = 0, 0.1, \cdots, 1$. As an illustration, we fit the data by a polynomial model with degrees p equal to $0, 1, \cdots, 10$. Note that when the degree of polynomial equals 10 (i.e., $n = p$), it yields an interpolation to the collected sample. Figure 5.1 (a) depicts the plots of true curve and the scatter plot of collected samples. The weighted function $w(\mathbf{x})$ here is set to be equal to 1 for computing the PE. Figure 5.1(b) displays the plot of prediction error versus the degrees of polynomial model fitted to the data. The prediction error reaches its minimum at $p = 9$. The fitted curve and fitted values at observed x's are also depicted in Figure 5.1(a), where the approximate polynomial is

$$\begin{aligned}
g(x) = {} & 0.9931 + 1.9600(x - 0.5) - 76.8838(x - 0.5)^2 - 152.0006(x - 0.5)^3 \\
& + 943.8565(x - 0.5)^4 + 1857.1427(x - 0.5)^5 - 3983.9332(x - 0.5)^6 \\
& - 7780.7937(x - 0.5)^7 + 5756.3571(x - 0.5)^8 + 11147.1698(x - 0.5)^9.
\end{aligned}$$

In many situations, computer experiments are computationally intensive. For instance, each run in Example 2 required 24 hours of simulation. In such

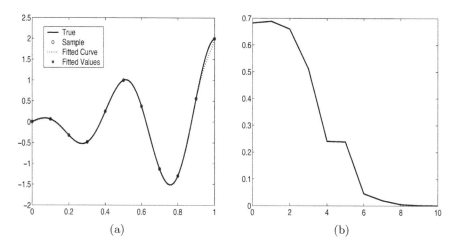

(a) (b)

FIGURE 5.1
Plots for Example 18. (a) is the plot of the true model, scatter plot of collected sample, fitted curve, and fitted values using the polynomial regression model. (b) is the plot of prediction error against the degree of polynomial fitted to the data. The prediction error reaches its minimum when $p = 9$.

situations, direct evaluation of the prediction error seems to be computationally impractical. To alleviate this problem, a general strategy is to estimate the prediction error of a metamodel g using the *cross validation* (CV) procedure. For $i = 1, \cdots, n$, let g_{-i} denote the metamodel based on the sample excluding $(\mathbf{x}_i, y_i)$. The cross validation score is defined as

$$\mathrm{CV}_n = \frac{1}{n} \sum_{i=1}^{n} \{f(x_i) - g_{-i}(\mathbf{x}_i)\}^2. \tag{5.2}$$

The cross validation score can give us a good estimate for the prediction error of g. This procedure is referred to as *leave-one-out cross validation* in the statistical literature.

When the sample size n is large and the process of building nonlinear metamodel is time consuming (e.g., Kriging and neural network models that will be introduced in Sections 5.4 and 5.6, respectively), using CV_1 scores becomes computationally too demanding because we need to build n metamodels. To reduce the computational burden, we may modify the cross validation procedure in the following way. For a pre-specified K, divide the sample into K groups with approximately equal sample size among the groups. Let $g_{(-k)}$ be the metamodel built on the sample excluding observations in the k-th group. Further let $\mathbf{y}_{(k)}$ and $\mathbf{g}_{(k)}$ be the vector consisting of observed values and predicted values for the k-th group using the $g_{(-k)}$, respectively. Then the K-fold *cross validation* score is defined as

$$\mathrm{CV}_K = \frac{1}{n} \sum_{k=1}^{K} (\mathbf{y}_{(k)} - \mathbf{g}_{(k)})'(\mathbf{y}_{(k)} - \mathbf{g}_{(k)}).$$

Thus, the cross validation score in (5.2) coincides with the n-fold cross validation score.

In most modeling procedures, the metamodel g depends on some regularization parameter (see the section below), say λ, to be selected. Therefore, the cross validation score depends on the regularization parameter and is denoted by $CV(\lambda)$; here CV can be either CV_1 or CV_K. Minimizing the cross validation score with respect to λ yields a data-driven method to select the regularization parameter. In other words,

$$\hat{\lambda} = \min_{\lambda} \mathrm{CV}(\lambda).$$

Here, CV can be either CV_n or CV_K. In the statistical literature, the minimization is carried out by searching over a set of grid values for λ. This data-driven approach to selecting the regularization parameter may be implemented in many modeling procedures for computer experiments. The theoretical properties of cross validation have been studied extensively in the statistical literature. See, for example, Li (1987).

5.1.2   Regularization

Most metamodels in the literature for modeling computer experiments have the following form:

$$g(\mathbf{x}) = \sum_{j=0}^{L} B_j(\mathbf{x})\beta_j, \tag{5.3}$$

where $\{B_0(\mathbf{x}), \cdots, B_L(\mathbf{x})\}$ is a set of basis functions defined on the experimental domain. For instance, the basis function may be the polynomial basis function, the spline basis function, radial basis function, or the basis function constructed by the covariance function in the Kriging model (see Section 1.6). We shall discuss these basis functions in detail later on.

Since outputs of computer experiments are deterministic, the construction of a metamodel is in fact an interpolation problem mathematically. Let us begin with some matrix notation. Denote

$$\begin{aligned} \mathbf{y} &= (y_1, \cdots, y_n)', \\ \boldsymbol{\beta} &= (\beta_0, \cdots, \beta_L)', \\ \mathbf{b}(\mathbf{x}) &= (B_0(\mathbf{x}), \cdots, B_L(\mathbf{x}))', \end{aligned} \tag{5.4}$$

and

$$\mathbf{B} = \begin{pmatrix} B_0(\mathbf{x}_1) & \cdots & B_L(\mathbf{x}_1) \\ B_0(\mathbf{x}_2) & \cdots & B_L(\mathbf{x}_2) \\ \cdots & \cdots & \cdots \\ B_0(\mathbf{x}_n) & \cdots & B_L(\mathbf{x}_n) \end{pmatrix}. \tag{5.5}$$

To interpolate the observed outputs $y_1, \cdots, y_n$ over the observed inputs $\{\mathbf{x}_1, \cdots, \mathbf{x}_n\}$ using the basis $B_0(\mathbf{x}), B_1(\mathbf{x}), \cdots$, we take L large enough such that following equation has a solution:

$$\mathbf{y} = \mathbf{B}\boldsymbol{\beta}. \tag{5.6}$$

There are many different ways to construct the basis functions. We might face a situation where Equation (5.6) becomes an undetermined system of equations, and therefore, there are many different solutions for $\boldsymbol{\beta}$ that interpolate the sampled data $\mathbf{y}$. This occurs when $L \geq n$ or when the columns of B are not linearly independent. To deal with this issue, statisticians choose a parameter set $\boldsymbol{\beta}$ that minimizes $\|\boldsymbol{\beta}\|^2$ under the constraint of (5.6). This method is referred to as the *normalized method of frame* in the statistical literature (Rao (1973) and Antoniadis and Fan (2001)). Donoho (2004*a,b*) showed that minimizing the ℓ_1-norm $\|\boldsymbol{\beta}\|_1 = \sum_{j=1}^{L} |\beta_j|$ under the constraint of (5.6) yields a very sparse solution, i.e., the resulting estimate of $\boldsymbol{\beta}$ has many zero elements.

Li (2002) established a connection between the normalized method of frame and penalized least squares for modeling computer experiments. Instead of dealing with the constraint minimization problem directly, we consider the following minimization formulation:

$$\|\boldsymbol{\beta}\|^2 + \lambda_0 \|\mathbf{y} - \mathbf{B}\boldsymbol{\beta}\|^2, \tag{5.7}$$

where $\| \cdot \|$ is the Euclidean norm and λ_0 is a Lagrange multiplier. Consider (5.7) as a penalized sum of least squares

$$\frac{1}{2}\|\mathbf{y} - \mathbf{B}\boldsymbol{\beta}\|^2 + \lambda \sum_{l=1}^{L} \beta_j^2, \tag{5.8}$$

where λ is referred to as a *regularization parameter* or *tuning parameter*. The solution of (5.8) can be expressed as

$$\hat{\boldsymbol{\beta}}_\lambda = (\mathbf{B}'\mathbf{B} + 2\lambda\mathbf{I}_L)^{-1}\mathbf{B}'\mathbf{y}, \tag{5.9}$$

where $\mathbf{I}_L$ is the identity matrix of order L. The solution of (5.9) is known as ridge regression in the statistical literature (Hoerl and Kennard (1970)), and λ corresponds to a *ridge parameter*. We can further obtain the following metamodel

$$g_\lambda(\mathbf{x}) = \mathbf{b}(\mathbf{x})'\hat{\boldsymbol{\beta}}_\lambda.$$

The regularization parameter λ can be selected to trade off the balance between bias and variance of the resulting metamodel using the cross validation procedure introduced in the previous section. In particular, taking λ to be zero yields an interpolation, which may not be the best predictor in terms of prediction error, as demonstrated in Example 18.

The connection between the normalized method of frame and penalized least squares allows us to extend the concepts of variable selection and model selection in statistics to modeling for computer experiments, because variable selection can be viewed as a type of regularization. In many applications, we may want to perform optimization using the computer model to find a point with maximum/minimum response over the experimental domain. In this situation, it is desirable to build a metamodel g which has good prediction capability over the experimental domain yet which also has a simple form. This motivates us to pursue a less complicated metamodel (i.e., more parsimonious model) by applying the variable selection approach. For this reason, variable selection may play an important role in modeling for computer experiments.

Considering other penalty functions, define a penalized least squares as

$$Q(\boldsymbol{\beta}) = \frac{1}{2}\|\mathbf{y} - \mathbf{B}\boldsymbol{\beta}\|^2 + n\sum_{j=0}^{L} p_\lambda(|\beta_j|), \tag{5.10}$$

where $p_\lambda(\cdot)$ is a pre-specified nonnegative penalty function, and λ is a regularization parameter, which may depend on n and can be chosen by a data-driven criterion, such as the cross validation procedure or a modification thereof. In principle, a larger value of λ yields a simpler model, i.e., a model with fewer terms (cf. Table 5.3). Minimizing the penalized least squares yields a penalized least squares estimator. Fan and Li (2001) demonstrated that with proper choices of penalty function, the resulting estimate can automatically reduce model complexity. Many traditional variable selection criteria in the literature can be derived from the penalized least squares with the L_0 penalty:

$$p_\lambda(|\beta|) = \frac{1}{2}\lambda^2 I(|\beta| \neq 0),$$

where $I(\cdot)$ is the indicator function.

Tibshirani (1996) proposed the LASSO algorithm, which can provide the solution of (5.10) with the L_1 penalty:

$$p_\lambda(|\beta|) = \lambda|\beta|.$$

The L_1 penalty is popular in the literature of support vector machines (see, for example, Smola and Schölkopf (1998).) Frank and Friedman (1993) considered the L_q penalty:

$$p_\lambda(|\beta|) = \lambda|\beta|^q, \quad q > 0.$$

The penalized least squares with the L_q penalty yields a "bridge regression." Fan and Li (2001) advocated the use of the smoothly clipped absolute deviation (SCAD) penalty, whose derivative is defined by

$$p'_\lambda(\beta) = \lambda\{I(\beta \le \lambda) + \frac{(a\lambda - \beta)_+}{(a - 1)\lambda}I(\beta > \lambda)\} \text{ for some } a > 2 \quad \text{and } \beta > 0$$

with $p_\lambda(0) = 0$ and $a = 3.7$. See Section A.4 for more discussions on variable selection criteria and procedures for linear regression models.

Throughout this chapter, we shall apply the above regularization methods to various metamodeling techniques.

5.2 Polynomial Models

Polynomial models have been widely used by engineers for modeling computer experiments. In polynomial models, the basis functions are taken as (see (1.23)):

$$\begin{aligned}
&B_0(\mathbf{x}) = 1, B_1(\mathbf{x}) = x_1, \cdots, B_s(\mathbf{x}) = x_s, \\
&B_{s+1}(\mathbf{x}) = x_1^2, \cdots, B_{2s}(\mathbf{x}) = x_s^2, \\
&B_{2s+1}(\mathbf{x}) = x_1 x_2, \cdots, B_{s(s+3)/2}(\mathbf{x}) = x_{s-1} x_s, \\
&\cdots.
\end{aligned} \tag{5.11}$$

Note that the number of polynomial basis functions dramatically increases with the number of input variables and the degree of the polynomial. Thus, to simplify the modeling process, lower-order polynomials such as the quadratic polynomial model or the centered quadratic model (1.24) with all first-order interactions have been the common choice in practice. Let $\{\mathbf{x}_1, \cdots, \mathbf{x}_n\}$ consist of a design, and $y_1, \cdots, y_n$ are their associated outputs. Using notation defined in (5.4) and (5.5), the least squares approach is to minimize

$$\sum_{i=1}^{n}\{y_i - \sum_{j=0}^{L} B_j(\mathbf{x}_i)\beta_j\}^2 = \|\mathbf{y} - \mathbf{B}\boldsymbol{\beta}\|^2,$$

which results in the least squares estimator

$$\hat{\boldsymbol{\beta}} = (\mathbf{B}'\mathbf{B})^{-1}\mathbf{B}'\mathbf{y},$$

if $\mathbf{B}$ has a full rank.

The polynomial basis functions in (5.11) are adapted from traditional polynomial regression in statistics. They are not an orthogonal basis, and may suffer the drawback of collinearity. To alleviate the problem, orthogonal polynomial bases have been introduced to model computer experiments in the

literature. An and Owen (2001) introduced quasi-regression for approximation of functions on the unit cube $[0,1]^s$. They further constructed the basis
function over the unit cube by taking tensor products of univariate basis function. Let $\phi_0(u) = 1$ for all $u \in [0,1]$. For integers $j \geq 1$, let $\phi_j(u)$s satisfy

$$\int_0^1 \phi_j(u)\,du = 0, \quad \int_0^1 \phi_j^2(u)\,du = 1, \quad \text{and} \quad \int_0^1 \phi_j(u)\phi_k(u)\,du = 0, \text{ for } j \neq k.$$

An s dimensional tensor product basis function over $\mathbf{x} = (x_1, \cdots, x_s)' \in [0,1]^s$
is then as follows

$$\Phi_{r_1,\cdots,r_s}(\mathbf{x}) = \prod_{j=1}^s \phi_{r_j}(x_j). \tag{5.12}$$

Let $B_0(\mathbf{x}) \equiv \Phi_{0,\cdots,0}(\mathbf{x}) = 1$, and any finite set of functions $\Phi_{r_1,\cdots,r_s}(\mathbf{x})$ can
serve to define the basis function $B_l(\mathbf{x})$ in (5.11).

The univariate orthogonal polynomials over $[0,1]$ can be constructed using
Legendre polynomials over $[-1/2, 1/2]$ by a location transformation. The first
few Legendre polynomials are:

$\phi_0(u) = 1$

$\phi_1(u) = \sqrt{12}(u - 1/2)$

$\phi_2(u) = \sqrt{180}\{(u - 1/2)^2 - \dfrac{1}{12}\}$

$\phi_3(u) = \sqrt{2800}\{(u - 1/2)^3 - \dfrac{3}{20}(u - 1/2)\}$

$\phi_4(u) = 210\{(u - 1/2)^4 - \dfrac{3}{14}(u - 1/2)^2 + \dfrac{3}{560}\}$

$\phi_5(u) = 252\sqrt{11}\{(u - 1/2)^5 - \dfrac{5}{18}(u - 1/2)^3 + \dfrac{5}{336}(u - 1/2)\}$

$\phi_6(u) = 924\sqrt{13}\{(u - 1/2)^6 - \dfrac{15}{44}(u - 1/2)^4 + \dfrac{5}{176}(u - 1/2)^2 - \dfrac{5}{14784}\}.$

These orthogonal polynomials together with their tensor products (5.12) can
be easily used to construct an orthogonal polynomial basis.

Due to the structure of the polynomial basis, as the number of variables
and the order of polynomials increase, the number of possible terms in the
polynomial basis grows rapidly. Hence, the number of possible candidate polynomial interpolators grows dramatically. As a result, the required number of
data samples also increases dramatically, which can be prohibitive for computationally expensive simulation models. Therefore, people usually limit the
model to only linear or up to lower-order models or models with fixed terms. In
practice, once a polynomial interpolator is selected, the second stage consists
of reducing the number of terms in the model following a selection procedure,
such as a stepwise selection based on C_p, AIC, BIC, or ϕ-criterion (see Section
A.4). The selected model usually has better prediction power, although it may
not exactly interpolate the observed data. Giglio et al. (2000) suggested a new

orthogonal procedure to select a parsimonious model. Bates et al. (2003) proposed a global selection procedure for polynomial interpolators based upon the concept of curvature of fitted functions. As discussed in the last section, variable selection can be regarded as a kind of regularization. We may directly employ the penalized least squares (5.10) to select a parsimonious model.

There are many case studies using a polynomial basis in the literature. See, for instance, Fang, Lin, Winker and Zhang (2000), An and Owen (2001), and Bates et al. (2003). We conclude this section by illustrating a new application of polynomial basis approximation.

Example 2 (Continued)

The computer model of interest simulates engine cylinder head and block joint sealing including the assembly process (e.g., head bolt rundown) as well as engine operating conditions (e.g., thermal and cylinder pressure cyclical loads due to combustion process). A computer experiment employing uniform design with 27 runs and 8 factors was conducted to optimize the design of the head gasket for sealing function. The 8 factors (see Table 5.1) are given below:

x_1: gasket thickness

x_2: number of contour zones

x_3: zone-to-zone transition

x_4: bead profile

x_5: coining depth

x_6: deck face surface flatness

x_7: load/deflection variation

x_8: head bolt force variation

The small number of runs is necessary due to the simulation setup complexity and excessive computing requirements. The objective of the design is to optimize the head gasket design factors $(x_1, \cdots, x_5)$ so that it minimizes the "gap lift" of the assembly as well as its sensitivity to manufacturing variation (x_6, x_7, x_8). The response y, the gap lift, is given in the last column of Table 5.1.

The main effect plots shown in Figure 5.2 indicate that gasket thickness (x_1) and surface flatness (x_6) are the most important factors affecting gap lift. One particular interest in this study is to set gasket design factors so that the effect of manufacturing variations (e.g., surface flatness) are minimized. The interaction plots are shown in Figure 5.3, from which it can be seen that the interaction between gasket thickness and surface flatness is strong; therefore, this suggests that one can select a gasket thickness such that the gap lift is insensitive to the surface flatness variation. The main effect plots in Figure 5.2 are used only for reference in this case. Very often, the main effect plots can provide us with very rich information.

TABLE 5.1
Engine Block and Head Joint Sealing Assembly Data

x_1	x_2	x_3	x_4	x_5	x_6	x_7	x_8	y
2	2	3	2	2	1	2	3	1.53
3	3	3	2	3	1	3	1	2.21
1	1	2	3	2	1	3	3	1.69
3	1	2	1	2	2	3	1	1.92
1	1	2	2	3	1	1	2	1.42
1	3	2	3	3	3	2	2	5.33
1	3	1	2	1	2	3	3	2.00
2	3	2	1	1	1	1	1	2.13
3	2	1	3	3	2	1	2	1.77
2	1	1	2	1	3	1	3	1.89
1	3	3	1	3	2	1	3	2.17
3	2	2	3	1	2	1	3	2.00
3	3	1	3	2	1	2	3	1.66
2	1	1	3	3	2	3	1	2.54
1	2	1	1	3	1	2	1	1.64
3	1	3	2	3	3	2	3	2.14
1	2	3	1	1	3	3	2	4.20
3	2	2	2	1	3	2	1	1.69
1	2	1	2	2	3	1	1	3.74
2	2	2	1	3	3	3	3	2.07
2	3	3	3	2	3	1	1	1.87
2	3	2	2	2	2	2	2	1.19
3	3	1	1	2	3	3	2	1.70
2	2	3	3	1	1	3	2	1.29
2	1	1	1	1	1	2	2	1.82
1	1	3	3	1	2	2	1	3.43
3	1	3	1	2	2	1	2	1.91

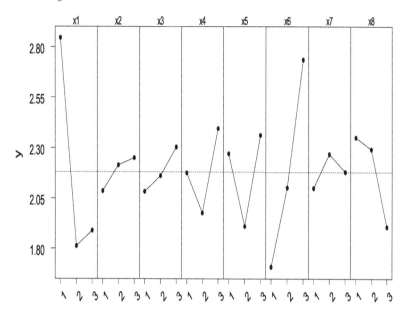

FIGURE 5.2
Main effect plots.

Consider the following quadratic polynomial model:

$$f(\mathbf{x}) = \beta_0 + \beta_1 x_1 + \cdots, \beta_8 x_8 + \beta_9 x_1^2 + \cdots + \beta_{16} x_8^2$$
$$+ \beta_{17} x_1 x_2 + \cdots + \beta_{44} x_7 x_8.$$

Note that we have only 27 runs. The above quadratic polynomial model has 45 parameters to estimate and will result in an over-fitting model. To select significant terms in the model, we first standardize all x-variables, and then apply traditional variable selection procedures to select significant terms. Table 5.2 lists the variables in the selected models. From Table 5.2, it can be concluded that *gasket thickness* and *desk face surface flatness* are important factors. From the selected model by the ϕ-criterion, x_8 and x_2 may have some effects. We also noticed that the coefficient for the interaction between x_1 and x_6 in all selected models is negative. This implies that the effect of surface flatness, a manufacturing variation variable, can be minimized by adjusting gasket thickness.

We next illustrate how to rank importance of variables using penalized least squares with the SCAD penalty (see Section 2.3.3). We first standardize all x-variables and then apply the penalized least squares with the SCAD penalty for the underlying polynomial model, and take the values of λ from a very large one to a very small one. The results are depicted in Table 5.3.

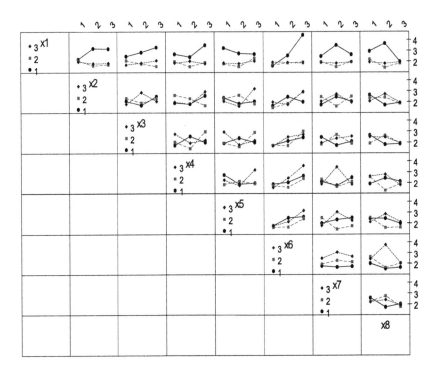

FIGURE 5.3
Interaction effect plots.

TABLE 5.2
Variable Selection

Criterion	Variables in the Selected Model
AIC	x_1^2, x_6^2, x_1x_2, x_1x_6
BIC	x_1^2, x_6^2, x_1x_6
RIC	x_6^2, x_1x_6
ϕ-criterion	x_1^2, x_6^2, x_8^2, x_1x_2, x_1x_6, x_2x_6

TABLE 5.3

Rank Importance of Variables

λ	Selected Model
0.4212	x_6
0.4012	x_6, x_1
0.1930	**x_6, x_1x_6**
0.1750	x_6, x_8^2, x_1x_6
0.1512	$x_6, x_8^2, x_1x_6, x_4x_5$
0.1185	$x_6^2, x_8^2, x_1x_6, x_4x_5$
0.1023	$x_8, x_6^2, x_1x_6, x_4x_5$
0.0975	$x_8, x_6^2, x_1x_6, x_3x_7, x_4x_5$
0.0928	$x_6^2, x_8^2, x_1x_6, x_3x_7, x_4x_5$
0.0884	$x_4^2, x_6^2, x_1x_6, x_4x_5, x_6x_7, x_7x_8$

From Table 5.3, when $\lambda = 0.4212$, only x_6 is selected. This implies that surface flatness is the most important variable. When $\lambda = 0.4012$, both x_6 and x_1 are selected. Therefore, the gasket thickness is also an important variable, next in importance to the surface flatness. Following Fan and Li (2001), the generalized cross validation (GCV, Craven and Wahba (1979)) method (see also Appendix) was employed to select the tuning parameter λ. For a certain class of estimators, the GCV is equivalent to the cross validation introduced in Section 5.1.1, but the GCV method requires much less computation. The GCV chooses $\lambda = 0.1930$; the corresponding row is highlighted by bold-face font in Table 5.3. With this value of λ, the interaction between x_1 and x_6 has significant impact. From Table 5.3, the rank of importance of variables seems to be

$$x_6, x_1, x_8, x_4, x_5, x_3, x_7, x_2.$$

With the chosen value of $\lambda = 0.1930$, the selected model is

$$\hat{y}(\mathbf{x}) = 1.1524 + 1.0678x_6 - 0.2736x_1x_6.$$

The negative interaction between gasket thickness and the surface flatness implies that the effect of surface flatness, a manufacturing variation variable, can be minimized by adjusting gasket thickness. We will further use this example to demonstrate how to rank important variable by using decomposition of sum of squares in Chapter 6.

5.3 Spline Method

Splines are frequently used in nonparametric regression in the statistical literature. The spline method mainly includes *smoothing splines* (Wahba (1990)), *regression splines* (Stone et al. (1997)), and *penalized splines* (Eilers and Marx (1996) and Ruppert and Carroll (2000)). Here we focus on regression splines, as an extension to polynomial models. We will introduce penalized splines in Section 7.3.

Example 18 (Continued)

For the purpose of illustration, we refit the data in Example 18 using a regression spline. Since the polynomial of order 9 is the best polynomial fit, in order to make the spline fit comparable to the polynomial fit, we consider a quadratic spline model with 10 degrees of freedom (i.e., terms):

$$\begin{aligned} f(x) &= \beta_0 + \beta_1 x + \beta_2^2 x^2 + \beta_3(x - \kappa_1)_+^2 + \beta_4(x - \kappa_2)_+^2 + \beta_5(x - \kappa_3)_+^2 \\ &+ \beta_6(x - \kappa_4)_+^2 + \beta_7(x - \kappa_5)_+^2 + \beta_8(x - \kappa_6)_+^2 + \beta_9(x - \kappa_7)_+^2, \quad (5.13) \end{aligned}$$

where $\kappa_i = i/8$ are termed as knots, $a_+ = a\,I(a \geq 0)$, and $I(\cdot)$ is the indicator function. The linear spline basis is depicted in Figure 5.4(a), from which we can see that the linear spline basis is a piecewise linear function. The resulting metamodel is given by

$$\begin{aligned} \hat{f}(x) &= 0.00001756 + 3.8806x - 32.6717x^2 + 37.4452(x - 0.125)_+^2 \\ &+ 52.1643(x - 0.25)_+^2 - 85.5367(x - 0.375)_+^2 - 67.4558(x - 0.50)_+^2 \\ &+ 166.3704(x - 0.625)_+^2 + 39.9065(x - 0.75)_+^2 - 181.3511(x - 0.875)_+^2. \end{aligned}$$

The resulting fit is depicted in Figure 5.4, from which the linear spline has a slightly better fit to data than the best polynomial fit. One may further pursue a better spline fit by using a higher order spline and/or adding more knots.

5.3.1 Construction of spline basis

The above example shows that splines can be used to approximate univariate functions as easily as the polynomial regression. The mechanism of constructing a multi-dimensional spline basis is also similar to that of polynomial regression. Let us begin with construction of a one-dimensional spline basis. For given knots, $\kappa_1, \cdots, \kappa_K$, the p-th order spline is defined as

$$s(x) = \beta_0 + \beta_1 x + \beta_2 x^2 + \cdots + \beta_p x^p + \sum_{k=1}^{K} \beta_{p+k}(x - \kappa_k)_+^p, \quad (5.14)$$

where $p \geq 1$ is an integer. It is not difficult to see that $s(x)$ given by (5.14) is a pth degree polynomial on each interval between two consecutive knots and has $p - 1$ continuous derivatives everywhere. The pth derivative of $s(\cdot)$ takes a jump of size β_{p+l} at the l-th knot, k_l. The spline basis used in (5.14) is a

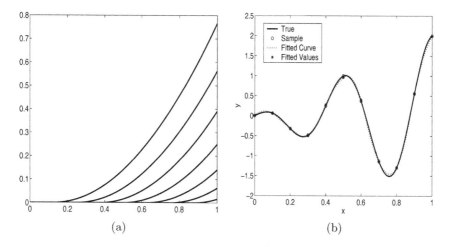

FIGURE 5.4

(a) Quadratic spline basis, and (b) plot of spline fit.

power basis and can be replaced by another spline basis, such as a B-spline (De Boor (1978))).

A spline basis in multi-dimensional functional space is constructed by using a tensor product. It is common to standardize the x-variables first such that the ranges for all x-variables are the same. Then take the same knots for each x-variables. For given fixed knots $\kappa_1, \cdots, \kappa_K$, denote

$$S_0(u) = 1, S_1(u) = u, \cdots, S_p(u) = u^p,$$
$$S_{p+1}(u) = (u - \kappa_1)_+^p, \cdots, S_{p+K}(u) = (u - \kappa_K)_+^p.$$

Then an s-dimensional tensor product basis function over $\mathbf{x} = (x_1, \cdots, x_s)'$ is

$$B_{r_1, \cdots, r_s}(\mathbf{x}) = \prod_{j=0}^{s} S_{r_j}(x_j). \qquad (5.15)$$

It is easy to see that $B_{0, \cdots, 0}(\mathbf{x}) = 1$, and any finite set of functions $B_{r_1, \cdots, r_s}(\mathbf{x})$, $r_1 = 0, \cdots, K + p, \cdots; r_s = 0, \cdots, K + p$, can form the basis function of a function space over $\mathbf{x}$. As discussed in Friedman (1991), the tensor product method presents us with a straightforward approach for extending one dimensional regression spline fitting to multidimensional model fitting. However, its implementation is difficult in practice because the number of coefficients to be estimated from data increases exponentially as the dimension s increases. Using a recursive partitioning approach, Friedman (1991) proposed an approach called *multivariate adaptive regression splines* (MARS). In MARS, the number of basis functions and the knot locations are adaptively determined by the data. Friedman (1991) presented a thorough discussion on MARS and gave an algorithm to create spline basis functions for MARS. Readers are referred to his paper for details.

Let $B_0(\mathbf{x}), \cdots, B_L(\mathbf{x})$ be a finite subset of $\{B_{r_1,\cdots,r_s}(\mathbf{x}) : r_1 = 0, \cdots, K + p; \cdots; r_s = 0, \cdots, K + p\}$. Consider a regression spline model:

$$f(\mathbf{x}) = \sum_{j=1}^{L} \beta_j B_j(\mathbf{x}).$$

When $L + 1 \leq n$, the least squares estimator for $(\beta_0, \cdots, \beta_L)'$ is

$$\hat{\boldsymbol{\beta}} = (\mathbf{B}'\mathbf{B})^{-1}\mathbf{B}'\mathbf{y},$$

where $\mathbf{B}$ is the associated design matrix (cf. Section 1.2). However, only when $N + 1 \geq n$, does the resulting metamodel exactly interpolate the observed sample. How to choose a good subset from the set of spline basis functions is an important issue. In a series of works by C.J. Stone and his collaborators (see, for example, Stone et al. (1997)), the traditional variable selection approaches were modified to select useful sub-bases. The regularization method introduced in Section 5.1.2 can be applied to the regression spline model directly. This is referred to as *penalized splines* in the statistical literature. Ruppert (2002) has explored the usefulness of penalized splines with the L_2 penalty.

5.3.2 An illustration

The borehole example has been analyzed using the polynomial regression model in Section 1.8. Here, the penalized spline approach is used to analyze the borehole data, as demonstrated by Li (2002). We first employ a $U_{30}(30^8)$ to generate a design with 30 experiments, and the outcomes and their corresponding design points are depicted in Table 5.4. Similar to analysis in Section 1.8, we take $x_1 = \log(r_w)$, $x_2 = \log(r)$, $x_3 = T_u$, $x_4 = T_l$, $x_5 = H_u$, $x_6 = H_l$, $x_7 = L$, and $x_8 = K_w$. Let z_i be the standardized variable of x_i. We construct quadratic splines $(z_i - \kappa_{ij})_+^2$ with the knots at the median of z_i, denoted by κ_{i2}, and lower and upper 20 percentiles of z_i, denoted by κ_{i1} and κ_{i3}. The values of the knots of z_i's are depicted in Table 5.5. Here a_+ means the positive part of a, i.e., $a_+ = aI(a > 0)$, and is taken before taking square in the construction of the quadratic spline basis. Like the analysis in Section 1.8, we take natural logarithm of outputs as response variable y. For each variable z_i, $i = 1, \cdots, 8$,

$$z_i, z_i^2, (z_i - \kappa_1)_+^2, (z_i - \kappa_2)_+^2, (z_i - \kappa_3)_+^2,$$

are introduced to the initial model. We also include an intercept and all interactions of z_i's in the initial model. Thus, there is a total of 68 x-variables in the

TABLE 5.4

Designs and Outputs

r_w	r	T_u	T_l	H_u	H_l	L	K_w	output
0.0617	45842	84957	106.3017	1108	762	1335	9990	30.8841
0.1283	35862	93712	81.6150	1092	790	1241	10136	126.2840
0.0950	10912	81456	111.5917	1016	794	1223	10063	51.6046
0.0883	22555	114725	67.5083	1008	778	1260	10048	44.7063
0.0583	14238	95464	71.0350	1068	810	1559	9983	17.6309
0.0683	4258	88460	88.6683	1080	766	1129	10005	40.7011
0.0817	27545	105970	109.8283	1088	802	1428	10034	41.9919
0.1350	19228	76203	69.2717	1032	738	1148	10151	146.8108
0.0717	40852	104218	74.5617	1000	746	1372	10012	29.8083
0.1050	39188	72701	101.0117	1064	814	1167	10085	74.3997
0.0750	20892	98965	99.2483	1036	754	1671	10019	29.8223
0.1383	17565	79704	93.9583	1100	782	1652	10158	116.6914
0.1183	32535	74451	108.0650	1076	722	1540	10114	101.7336
0.1417	44178	83207	72.7983	1060	706	1447	10165	154.9332
0.1083	12575	112974	86.9050	1104	730	1484	10092	93.2778
0.1117	5922	69199	63.9817	1056	774	1409	10100	78.5678
0.1017	2595	86708	83.3783	992	714	1633	10078	55.4821
0.1150	49168	97215	90.4317	1020	726	1204	10107	101.7270
0.0783	24218	63945	76.3250	1096	718	1279	10027	56.9115
0.1217	7585	107721	78.0883	1040	806	1353	10121	80.7530
0.0850	47505	67447	85.1417	1044	798	1615	10041	34.6025
0.0983	37525	100717	65.7450	1084	742	1596	10070	65.1636
0.0650	9248	70949	102.7750	1012	734	1521	9997	24.2095
0.0550	30872	109472	95.7217	1052	710	1185	9975	27.3042
0.0517	34198	77954	79.8517	1024	786	1465	9968	13.5570
0.1450	932	102468	104.5383	1072	750	1297	10173	165.6246
0.0917	15902	91961	115.1183	1048	702	1391	10056	65.8352
0.1250	29208	65697	97.4850	996	758	1316	10129	89.2366
0.1483	25882	90211	92.1950	1004	818	1503	10180	86.2577
0.1317	42515	111222	113.3550	1028	770	1577	10143	89.7999

TABLE 5.5

Knots for Regression Spline

Variable	z_1	z_2	z_3	z_4	z_5	z_6	z_7	z_8
κ_1	−1.042	−0.7144	−1.0564	−1.0564	−1.0564	−1.0564	−1.0564	−1.0564
κ_2	0.1439	0.3054	0	0	0	0	0	0
κ_3	1.0072	0.8152	1.0564	1.0564	1.0564	1.0564	1.0564	1.0564

initial model. Since we only conduct 30 experiments, the initial model is over-parameterized. Penalized least squares with the SCAD penalty (Section A.4) is applied for this example. The estimated λ is $1.1923 * 10^{-5}$. The resulting coefficients are depicted in Table 5.6. We further estimate the MSE via randomly generating 10,000 new samples, and the estimated MSE is 0.1112, which is much smaller than the one obtained by polynomial regression in Section 1.8.

TABLE 5.6
Estimate of Penalized Least Squares Estimates

Variable	Est. Coeff.	Variable	Est. Coeff.	Variable	Est. Coeff.
intercept	4.0864	z_7^2	0.0067	$(z_3 - k_1)_+^2$	0.0021
z_1	0.6289	$z_1 z_2$	0.0005	$(z_4 - k_1)^2$	-0.0004
z_3	-0.0041	$z_1 z_4$	0.0024	$(z_2 - k_2)_+^2$	-0.0002
z_4	0.0010	$z_2 z_3$	-0.0005	$(z_5 - k_2)_+^2$	-0.0127
z_5	0.1231	$z_4 z_6$	0.0007	$(z_7 - k_2)_+^2$	0.0014
z_6	-0.1238	$z_5 z_6$	0.0165	$(z_8 - k_2)_+^2$	0.0275
z_7	-0.1179	$z_5 z_7$	0.000035	$(z_1 - k_3)_+^2$	-0.0174
z_5^2	-0.0095	$z_6 z_7$	-0.0011	$(z_2 - k_3)_+^2$	-0.0014
z_6^2	-0.0072	$(z_2 - k_1)_+^2$	-0.0007		

5.3.3 Other bases of global approximation

Polynomial basis and spline basis are the most popular bases in modeling computer experiments. Other bases for global approximation of an unknown function have also appeared in the literature. For example, the Fourier basis can be used to approximate periodic functions. It is well known that

$$1, \cos(2\pi x), \sin(2\pi x), \cdots, \cos(2k\pi x), \sin(2k\pi x), \cdots,$$

forms an orthogonal basis for a functional space over $[0, 1]$. Using the method of tensor product, we may construct an s-dimensional basis as follows:

$$B_{r_1, \cdots, r_{2s}}(\mathbf{x}) = \prod_{j=1}^{s} \cos(2r_1\pi x_1)\sin(2r_2\pi x_1)\cdots\cos(2(r_{2s-1}\pi x_s)\sin(2r_{2s}\pi x_s)$$

for $r_1, \cdots, r_{2s} = 0, 1, \cdots$. In practice, the following Fourier regression model is recommended:

$$\beta_0 + \sum_{i=1}^{s}\sum_{j=1}^{m}\{\alpha_{ij}\cos(2j\pi x_i) + \beta_{ij}\sin(2j\pi x_i)\}.$$

The Fourier basis has been utilized for modeling computer experiments in Bates et al. (1996). Motivated by Fang and Wang (1994), Bates et al. (1996) further showed that the good lattice point method (see Section 3.3.3) forms a *D*-optimal design for the Fourier basis.

5.4 Gaussian Kriging Models

The Kriging method was proposed by a South African geologist, D.G. Krige, in his master's thesis (Krige (1951)) on analyzing mining data. His work was further developed by many other authors. The Gaussian Kriging method was proposed by Matheron (1963) for modeling spatial data in geostatistics. See Cressie (1993) for a comprehensive review of the Kriging method. The Kriging approach was systematically introduced to model computer experiments by Sacks, Welch, Mitchell and Wynn (1989). Since then, Kriging models have become popular for modeling computer experiment data sets. Suppose that $\mathbf{x}_i$, $i = 1, \cdots, n$ are design points over an s-dimensional experimental domain T, and $y_i = y(\mathbf{x}_i)$ is the associated output to $\mathbf{x}_i$. The Gaussian Kriging model is defined as

$$y(\mathbf{x}) = \sum_{j=0}^{L} \beta_j B_j(\mathbf{x}) + z(\mathbf{x}), \tag{5.16}$$

where $\{B_j(\mathbf{x}), j = 1, \cdots, L\}$ is a chosen basis over the experimental domain, and $z(\mathbf{x})$ is a random error. Instead of assuming the random error is independent and identically distributed, it is assumed that $z(\mathbf{x})$ is a Gaussian process (cf. Appendix A.2.2 for a definition and Section 5.5.1 for basic properties) with zero mean, variance σ^2, and correlation function

$$r(\boldsymbol{\theta}; \mathbf{s}, \mathbf{t}) = \text{Corr}(z(\mathbf{s}), z(\mathbf{t})). \tag{5.17}$$

Here $r(\boldsymbol{\theta}; \mathbf{s}, \mathbf{t})$ is a pre-specified positive definite bivariate function of $z(\mathbf{s})$ and $z(\mathbf{t})$. Model (5.16) is referred to as the *universal Kriging model*. In the literature,

$$y(\mathbf{x}) = \mu + z(\mathbf{x})$$

is referred to as the *ordinary Kriging model*, which is the most commonly used Kriging model in practice.

A natural class for the correlation function of z, and hence of y, is the stationary family where

$$r(\boldsymbol{\theta}; \mathbf{s}, \mathbf{t}) \equiv r(\boldsymbol{\theta}; |\mathbf{s} - \mathbf{t}|).$$

For instance, the following function

$$r(\boldsymbol{\theta}; \mathbf{s}, \mathbf{t}) = \exp\{-\sum_{k=1}^{s} \theta_k |s_k - t_k|^q\}, \qquad \text{for} \quad 0 < q \le 2$$

or

$$r(\boldsymbol{\theta}; \mathbf{s}, \mathbf{t}) = \exp\{-\theta \sum_{k=1}^{s} |s_k - t_k|^q\}, \qquad \text{for} \quad 0 < q \le 2.$$

is popular in the literature. The case $q = 1$ is the product of Ornsterin-Uhenbeck processes. The case $q = 2$ gives a process with infinitely differentiable paths and is useful when the response is analytic. We will further introduce some other correlation functions in Section 5.5.

5.4.1   Prediction via Kriging

It is common to expect that a predictor $g(\mathbf{x})$ of $f(\mathbf{x})$ or $\hat{y}(\mathbf{x})$ of $y(\mathbf{x})$ satisfies some properties:

DEFINITION 5.1   (a) A predictor $\hat{y}(\mathbf{x})$ of $y(\mathbf{x})$ is a *linear predictor* if it has the form $\hat{y}(\mathbf{x}) = \sum_{i=1}^{n} c_i(\mathbf{x}) y_i$.

(b) A predictor $\hat{y}(\mathbf{x})$ is an *unbiased predictor* if $E\{\hat{y}(\mathbf{x})\} = E\{y(\mathbf{x})\}$.

(c) A predictor $\hat{y}(\mathbf{x})$ is the *best linear unbiased predictor* (BLUP) if it has the minimal mean squared error (MSE), $E\{\hat{y}(\mathbf{x}) - y(\mathbf{x})\}^2$, among all linear unbiased predictors.

The Gaussian Kriging approach essentially is a kind of linear interpolation built on the following property of the multivariate normal distribution. Let $\mathbf{z} \sim N_{n+1}(\mathbf{0}, \boldsymbol{\Sigma})$, and partition $\mathbf{z}$ into $(z_1, \mathbf{z}_2)'$, where z_1 is univariate and $\mathbf{z}_2$ is n-dimensional. Then, given $\mathbf{z}_2$, the conditional expectation of z_1 is

$$E(z_1|\mathbf{z}_2) = \boldsymbol{\Sigma}_{12}\boldsymbol{\Sigma}_{22}^{-1}\mathbf{z}_2, \tag{5.18}$$

where $\boldsymbol{\Sigma}_{12} = \text{Cov}(z_1, \mathbf{z}_2)$ and $\boldsymbol{\Sigma}_{22} = \text{Cov}(\mathbf{z}_2)$. Equation (5.18) implies that for a given observed value of $\mathbf{z}_2$, a prediction to z_1 is

$$\hat{z}_1 = \boldsymbol{\Sigma}_{12}\boldsymbol{\Sigma}_{22}^{-1}\mathbf{z}_2.$$

This is a linear combination of components of $\mathbf{z}_2$.

It can be shown that $\hat{z}_1$ is a linear unbiased predictor. Under the normality assumption, it can be further shown that $\hat{z}_1$ is the BLUP. To facilitate the normality assumption, the Gaussian Kriging approach views $z(\mathbf{x}_i) = y(x_i) - \sum_{j=0}^{L} \beta_j B_j(\mathbf{x}_i)$, $i = 1, \cdots, n$, the residuals of deterministic output, as the realization of a Gaussian process $z(\mathbf{x})$. Thus, for any new $\mathbf{x}^* \in T$ (that is, $\mathbf{x}^*$ is different from $\mathbf{x}_1, \cdots, \mathbf{x}_n$), $(z(\mathbf{x}^*), z(\mathbf{x}_1), \cdots, z(\mathbf{x}_n))'$ follows an $n + 1$-dimensional normal distribution. Therefore, a linear predictor for $z(\mathbf{x}^*)$ can easily be obtained by using (5.18), and is denoted by $\hat{z}(\mathbf{x}^*)$. Furthermore, the predictor for $y(\mathbf{x}^*)$ can be obtained by using (5.19) below.

Let $R(\boldsymbol{\theta})$ be an $n \times n$ matrix whose (i, j)-element is $r(\boldsymbol{\theta}; \mathbf{x}_i, \mathbf{x}_j)$. Under the normality assumption on $z(\mathbf{x})$, the BLUP of $y(\mathbf{x})$ can be written as

$$\hat{y}(\mathbf{x}) = \mathbf{b}(\mathbf{x})\hat{\boldsymbol{\beta}} + \mathbf{r}'(\mathbf{x})R^{-1}(\boldsymbol{\theta})(\mathbf{y} - \mathbf{B}\hat{\boldsymbol{\beta}}), \tag{5.19}$$

where $\mathbf{r}(\mathbf{x}) = (r(\boldsymbol{\theta}; \mathbf{x}_1, \mathbf{x}), \cdots, r(\boldsymbol{\theta}; \mathbf{x}_n, \mathbf{x}))'$, and other notations given in (5.4) and (5.5). Furthermore, the MSE of the BLUP is

$$\sigma^2 \left[1 - (\mathbf{b}'(\mathbf{x}), \mathbf{r}'(\mathbf{x})) \begin{pmatrix} \mathbf{0} & \mathbf{B}' \\ \mathbf{B} & R(\boldsymbol{\theta})^{-1} \end{pmatrix} \begin{pmatrix} \mathbf{b}(\mathbf{x}) \\ \mathbf{r}(\mathbf{x}) \end{pmatrix} \right].$$

Equation (5.18) is a critical assumption of the Gaussian Kriging approach to model computer experiments. This assumption can be satisfied not only by normal distributions but also by elliptical distributions, such as multivariate t-distribution and scale mixture of normal distributions. See Fang, Kotz and Ng (1990) for properties of elliptical distributions.

5.4.2 Estimation of parameters

From the normality assumption of the Gaussian Kriging model, the density of $\mathbf{y}$ is

$$(2\pi\sigma^2)^{-n/2} |R(\boldsymbol{\theta})|^{-1/2} \exp \left\{ -\frac{1}{2\sigma^2} (\mathbf{y} - \mathbf{B}\boldsymbol{\beta})' R(\boldsymbol{\theta})^{-1} (\mathbf{y} - \mathbf{B}\boldsymbol{\beta}) \right\}.$$

Then, after dropping a constant, the log-likelihood function of the collected data equals

$$\ell(\boldsymbol{\beta}, \sigma^2, \boldsymbol{\theta}) = -\frac{n}{2} \log(\sigma^2) - \frac{1}{2} \log |R(\boldsymbol{\theta})| - \frac{1}{2\sigma^2} (\mathbf{y} - \mathbf{B}\boldsymbol{\beta})' R^{-1}(\boldsymbol{\theta})(\mathbf{y} - \mathbf{B}\boldsymbol{\beta}). \tag{5.20}$$

Maximizing the log-likelihood function yields the maximum likelihood estimate of $(\boldsymbol{\beta}, \sigma^2, \boldsymbol{\theta})$. In practice, simultaneous maximization over $(\boldsymbol{\beta}, \sigma^2, \boldsymbol{\theta})$ is unstable because $R(\boldsymbol{\theta})$ may be nearly singular and σ^2 could be very small. Furthermore, the parameters $\boldsymbol{\beta}$ and $\boldsymbol{\theta}$ play different roles: $\boldsymbol{\beta}$ is used to model overall trend, while $\boldsymbol{\theta}$ is a smoothing parameter vector. It is desirable to estimate them separately. Since

$$\frac{\partial^2 \ell(\boldsymbol{\beta}, \sigma^2, \boldsymbol{\theta})}{\partial \boldsymbol{\beta} \partial \sigma^2} = -\frac{1}{2\sigma^4} \mathbf{B}' R^{-1}(\boldsymbol{\theta})(\mathbf{y} - \mathbf{B}\boldsymbol{\beta}),$$

$$\frac{\partial^2 \ell(\boldsymbol{\beta}, \sigma^2, \boldsymbol{\theta})}{\partial \boldsymbol{\beta} \partial \boldsymbol{\theta}} = \frac{1}{2\sigma^2} \mathbf{B}' \frac{\partial \mathbf{R}^{-1}(\boldsymbol{\theta})}{\partial \boldsymbol{\theta}} (\mathbf{y} - \mathbf{B}\boldsymbol{\beta}),$$

and $E(\mathbf{y}) = \mathbf{B}\boldsymbol{\beta}$,

$$E \left\{ \frac{\partial^2 \ell(\boldsymbol{\beta}, \sigma^2, \boldsymbol{\theta})}{\partial \boldsymbol{\beta} \partial \sigma^2} \right\} = \mathbf{0}, \quad \text{and} \quad E \left\{ \frac{\partial^2 \ell(\boldsymbol{\beta}, \sigma^2, \boldsymbol{\theta})}{\partial \boldsymbol{\beta} \partial \boldsymbol{\theta}} \right\} = \mathbf{0}.$$

This implies that the maximum likelihood estimator of β is asymptotically independent of that of $(\sigma^2, \boldsymbol{\theta})$ because the Fisher information matrix,

$$E\left\{\frac{\partial^2 \ell(\beta, \sigma^2, \boldsymbol{\theta})}{\partial \beta \partial(\sigma^2, \boldsymbol{\theta})}\right\}$$

is block-diagonal. This allows one to estimate β and $(\sigma^2, \boldsymbol{\theta})$ separately and iteratively by using the Fisher scoring algorithm or using the Newton-Raphson algorithm and ignoring the off-block-diagonal matrix.

It has been empirically observed that the prediction based on the simultaneous maximization over $(\beta, \sigma^2, \boldsymbol{\theta})$ performs almost the same as that which relies on the estimation of β and $(\sigma^2, \boldsymbol{\theta})$ separately. This is consistent with the above theoretical analysis. In practice, we may estimate β and $(\sigma^2, \boldsymbol{\theta})$ iteratively in the following way.

Initial value of $\boldsymbol{\theta}$: As studied by Bickel (1975) in other statistical settings, with a good initial value, the one-step estimator may be as efficient as the full iteration one. In practice, one may stop at any step during the course of iterations. It is clear that a good initial value for β is the least squares estimator. This is equivalent to setting the initial value of $\boldsymbol{\theta}$ to be $\mathbf{0}$.

The maximum likelihood estimate of β: For a given $\boldsymbol{\theta}$, the maximum likelihood estimate of β is

$$\hat{\beta}_{\mathrm{mle}} = (\mathbf{B}'R^{-1}(\boldsymbol{\theta})\mathbf{B})^{-1}\mathbf{B}'R^{-1}(\boldsymbol{\theta})\mathbf{y} \tag{5.21}$$

which can be derived via minimizing

$$(\mathbf{y} - \mathbf{B}\beta)'R^{-1}(\boldsymbol{\theta})(\mathbf{y} - \mathbf{B}\beta)$$

with respect to β. This formulation is referred to as the generalized least squares criterion in the literature since

$$E(\mathbf{y}) = \mathbf{B}\beta \quad \text{and} \quad \mathrm{Cov}(\mathbf{y}) = \sigma^2 R(\boldsymbol{\theta}).$$

Estimation of σ^2: The maximum likelihood estimator for σ^2 also has an expressive form. By setting $\partial \ell(\beta, \sigma^2, \boldsymbol{\theta})/\partial \sigma^2 = 0$, it follows that for a given $\boldsymbol{\theta}$,

$$\hat{\sigma}^2 = n^{-1}(\mathbf{y} - \mathbf{B}\hat{\beta})'R^{-1}(\boldsymbol{\theta})(\mathbf{y} - \mathbf{B}\hat{\beta}), \tag{5.22}$$

which is a biased estimator for σ^2, although it is close to unbiased as n increases. An unbiased estimator for σ^2 is

$$\hat{\sigma}^2 = (n - L - 1)^{-1}(\mathbf{y} - \mathbf{B}\hat{\beta})'R^{-1}(\boldsymbol{\theta})(\mathbf{y} - \mathbf{B}\hat{\beta}),$$

Estimation of $\boldsymbol{\theta}$: The maximum likelihood estimator for $\boldsymbol{\theta}$ does not have a closed form. The Newton-Raphson algorithm or Fisher scoring algorithm may be used to search for the solution. Before we continue, we apply the Gaussian Kriging model for the data set in Example 18.

Algorithm for Estimation of Gaussian Kriging Model:

Step 1: Set the initial value of $\boldsymbol{\beta}$ to be $(\mathbf{B}'\mathbf{B})^{-1}\mathbf{B}\mathbf{y}$, the least squares estimator of $\boldsymbol{\beta}$;

Step 2: For a given $\boldsymbol{\beta}$, we update $(\sigma^2, \boldsymbol{\theta})$ by using (5.22) and solving the following equation

$$\partial\ell(\boldsymbol{\beta}, \sigma^2, \boldsymbol{\theta})/\partial\boldsymbol{\theta} = 0.$$

This step requires a numerical iteration algorithm, such as the Newton-Raphson algorithm or Fisher scoring algorithm, to solve the above equation. Fortunately, these algorithms are rather simple to implement.

Step 3: For a given $\boldsymbol{\theta}$, update $\boldsymbol{\beta}$ using (5.21);

Step 4: Iterate Step 2 and Step 3 until it converges.

Example 18 (Continued) As a demonstration, we refit the data in Example 18 using the following Gaussian Kriging model:

$$y(x) = \mu + z(x),$$

where $z(x)$ is a Gaussian process with mean zero and covariance

$$\mathrm{Cov}\{z(s), z(t)\} = \sigma^2 \exp\{-\theta|s - t|^2\}.$$

The maximum likelihood estimates for (μ, θ, σ^2) are $(\hat{\mu}, \hat{\theta}, \hat{\sigma}^2) = (-0.6745, 10, 14.9443)$. Furthermore, $\hat{\mathbf{b}} = R^{-1}(\hat{\theta})(\mathbf{y} - \mu\mathbf{1}_n) = (26.6721, -103.5381, 167.6298, -76.4172, -219.0115, 589.1082, -868.8839, 970.2693, -822.4150, 459.6337)$. The resulting prediction is depicted in Figure 5.5, from which we can see that the Gaussian Kriging exactly interpolates the observed sample. The predicted curve and the true curve are almost identical in this example. Compared with the polynomial regression model and regression spline, the Gaussian Kriging model provides us with a better prediction in this case.

Remark 5.1 As a natural alternative approach to the maximum likelihood estimate, one may use the restricted maximum likelihood (REML, Patterson and Thompson (1971)) method to estimate the parameters involving the covariance matrix. The REML is also called the residual or modified maximum likelihood in the literature. After dropping a constant, the logarithm of REML for model (5.16) is given by

$$\ell_r(\boldsymbol{\beta}, \sigma^2, \boldsymbol{\theta}) = \frac{(n - L - 1)}{2} \log \sigma^2 - \frac{1}{2} \log |R(\boldsymbol{\theta})| - \frac{1}{2} \log |\mathbf{B}'R^{-1}(\boldsymbol{\theta})\mathbf{B}|$$
$$- \frac{1}{2\sigma^2}(\mathbf{y} - \mathbf{B}\boldsymbol{\beta})'R^{-1}(\boldsymbol{\theta})(\mathbf{y} - \mathbf{B}\boldsymbol{\beta}).$$

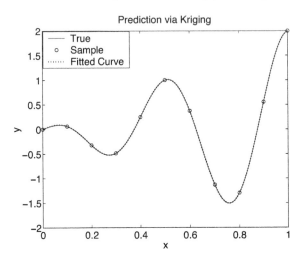

FIGURE 5.5
Plot of Gaussian Kriging model fit.

Maximizing $\ell_r(\boldsymbol{\beta}, \sigma^2, \boldsymbol{\theta})$ yields a REML estimate for the unknown parameters. The algorithm to optimize $\ell_r(\boldsymbol{\beta}, \sigma^2, \boldsymbol{\theta})$ is parallel to that for $\ell(\boldsymbol{\beta}, \sigma^2, \boldsymbol{\theta})$.

In practice, one of the serious problems with the maximum likelihood estimate of $\boldsymbol{\theta}$ is that the resulting estimate may have a very large variance because the likelihood function near the optimum is flat. Li and Sudjianto (2005) proposed a penalized likelihood approach to deal with this problem. For simplicity, we refer to the Gaussian Kriging approach with penalized likelihood as a penalized Kriging approach. We illustrate the motivation of the penalized Kriging approach via the following simple example, demonstrated in Li and Sudjianto (2005).

Example 19 Consider the following one-dimensional function:

$$y = \sin(x).$$

Let the sample data be $\{x_i = i : i = 0, 2, \cdots, 10\}$. Consider the following Gaussian Kriging model to fit the data:

$$y(x) = \mu + z(x),$$

where $z(x)$ is a Gaussian process with mean zero and covariance:

$$\text{Cov}(z(s), z(t)) = \sigma^2 \exp\{-\theta|s - t|^2\}.$$

For a given θ, the maximum likelihood estimate for μ and σ^2 can be easily computed. We can further compute the profile likelihood function $\ell(\theta)$, which equals the maximum of the likelihood function over μ and σ^2 for any given θ. The corresponding Gaussian logarithm of profile likelihood (log-likelihood, for short) function $\ell(\theta)$ versus θ is depicted in Figure 5.6(a), from which we can see that the likelihood function becomes almost flat for $\theta \geq 1$. The prediction based on the Gaussian Kriging model is displayed in Figure 5.6(b), which shows that the prediction becomes very erratic when x is not equal to the sampled data.

We now consider the REML method. The corresponding logarithm of the profile restricted likelihood function versus θ is depicted in Figure 5.6(c). From the figure we can see that the shape of the profile restricted likelihood function in this example is the same as that of Figure 5.6(a), and that it achieves its maximum at $\theta = 3$ and becomes almost flat for $\theta \geq 1$. The prediction based on the REML is displayed in Figure 5.6(c). The prediction is the same as that in Figure 5.6(b) because $\hat{\theta} = 3$, which is the same as that obtained by the ordinary likelihood approach. The prediction based on REML becomes very erratic when x is not equal to the sample data.

To avoid erratic behavior, Li and Sudjianto (2005) considered a penalized likelihood approach as follows. Define a penalized likelihood as

$$Q(\boldsymbol{\beta}, \sigma^2, \boldsymbol{\theta}) = \ell(\boldsymbol{\beta}, \sigma^2, \boldsymbol{\theta}) + n \sum_{j=1}^{s} p_\lambda(|\theta_j|),$$

where $\ell(\boldsymbol{\beta}, \sigma^2, \boldsymbol{\theta})$ is defined in (5.18), and $p_\lambda(\cdot)$ is a penalty function with a regularization parameter λ. For investigating the performance of the penalized Kriging method, a penalized log-likelihood function with the SCAD penalty is depicted in Figure 5.6(e), and its corresponding prediction is displayed in Figure 5.6(f). Figure 5.6(e) clearly shows that the penalized likelihood function is not flat around the optimum. The curvature around the optimum implies that the resulting estimate for θ possesses smaller standard error. From Figure 5.6(f), the prediction and the true curve are almost identical. Readers are referred to Li and Sudjianto (2005) for theoretical analysis of the penalized likelihood approach.

To confirm the model selected by penalized Kriging, we take a much larger sample $\{x_i = i : i = 0, 0.5, 1, \cdots, 10\}$. The corresponding likelihood function is depicted in Figure 5.6(g) and the prediction is shown in Figure 5.6(h). Comparing Figure 5.6(e) with Figure 5.6(g), the locations of the maximum likelihood estimate and the penalized maximum likelihood estimate are very close. Furthermore, Figure 5.6(h) confirms that the predictions yielded by the penalized Kriging method are quite accurate.

Model selection techniques can be employed to estimate the parameter $\boldsymbol{\theta}$, because it can be viewed as a smoothing parameter vector (see Section 5.7 for more discussions). In the literature, the cross validation procedure described in Section 5.1.1 has been utilized to select the smoothing parameter vector. However, it can be very computationally expensive because the cross validation procedure has to search over s-dimensional grid points. A useful way

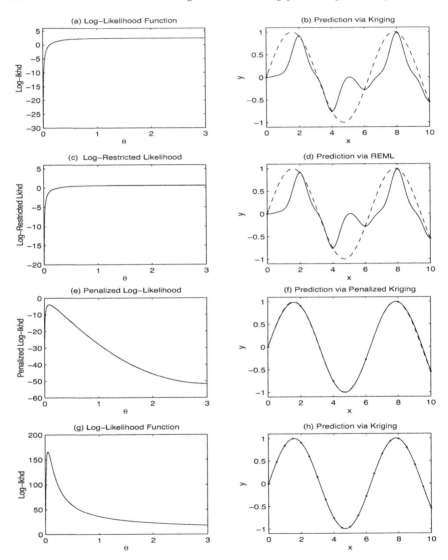

FIGURE 5.6

Kriging and penalized Kriging. (a) and (g) are the log-likelihood functions
for Kriging with sample size $n = 6$ and 21, respectively. (b) and (h) are
prediction via Kriging with sample size $n = 6$ and 21, respectively. (c) is the
log-restricted likelihood function for Kriging with $n = 6$. (d) is the prediction
via Kriging with $n = 6$ using the REML method. (e) is the penalized log-
likelihood function for Kriging with $n = 6$. (f) is the prediction via penalized
Kriging with $n = 6$. In (b), (d), (f) and (h), the solid line stands for prediction,
the dashed line stands for the true curve, and dots stands for prediction at

to reduce computational burden is to set $\theta_j = \sigma_j \theta$, where σ_j is the sample standard deviation of the j-component of $\mathbf{x}$. This indeed reduces the Kriging method to a radial basis function method. See Section 5.6 for a thorough discussion on how to implement the radial basis function method to model computer experiments.

5.4.3 A case study

Total vehicle customer satisfaction depends greatly on the customer's level of satisfaction with the vehicle's engine. The noise, vibration, and harshness (NVH) characteristics of the vehicle and engine are critical elements of customer dissatisfaction. Piston slap is an unwanted engine noise caused by piston secondary motion. de Luca and Gerges (1996) gave a comprehensive review of the piston slap mechanism and experimental piston slap analysis, including noise source analysis and parameters influencing piston slap. Since then, with the advent of faster, more powerful computers, much of the piston slap study has shifted from experimental analysis to analytical analysis for both the power cylinder design phase and for piston noise troubleshooting. Thus, it is desirable to have an analytical model to describe the relationship between the piston slap noise and its covariates, such as piston skirt length, profile, and ovality.

We first give a brief description of this study. A detailed and thorough explanation of this study can be found in Hoffman et al. (2003). Piston slap is a result of piston secondary motion, that is, the departure of the piston from the nominal motion prescribed by the slider crank mechanism. The secondary motion is caused by a combination of transient forces and moments acting on the piston during engine operation and the presence of clearances between the piston and the cylinder liner. This combination results in both a lateral movement of the piston within the cylinder and a rotation of the piston about the piston pin, and it causes the piston to impact the cylinder wall at regular intervals. These impacts may result in the objectionable engine noise.

For this study, the power cylinder system is modeled using the multi-body dynamics code ADAMS/Flex that includes a finite element model. The piston, wrist pin, and connecting rod were modeled as flexible bodies, where flexibility is introduced via a model superposition. Boundary conditions for the flexible bodies are included via a Craig-Bampton component mode synthesis. The crankshaft is modeled as a rigid body rotating with a constant angular velocity. In addition, variation in clearance due to cylinder bore distortion and piston skirt profile and ovality are included in the analysis.

We take the piston slap noise as the output variable and have the following input variables:

x_1 = Clearance between the piston and the cylinder liner,
x_2 = Location of Peak Pressure,
x_3 = Skirt length,

x_4 = Skirt profile,
x_5 = Skirt ovality,
x_6 = Pin offset.

TABLE 5.7
Piston Slap Noise Data

Run	x_1	x_2	x_3	x_4	x_5	x_6	y
1	71.00	16.80	21.00	2.00	1.00	0.98	56.75
2	15.00	15.60	21.80	1.00	2.00	1.30	57.65
3	29.00	14.40	25.00	2.00	1.00	1.14	53.97
4	85.00	14.40	21.80	2.00	3.00	0.66	58.77
5	29.00	12.00	21.00	3.00	2.00	0.82	56.34
6	57.00	12.00	23.40	1.00	3.00	0.98	56.85
7	85.00	13.20	24.20	3.00	2.00	1.30	56.68
8	71.00	18.00	25.00	1.00	2.00	0.82	58.45
9	43.00	18.00	22.60	3.00	3.00	1.14	55.50
10	15.00	16.80	24.20	2.00	3.00	0.50	52.77
11	43.00	13.20	22.60	1.00	1.00	0.50	57.36
12	57.00	15.60	23.40	3.00	1.00	0.66	59.64

Since each computer experiment requires intensive computational resources, uniform design (cf. Chapter 3) was employed to plan a computer experiment with 12 runs. The collected data are displayed in Table 5.7. The ultimate goal of the study is to perform robust design optimization (Hoffman et al. (2003)) to desensitize the piston slap noise from the source of variability (e.g., clearance variation). To accomplish this goal, the availability of a good metamodel is a necessity. A Gaussian Kriging model is employed to construct a metamodel as an approximation to the computationally intensive analytical model. In this discussion, we only focus on the development of the metamodel. Interested readers should consult Hoffman et al. (2003) and Du et al. (2004) for the probabilistic design optimization study.

In this example, we consider the following Gaussian Kriging model:

$$y(\mathbf{x}_i) = \mu + z(\mathbf{x}_i),$$

where $z(\mathbf{x})$ is a Gaussian process with zero mean, variance σ^2, and correlation function between $\mathbf{x}_i$ and $\mathbf{x}_j$:

$$r(\mathbf{x}_i, \mathbf{x}_j) = \exp\left\{ -\sum_{k=1}^{s} \theta_k |x_{ik} - x_{jk}|^2 \right\}.$$

We first conduct some preliminary analysis using correlation functions with a single parameter, i.e., $\theta_k = \theta/\sigma_k$, in order to quickly gain a rough picture

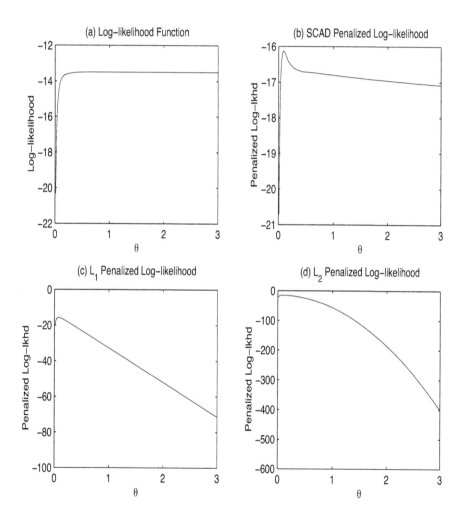

FIGURE 5.7

Log-likelihood and penalized log-likelihood of the case study example with $n = 12$. (a) is log-likelihood function. (b), (c), and (d) are penalized log-likelihood functions with the SCAD, L_1, and L_2 penalty, respectively.

of the logarithm of the profile likelihood function (log-likelihood, for short).
In our implementation, we intuitively set $\theta_k = \theta/\sigma_k$, where σ_k stands for the
sample standard deviation of the k-th component of $\mathbf{x}_i, i = 1, \cdots, n$. This
choice of θ_k allows us to plot the log-likelihood function $l(\theta)$ against θ. Plots
of the log-likelihood function, and the penalized log-likelihood functions with
the SCAD, the L_1 and the L_2 penalties, where $\lambda = 0.2275 (= 0.5\sqrt{\log(n)/n})$
and $n = 12$, are depicted in Figure 5.7, from which we can see that the log-
likelihood function is flat when the log-likelihood function near its optimum
$(\hat{\theta} = 3)$ is flat. This creates the same problem as that in Example 19 when
the sample size equals 6. The three penalized log-likelihood functions near
its optimum are not flat. Although the three penalty functions are quite
different, and the shape of the corresponding penalized likelihood functions
look very different, their resulting penalized maximum likelihood estimates for
θ under the constraint $\theta_k = \theta/\sigma_k$ and $\theta_j \geq 0$ are very close. From the shape
of the penalized log-likelihood functions, the resulting estimate of penalized
likelihood with the SCAD penalty may be more efficient than the other two.
This preliminary analysis gives us a rough picture of log-likelihood function
and the penalized likelihood function.

TABLE 5.8
Penalized Maximum Likelihood Estimate

Parameter	MLE	SCAD	L_1	L_2
$\hat{\lambda}$		0.1100	0.1300	0.0600
$\hat{\mu}$	56.7275	56.2596	56.5177	56.5321
$\hat{\sigma}^2$	3.4844	4.1170	3.6321	3.4854
$\hat{\theta}_1$	0.1397	0.8233E-3	0.1670E-2	0.3776E-2
$\hat{\theta}_2$	1.6300	0.18570E-6	0.1418E-3	0.2433E-1
$\hat{\theta}_3$	2.4451	0.4269E-1	0.5779	0.2909
$\hat{\theta}_4$	4.0914	0.5614E-6	0.2022E-3	0.3264E-1
$\hat{\theta}_5$	4.0914	0.3027E-5	0.1501	0.9798E-1
$\hat{\theta}_6$	12.2253	4.6269	0.1481E-1	0.2590
MSE	2.9953	2.1942	4.3067	3.1087
MAR	1.3375	1.0588	1.4638	1.3114

The leave-one-out cross validation procedure (cf. Section 5.2) was used to
estimate the tuning parameter λ. The resulting estimate of λ equals 0.1100,
0.1300, and 0.06 for the SCAD, the L_1, and the L_2 penalties, respectively.
The resulting estimates of μ, σ^2, θ_js are depicted in Table 5.8. The four
estimates for μ are very close, but the four penalized likelihood estimates for
σ^2 and θ_js are quite different. Comparisons below recommend the resulting
estimate of the penalized likelihood with the SCAD penalty.

Denote by $\hat{\mathbf{b}} = R^{-1}(\hat{\boldsymbol{\theta}})(\mathbf{y} - \mathbf{1}_n\hat{\mu})$ the best linear unbiased predictor for the response variable at input variable $\mathbf{x}$,

$$\hat{y}(\mathbf{x}) = \hat{\mu} + \mathbf{r}(\mathbf{x})\hat{\mathbf{b}},$$

where $\mathbf{r}(\mathbf{x})$ was defined in Section 5.4.1.

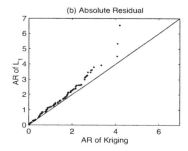

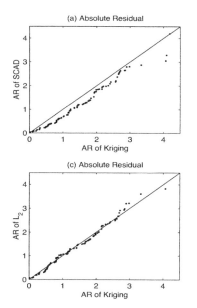

FIGURE 5.8

Plots of absolute residuals.

To assess the performance of the penalized Gaussian Kriging approach, we conduct another computer experiment with 100 runs at new sites, and further compute the median of absolute residuals (MAR), defined as

$$\text{MAR} = \text{median}\{|y(\mathbf{x}_i) - \hat{y}(\mathbf{x}_i)| : i = 1, \cdots, 100\}.$$

The MAR for the ordinary Kriging method is 1.3375, and MARs equals 1.0588, 1.4638, and 1.3114 for the penalized Kriging method with the SCAD, the L_1, and the L_2 penalties, respectively. The metamodel obtained by the penalized Kriging with the SCAD penalty outperforms the other three metamodels. We now plot the sorted absolute residuals from penalized Kriging versus the absolute residuals from Kriging in Figure 5.8, from which we can see that the penalized Kriging with the SCAD uniformly improves the ordinary Kriging model. The penalized Kriging with the L_2 penalty has almost the same performance as that of the ordinary Kriging model, while the penalized Kriging with the L_1 penalty does not perform well in this case.

To understand the behavior of the penalized Kriging method when the sample size is moderate, we apply the penalized Kriging method to the new sample with 100 runs. Again, let $\theta_j = \theta/\sigma_j$, and plot the log-likelihood against θ in Figure 5.9, from which we can see that the shape of the log-likelihood function is the same as that of the penalized log-likelihood function with the SCAD penalty. We further compute the maximum likelihood estimate for all of the parameters θ_j. The selected λ values are 0.18, 0.105, and 0.18, for the SCAD, L_1, and L_2 penalties, respectively. The resulting estimate is listed in Table 5.9, from which we found that all of these estimates are close, as expected.

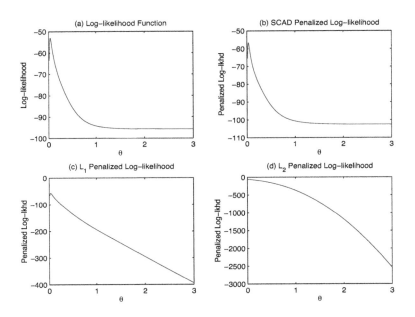

FIGURE 5.9

Log-likelihood and penalized log-likelihood when $n = 100$. (a) is the log-likelihood function. (b), (c) and (d) are the penalized log-likelihood functions with the SCAD, L_1, and L_2 penalties, respectively.

TABLE 5.9
Penalized Maximum Likelihood Estimate When $n = 100$

Parameter	MLE	SCAD	L_1	L_2
$\hat{\theta}_1$	0.4514E-4	0.3971E-4	0.3943E-4	0.5858E-4
$\hat{\theta}_2$	0.5634E-3	0.5192E-3	0.5204E-3	0.6519E-3
$\hat{\theta}_3$	0.3150E-5	0.2602E-5	0.2618E-5	0.4261E-5
$\hat{\theta}_4$	0.2880	0.2752	0.2765	0.3003
$\hat{\theta}_5$	0.2939E-1	0.2593E-1	0.2590E-1	0.3641E-1
$\hat{\theta}_6$	0.1792	0.1515	0.1548	0.2162

5.5 Bayesian Approach

Bayesian interpolation has a long history (see Diaconis (1988) for an interesting account of a method published by Poincaré in 1896). Kimeldorf and Wahba (1970) and Wahba (1978) established the connection between Bayesian interpolation and smoothing splines. Bayesian interpolation was introduced to model computer experiments by Currin et al. (1991) and Morris et al. (1993). These authors provided a Bayesian formulation for Gaussian Kriging models. Morris et al. (1993) focus on computer models that can provide not only the response but also its first partial derivatives.

5.5.1 Gaussian processes

Gaussian processes have played an important role in Bayesian interpolation. MacKay (1998) described how inferences for a Gaussian process work and how they can be implemented with finite computational resources. In this section, we briefly introduce some basic properties of Gaussian processes.

Let $\{Y(\mathbf{x}), \mathbf{x} \in T\}$ be a Gaussian process with the mean, variance, and covariance functions as

$$\mu(\mathbf{x}) = E\{Y(\mathbf{x})\}, \quad \sigma^2(\mathbf{x}) = \text{Var}\{Y(\mathbf{x})\},$$

and for any $\mathbf{s}, \mathbf{t} \in T$,

$$\sigma(\mathbf{s}, \mathbf{t}) = \text{Cov}(Y(\mathbf{s}), Y(\mathbf{t})),$$

respectively. When $\sigma(\mathbf{s}, \mathbf{t})$ is entirely a function of $\mathbf{s} - \mathbf{t}$, the process $\{Y(\mathbf{x}), \mathbf{x} \in T\}$ is called stationary.

For the sake of further discussion in Section 5.5.3, denote

$$Y_j(\mathbf{x}) = \frac{\partial Y(\mathbf{x})}{\partial x_j},$$

the derivative of $Y(\mathbf{x})$ with respect to x_j. The general derivative process is defined as

$$Y^{(a_1,\cdots,a_s)}(\mathbf{x}) = \frac{\partial^{a_1+\cdots+a_s}}{\partial x_1^{a_1}\cdots\partial x_s^{a_s}}Y(\mathbf{x})$$

for $a_j \geq 0$, $j = 1,2,\cdots,s$. We summarize the properties of the derivative process in the following theorem.

Theorem 11 *If $Y(\mathbf{x})$ is a Gaussian process with mean function $\mu(\mathbf{x})$ covariance function $\sigma(\mathbf{s},\mathbf{t})$, then $Y^{(a_1,\cdots,a_s)}(\mathbf{x})$ is a Gaussian process with mean function and covariance function given by*

$$E\{Y^{(a_1,\cdots,a_s)}(\mathbf{x})\} = \mu^{(a_1,\cdots,a_s)}(\mathbf{x}) \tag{5.23}$$

and

$$\mathrm{Cov}\{Y^{(a_1,\cdots,a_s)}(\mathbf{s}), Y^{(b_1,\cdots,b_s)}(\mathbf{t})\} = \sigma^{(a_1,\cdots,a_s,b_1,\cdots,b_s)}(\mathbf{s},\mathbf{t}). \tag{5.24}$$

Moreover, $(Y(\mathbf{x}), Y_1(\mathbf{x}), \cdots, Y_s(\mathbf{x}))$ is $(s+1)$-dimensional Gaussian processes with mean function and covariance function given in (5.23) and (5.24), respectively (cf. Section 1 of Morris et al. (1993)).

5.5.2 Bayesian prediction of deterministic functions

Suppose that $\{\mathbf{x}_i, i = 1,\cdots,n\}$ are design points over an s-dimensional experimental domain T, and $y_i = y(\mathbf{x}_i)$ is the associated output to $\mathbf{x}_i$. In the context of Bayesian interpolation, the prior "knowledge" about the unknown function $y(\mathbf{x})$, $\mathbf{x} \in T$, is taken to be the Gaussian process $Y = \{Y(\mathbf{x}), \mathbf{x} \in T\}$, such that for every finite set of sites $U = \{\mathbf{u}_1,\cdots,\mathbf{u}_N\} \subset T$, the random vector $\mathbf{Y}_U = (Y_{\mathbf{u}_1},\cdots,Y_{\mathbf{u}_N})'$ is multivariate normal with mean vector $E(\mathbf{Y}_U) = (\mu(\mathbf{u}_1),\cdots,\mu(\mathbf{u}_N))' \equiv \boldsymbol{\mu}_U$ and with covariance matrix $\mathrm{Cov}(\mathbf{Y}_U) \equiv \Sigma_{UU}$, an $N \times N$ positive definite matrix with (i,j)-element $\sigma(\mathbf{u}_i, \mathbf{u}_j)$. With this prior specification, it is well known that the posterior process, given the vector of observed responses $\mathbf{Y}_O \equiv (y_1,\cdots,y_n)^T$, is also a Gaussian process. Furthermore, the posterior mean and covariance at any finite set of sites $U \subset T$ have the following closed forms:

$$\boldsymbol{\mu}_{U|\mathbf{Y}_O} = E(\mathbf{Y}_U|\mathbf{Y}_O) = \boldsymbol{\mu}_U + \mathrm{Cov}(\mathbf{Y}_U, \mathbf{Y}_O)\,\mathrm{Cov}(\mathbf{Y}_O)^{-1}\{\mathbf{Y}_O - E(\mathbf{Y}_O)\}, \tag{5.25}$$

and

$$\begin{aligned}\Sigma_{UU|\mathbf{Y}_O} &= \mathrm{Cov}(\mathbf{Y}_U, \mathbf{Y}_U|\mathbf{Y}_O)\\ &= \Sigma_{UU} - \mathrm{Cov}(\mathbf{Y}_U, \mathbf{Y}_O)\,\mathrm{Cov}(\mathbf{Y}_O)^{-1}\,\mathrm{Cov}(\mathbf{Y}_O, \mathbf{Y}_U).\end{aligned} \tag{5.26}$$

Therefore, a prediction $\hat{y}(\mathbf{x})$ for $Y(\mathbf{x})$ is the posterior mean, which is also the optimal condition under the squared loss in the context of decision theory (Currin et al. (1991)). So,

$$\hat{y}(\mathbf{x}) = \mu_{\mathbf{x}|\mathbf{Y}_O} = \mu(\mathbf{x}) + \mathrm{Cov}(Y(\mathbf{x}), \mathbf{Y}_O)\,\mathrm{Cov}(\mathbf{Y}_O)^{-1}\{\mathbf{Y}_O - E(\mathbf{Y}_O)\} \tag{5.27}$$

which is the most popular choice of prediction function derived from Gaussian prior processes.

The elements of $\text{Cov}(Y(\mathbf{x}), \mathbf{Y}_O)$ can be viewed as functions of $\mathbf{x}$, and they further form a set of bases consisting of n functions of $\mathbf{x}$. The basis functions follow automatically from the choice of prior process Y and design D and do not need to be chosen by the data analyst. From an approximation theoretical point of view, (5.27) is the unique interpolating function in the spanned space of the n basis functions. Moreover, the prediction $\hat{y}(\mathbf{x})$ can also be viewed as a minimal norm interpolant. See, for example, Micchelli and Wahba (1981), Sack and Ylvisaker (1985), and Currin et al. (1991).

In practice, it is usual to consider only stationary prior processes. That is, the prior mean and variance are assumed to be constant for all sites $\mathbf{x} \in T$: $\mu(\mathbf{x}) = \mu$ and $\sigma^2(\mathbf{x}) = \sigma^2$, and prior correlation is assumed to depend on only the distance between two sites:

$$\text{Corr}(\mathbf{Y_s}, \mathbf{Y_t}) = r(\boldsymbol{\theta}; \|\mathbf{s} - \mathbf{t}\|),$$

where $r(\boldsymbol{\theta}; 0) = 1$, and $\boldsymbol{\theta}$ is a unknown parameter vector to be estimated by the collected data. The correlation function can be easily derived using the *product correlation rule*, defined by

$$r(\boldsymbol{\theta}; \|\mathbf{s} - \mathbf{t}\|) = \prod_{k=1}^{s} r_k(\theta_k; |s_k - t_k|),$$

where $r_k(\theta_k; \cdot)$ is a univariate correlation function. The correlation function

$$r(\boldsymbol{\theta}; \|\mathbf{s} - \mathbf{t}\|) = \exp\{-\sum_{k=1}^{s} \theta_k |s_k - t_k|^q\}$$

can be derived by taking

$$r_k(\theta_k; |s_k - t_k|) = \exp\{-\theta_k |s_k - t_k|^q\}$$

which is referred to as the *exponential correlation function*.

Denote l_k to be the range of the k-coordinate of input vector over T, $k = 1, \cdots, s$. The *linear correlation function* is defined as

$$r_k(\theta_k; |s_k - t_k|) = 1 - \frac{1}{\theta_k l_k} |s_k - t_k|, \quad \text{for} \quad \frac{1}{2} < \theta_k < \infty.$$

The *non-negative linear correlation function* is defined as

$$r_k(\theta_k; |s_k - t_k|) = \begin{cases} 1 - \frac{1}{\theta_k l_k} |s_k - t_k|, & \text{when} \quad |s_k - t_k| < \theta_k l_k, \\ 0, & \text{when} \quad |s_k - t_k| \geq \theta_k l_k, \end{cases}$$

Under the assumption of stationarity, equation (5.27) becomes

$$\hat{y}(\mathbf{x}) = \mu_{\mathbf{x}|\mathbf{y}_O} = \mu + \text{Cov}(Y(\mathbf{x}), \mathbf{Y}_O)\, \text{Cov}(\mathbf{Y}_O)^{-1}\{\mathbf{Y}_O - \mu \mathbf{1}_n\}, \tag{5.28}$$

which is the same as the one derived from the ordinary Gaussian Kriging model (5.16) and (5.17).

A natural way to eliminate μ and σ^2 from the prior process would be to assign them standard noninformative prior distributions. However, Currin et al. (1991) suggested estimating μ and σ^2 by their maximum likelihood estimate since the maximum likelihood estimate is the most reliable compared with various kinds of cross validation they have tried (see Currin et al. (1991)).

The estimation procedure described in Section 5.4.2 can be directly used to search the maximum likelihood estimate for $(\mu, \sigma^2, \boldsymbol{\theta})$ under the assumption of stationarity. The resulting metamodel based on the Bayesian method is exactly the same as the ordinary Gaussian Kriging model (5.16) and (5.17). The Bayesian formulation provides us the insights into why the ordinary Gaussian Kriging model works well in practice. Under the Bayesian formulation, we can easily incorporate other information into analysis of computer experiments. Specifically, we introduce how to utilize derivatives of the response.

5.5.3 Use of derivatives in surface prediction

In some situations, computer models can provide not only the response $y(\mathbf{x})$ but also first partial derivatives $y_j(\mathbf{x}) = \partial y(\mathbf{x})/\partial x_j$, $j = 1, \cdots, s$. Inspired by several applications and strong scientific interests, Morris et al. (1993) proposed a Bayesian modeling procedure which can simultaneously model the response and its derivatives. In what follows, we introduce their modeling procedure in detail.

Suppose that $\mathbf{x}_1, \cdots, \mathbf{x}_n$ are design points over the s-dimensional experimental domain T, and at each $\mathbf{x}_i$, $\mathbf{y}_i \equiv (y(\mathbf{x}_i), y_1(\mathbf{x}_i), \cdots, y_s(\mathbf{x}_i))'$ is the associated output of the true model, where $y_j(\mathbf{x}_i)$ is the partial derivative of $y(\mathbf{x}_i)$ with respect to x_j. Denote the vector consisting of all outputs by

$$\mathbf{y} = (\mathbf{y}_1', \cdots, \mathbf{y}_n')'.$$

Denote further $\mathbf{Y}_i \equiv (Y(\mathbf{x}_i), Y_1(\mathbf{x}_i), \cdots, Y_s(\mathbf{x}_i))'$. As suggested by Morris et al. (1993), the prior uncertainty about the vector $\mathbf{y}$ is represented by treating $\mathbf{y}$ as a realization of the random normal $n(s+1)$-vector

$$\mathbf{Y} = (\mathbf{Y}_1', \cdots, \mathbf{Y}_n')' \tag{5.29}$$

with mean $\boldsymbol{\mu}$ and covariance matrix Σ given by (5.23) and (5.24), respectively.

It is well known that the posterior process, denoted by $Y^*(\mathbf{x})$, is also a Gaussian process with mean function

$$\mu^*(\mathbf{x}) = E\{Y^*(\mathbf{x})|\mathbf{y}\} = \mu(\mathbf{x}) + \text{Cov}(Y(\mathbf{x}), \mathbf{Y})\Sigma^{-1}(\mathbf{y} - \boldsymbol{\mu})' \tag{5.30}$$

and covariance function

$$\begin{aligned} K^*(\mathbf{s}, \mathbf{t}) &= \text{Cov}\{Y^*(\mathbf{s}), Y^*(\mathbf{t})|\mathbf{y}\} \\ &= \sigma(\mathbf{s}, \mathbf{t}) - \text{Cov}\{Y(\mathbf{s}), \mathbf{Y}\}\Sigma^{-1}\text{Cov}\{\mathbf{Y}, Y(\mathbf{t})\}. \end{aligned} \tag{5.31}$$

Equations (5.30) and (5.31) are the posterior mean and covariance function for a general Gaussian process. The advantage to the use of stochastic processes as priors for $y(\mathbf{x})$ is that the variability of the posterior processes $Y^*(\mathbf{x})$, as expressed by the posterior covariance function $K^*(\mathbf{s}, \mathbf{t})$ in (5.31), can be used to provide measures of uncertainty, and design can be sought to minimize the expected uncertainty in some sense. See Morris et al. (1993) for details. Under the assumption of stationarity, the above formulas can be simplified. Notice that

$$\mu(\mathbf{x}) \equiv \mu$$

and

$$\sigma(\mathbf{s}, \mathbf{t}) = \sigma^2 r(|\mathbf{s} - \mathbf{t}|),$$

where σ^2 is a constant, $r(\cdot)$ is a correlation function that depends only on the distance between $\mathbf{s}$ and $\mathbf{t}$, and $|\mathbf{a}|$ stands for a vector with elements $|a_j|$. In practice, $r(\cdot)$ is often taken to have the following product correlation form:

$$r(|\mathbf{s} - \mathbf{t}|) = \prod_{k=1}^{s} r_j(s_j - t_j),$$

where r_js usually are chosen from a parametric family of suitably differentiable correlation functions on the real line.

Due to the stationarity, $\mu_k(\mathbf{x}) = E\{Y_k(\mathbf{x})\} = 0$, and we further have

$$\text{Cov}(Y^{(a_1, \cdots, a_s)}(\mathbf{s}), Y^{(b_1, \cdots, b_s)}(\mathbf{t})) = \sigma^2 (-1)^{\sum a_j} \prod_{k=1}^{s} r_k^{(a_j + b_j)}(s_j - t_j). \quad (5.32)$$

Note that here we consider only the first derivative, which for each of a_j and b_j is either 0 or 1.

The choice of correlation functions $r_j(\cdot)$ must be twice differentiable. However, notice that the exponential correlation function

$$r_k(\theta_k, s_k - t_k) = \exp\{-\theta_k |s_k - t_k|^q\}, \quad 0 < q \le 2,$$

is twice differentiable only in the case of $q = 2$. Both the linear correlation function and the non-negative linear correlation function are not differentiable at some places. One may, however, use a cubic spline to construct a correlation function that is differentiable everywhere. Define a cubic correlation function as

$$r_k(\theta_k, |s_k - t_k|) = 1 - \frac{\theta_{k1}}{2l_k^2}|s_k - t_k|^2 + \frac{\theta_{k1}}{6l_k^3}|s_k - t_k|^3,$$

where θ_{k1} and θ_{k2} are positive parameters that satisfy $\theta_{k2} \le 2\theta_{k1}$ and $\theta_{k2} - 6\theta_{k1}\theta_{k2} + 12\theta_{k1}^2 \le 24\theta_{k2}$. Define further the non-negative cubic correlation function

$$
r_k(\theta_k, |s_k - t_k|) = \begin{cases} 1 - 6\left(\dfrac{|s_k - t_k|}{\theta_k l_k}\right)^2 + 6\left(\dfrac{|s_k - t_k|}{\theta_k l_k}\right)^3, & \text{if } \dfrac{|s_k - t_k|}{l_k} < \dfrac{\theta}{2}, \\ 2\left(1 - \dfrac{|s_k - t_k|}{\theta_k l_k}\right)^3, & \text{if } \dfrac{\theta}{2} \le \dfrac{|s_k - t_k|}{l_k} < \theta, \\ 0, & \text{if } \dfrac{|s_k - t_k|}{l_k} \ge \theta, \end{cases}
$$

where $\theta_k > 0$.

We now discuss how to estimate the unknown parameters presented in the model. As usual, the correlation function is fitted by a family of parametric models indexed by the parameter vector $\boldsymbol{\theta}$. Morris et al. (1993) suggest estimating $(\mu, \sigma^2, \boldsymbol{\theta})$ using the maximum likelihood estimation. Let $R(\boldsymbol{\theta})$ be the correlation matrix of $\mathbf{Y}$ in (5.31), consisting of all outputs. Furthermore, denote

$$
\mathbf{r}(\boldsymbol{\theta}) = \text{Corr}(\mathbf{y}, Y(\mathbf{x})) = (r(\boldsymbol{\theta}; |\mathbf{x} - \mathbf{x}_1|), \cdots, r(\boldsymbol{\theta}; |\mathbf{x} - \mathbf{x}_n|))'.
$$

After dropping a constant, the log-likelihood of the collected data is

$$
\ell(\mu, \sigma^2, \boldsymbol{\theta}) = -\frac{1}{2}n(s+1)\log(\sigma^2) - \frac{1}{2}\log|R(\boldsymbol{\theta})|
$$
$$
- \frac{1}{2\sigma^2}(\mathbf{y} - \mu\mathbf{e})'R^{-1}(\boldsymbol{\theta})(\mathbf{y} - \mu\mathbf{e}), \tag{5.33}
$$

where $\mathbf{e}$ is an $n(s+1)$ binary vector with 1 in position $(i-1)(s+1)+1$, $i = 1, \cdots, n$, i.e., in each position corresponding to the mean of some $Y(\mathbf{x}_i)$, and 0 everywhere else. The log-likelihood function has the same form as that in (5.20). The estimation procedure described in Section 5.4.2 can also be utilized here. In particular, for fixed $\boldsymbol{\theta}$, the maximum likelihood estimate for (μ, σ^2) is

$$
\hat{\mu}(\boldsymbol{\theta}) = \frac{\mathbf{e}'R^{-1}(\boldsymbol{\theta})\mathbf{y}}{\mathbf{e}'R^{-1}(\boldsymbol{\theta})\mathbf{e}}
$$

and

$$
\sigma^2(\boldsymbol{\theta}) = \frac{1}{n(s+1)}(\mathbf{y} - \hat{\mu}(\boldsymbol{\theta})\mathbf{e})'R^{-1}(\boldsymbol{\theta})(\mathbf{y} - \hat{\mu}(\boldsymbol{\theta})\mathbf{e}).
$$

One may employ the cross validation procedure to select an optimal $\boldsymbol{\theta}$. But since $\boldsymbol{\theta}$ is multi-dimensional, the computation could be expensive. A popular way to determine $\boldsymbol{\theta}$ is to maximize $\ell(\hat{\mu}(\boldsymbol{\theta}), \hat{\sigma}^2(\boldsymbol{\theta}), \boldsymbol{\theta})$, which is called a *profile likelihood*, and therefore the resulting estimator for $\boldsymbol{\theta}$ is referred to as the *profile likelihood estimator*. Of course, the penalized likelihood approach described in Section 5.4.2 is also applicable to the current setting.

As usual, one may directly use the posterior mean to predict the response at a new site $\mathbf{x}$. Under the assumption of stationarity, the prediction of $Y(\mathbf{x})$ is

$$
\hat{y}(\mathbf{x}) = \mu + \mathbf{r}(\hat{\boldsymbol{\theta}})'R^{-1}(\hat{\boldsymbol{\theta}})(\mathbf{y} - \mu(\hat{\boldsymbol{\theta}})\mathbf{e}). \tag{5.34}
$$

This will be implemented in the next section.

5.5.4 An example: borehole model

The borehole model (see Example 8) is a classical example in the literature of modeling computer experiments. This example was first studied by Worley (1987). Then it was used to illustrate how to use derivatives in surface prediction in Morris et al. (1993). Recently, this example has been revisited by An and Owen (2001) and Fang and Lin (2003). Section 1.8 presents a detailed description of this example. Here we summarize the demonstration of Morris et al. (1993).

For simplicity of presentation, let $x_1, \cdots, x_8$ be the scaled version of r_w, r, T_u, H_u, T_l, H_l, L and K_w, respectively. That is, for example,

$$x_1 = (r_w - 0.05)/(0.15 - 0.05).$$

Therefore, the ranges of x_is are the interval $[0, 1]$. As illustrated in Morris et al. (1993), in order to produce a somewhat more nonlinear, nonadditive function, the range of K_w is extended to $[1500, 15000]$. To demonstrate the concepts defined in Section 5.5.1, fix $x_2, \cdots, x_7$ to be zero, and view $y(\mathbf{x})$ as a response of x_1 and x_8. The 3-D plots and contour plots of the response over x_1 and x_8 are depicted in Figure 5.10(a) and (b), which were reconstructed from Morris et al. (1993).

The data, $(y(\mathbf{x}_i), y_1(\mathbf{x}_i), y_8(\mathbf{x}_i))$, $i = 1, 2, 3$, are depicted in Table 5.10. Represent the output as a 9×1 vector $\mathbf{y} = (3.0489, 12.1970, \cdots, 244.4854)'$.

TABLE 5.10
Data for Borehole Example

$\mathbf{x}$	x_1	x_8	$y(\mathbf{x})$	$y_1(\mathbf{x})$	$y_8(\mathbf{x})$
$\mathbf{x}_1$	0.0000	0.0000	3.0489	12.1970	27.4428
$\mathbf{x}_2$	0.2680	1.0000	71.6374	185.7917	64.1853
$\mathbf{x}_3$	1.0000	0.2680	93.1663	123.6169	244.4854

In this example, let us take the correlation function to be the exponential correlation function. Let $r_j(\theta_j; |s_j - t_j|) = \exp\{-\theta_j |s_j - t_j|^2\}$ and

$$\mathbf{r}(\boldsymbol{\theta}; \|\mathbf{s} - \mathbf{t}\|) = \prod_{j=1}^{s} r_j(\theta_j; |s_j - t_j|^2).$$

To calculate the correlation function of the derivative processes of $Y(\mathbf{x})$ by

Theorem 11, we first calculate the derivatives of $\mathbf{r}(\boldsymbol{\theta}; |\mathbf{s} - \mathbf{t}|)$. Denote

$$r_{00}(\mathbf{s}, \mathbf{t}) = r(\boldsymbol{\theta}; |\mathbf{s} - \mathbf{t}|),$$
$$r_{10}(\mathbf{s}, \mathbf{t}) = r^{(10000000,00000000)}(\boldsymbol{\theta}; |\mathbf{s} - \mathbf{t}|) = -2\theta_1(s_1 - t_1)r(\boldsymbol{\theta}; |\mathbf{s} - \mathbf{t}|),$$
$$r_{01}(\mathbf{s}, \mathbf{t}) = r^{(00000000,10000000)}(\boldsymbol{\theta}; |\mathbf{s} - \mathbf{t}|) = -2\theta_1(t_1 - s_1)r(\boldsymbol{\theta}; |\mathbf{s} - \mathbf{t}|),$$
$$r_{11}(\mathbf{s}, \mathbf{t}) = r^{(10000000,10000000)}(\boldsymbol{\theta}; |\mathbf{s} - \mathbf{t}|) = (2\theta_1 - 4\theta_1^2(s_1 - t_1)^2)r(\boldsymbol{\theta}; |\mathbf{s} - \mathbf{t}|),$$
$$r_{80}(\mathbf{s}, \mathbf{t}) = r^{(00000001,00000000)}(\boldsymbol{\theta}; |\mathbf{s} - \mathbf{t}|) = -2\theta_8(s_8 - t_8)r(\boldsymbol{\theta}; |\mathbf{s} - \mathbf{t}|),$$
$$r_{08}(\mathbf{s}, \mathbf{t}) = r^{(00000000,00000001)}(\boldsymbol{\theta}; |\mathbf{s} - \mathbf{t}|) = -2\theta_8(t_8 - s_8)r(\boldsymbol{\theta}; |\mathbf{s} - \mathbf{t}|),$$
$$r_{88}(\mathbf{s}, \mathbf{t}) = r^{(00000001,00000001)}(\boldsymbol{\theta}; |\mathbf{s} - \mathbf{t}|) = (2\theta_8 - 4\theta_8^2(s_8 - t_8)^2)r(\boldsymbol{\theta}; |\mathbf{s} - \mathbf{t}|),$$
$$r_{18}(\mathbf{s}, \mathbf{t}) = r^{(10000000,00000001)}(\boldsymbol{\theta}; |\mathbf{s} - \mathbf{t}|)$$
$$= -4\theta_1\theta_8(s_1 - t_1)(s_8 - t_8)r(\boldsymbol{\theta}; |\mathbf{s} - \mathbf{t}|),$$
$$r_{81}(\mathbf{s}, \mathbf{t}) = r^{(10000000,00000001)}(\boldsymbol{\theta}; |\mathbf{s} - \mathbf{t}|)$$
$$= -4\theta_1\theta_8(s_1 - t_1)(s_8 - t_8)r(\boldsymbol{\theta}; |\mathbf{s} - \mathbf{t}|).$$

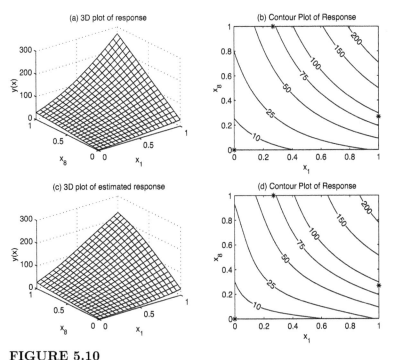

FIGURE 5.10
The 3-D plot and contour plot for the borehole example.

One may further compute Σ, the covariance matrix of $\mathbf{Y}$, as $\Sigma = \sigma^2 R(\boldsymbol{\theta})$, and the elements of $R(\boldsymbol{\theta})$ can be directly computed from the above expressions.

The log-likelihood function (5.33) achieves its maximum at

$$\hat{\mu} = 69.15, \ \hat{\sigma}^2 = 135.47, \ \hat{\theta}_1 = 0.429, \ \hat{\theta}_8 = 0.467.$$

Directly implementing (5.34) in this example, the prediction at a new site $\mathbf{x}$ is given by

$$\hat{y}(\mathbf{x}) = 69.15 + \mathbf{r}(\hat{\boldsymbol{\theta}})'\hat{\beta},$$

where $\hat{\mathbf{b}} = R^{-1}(\hat{\boldsymbol{\theta}})(\mathbf{y} - 69.15\mathbf{e})$, and

$$\hat{\mathbf{b}} = 10^3 \times (1.369, 0.579, 0.317, -0.914, 0.006, 0.418, -0.460, 0.453, 0.674)'.$$

The 3-D plot and the contour plot of the resulting metamodel are displayed in Figure 5.10(c) and (d), from which we can see the resulting metamodel provides us with a good prediction over the unobserved sites, although there are only three runs.

Morris et al. (1993) further tried four different experimental designs: Latin hypercube design, maximin design, maximum Latin hypercube design and modified maximin design, each having 10 runs, and provided a detailed analysis based on the 10 runs data. They reported that the maximization $\boldsymbol{\theta} \in [0, 1]^8$ is very challenging. Interested readers are referred to Section 4 of their paper.

5.6 Neural Network

The term *neural network* has evolved to encompass a large class of models and "learning" (i.e., parameter estimation) methods (Hassoun (1995), Bishop (1995), Haykin (1998), and Hagan et al. (1996)). Neural networks are composed of simple elements operating in parallel based on a network function that is determined largely by the connections between elements. We can train a neural network to perform a particular function by adjusting the values of the connections weights (i.e., parameters) between elements. The neuron model and the architecture of a neural network describe how a network transforms its input into an output. This mapping of inputs to outputs can be viewed as a non-parametric regression computation. Training in neural networks is synonymous with model building and parameter estimation. In this section, we will consider two popular types of neural network models for performing regression tasks known as *multi-layer perceptron (MLP)* and radial basis function (RBF) networks. The field of neural networks has a history of five decades and has gained widespread applications in the past fifteen years. Rumelhart et al. (1986) gave an systematic study with references therein.

Example 18 (Continued) For the purpose of illustration, we train MLP and RBF networks using the data set in Example 18. The predictions using MLP with 7 hidden units and RBF with 7 basis functions are depicted in Figure 5.11. We can see that both MLP and RBF networks have good performance. The selection of appropriate size of networks (i.e., the number of hidden units in MLP and the number of basis functions in RBF) will be further discussed in this section.

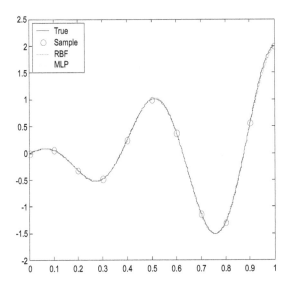

FIGURE 5.11
Plot of neural network model fit.

5.6.1   Multi-layer perceptron networks

A *single neuron* or *perceptron* that consists of inputs, weights and output (see Figure 5.12) performs a series of linear and non-linear mapping as follows:

$$v = \sum_{i=1}^{s} w_i x_i + w_0,$$

and

$$b(v) = \frac{1}{1 + e^{-\lambda v}} \quad \text{or} \quad b(v) = \tanh(\lambda v),$$

where x_i are the inputs, and w_i are the corresponding weights or parameters of the model, and $b(v)$ is the activation function or transfer function usually chosen to have a logistic-type of function. The steepness of the logistic function, λ, is typically set equal to 1.

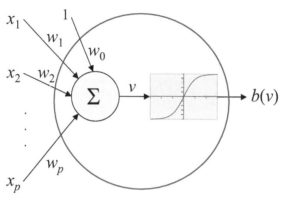

FIGURE 5.12

A neuron model with x_i as inputs, w_i as weights or parameters, and $b(v)$ as the activation function output.

A multi-layer perceptron (MLP) network (see Figure 5.13), that consists of input, hidden, and output layers with nonlinear and linear activation functions in the hidden and output layers, respectively, approximates inputs and outputs as follows:

$$\hat{y} = \sum_{j=1}^{d} \beta_j b_j(v_j) + \beta_0,$$

where d is a pre-specified integer, β_j is the weight connection between the output and the jth component in the hidden layer, and $b_j(v_j)$ is the output of the jth unit in the hidden layer,

$$b_j(v_j) = \frac{1}{1 + e^{-\lambda v_j}}, \quad \text{or} \quad b_j(v_j) = tanh(\lambda v_j),$$

and

$$v_j = \sum_{i=1}^{s} w_{ji} x_i + w_{j0},$$

where w_{ji} is the weight connection between the jth component in the hidden layer and the ith component of the input.

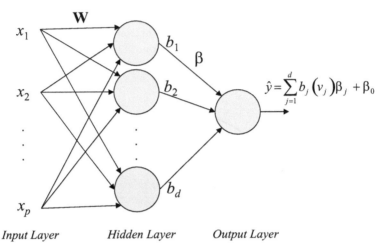

FIGURE 5.13
A three-layer MLP network.

Cybenko (1989) showed that a single hidden layer network employing a sigmoid type (as well as other types) activation function is a universal function approximator.

Theorem 12 *Let $b(.)$ be a continuous sigmoid-type function. Then for any continuous real-valued function f on $[0,1]^s$ (or any other compact subset of R^s) and $\varepsilon > 0$, there exist vectors $\mathbf{w}_1, \mathbf{w}_2, \cdots, \mathbf{w}_s$, $\boldsymbol{\beta}$ and a parameterized function $g(\mathbf{x}, \mathbf{W}, \boldsymbol{\beta}) : [0,1]^s \mapsto R$ such that*

$$| g(\mathbf{x}, \mathbf{W}, \boldsymbol{\beta}) - f(\mathbf{x}) |< \epsilon,$$

for all $\mathbf{x} \in [0,1]^s$, where

$$g(\mathbf{x}, \mathbf{W}, \boldsymbol{\beta}) = \sum_{j=1}^{d} \beta_j b_j(\mathbf{w}_j^T \mathbf{x}_i + w_{j0}),$$

and $\mathbf{w}_j \in R^s, \mathbf{w}_0 \in R^d$, $\mathbf{W} = (\mathbf{w}_1, \mathbf{w}_2, \cdots, \mathbf{w}_d)$, $\mathbf{w}_0 = (w_{10}, w_{20}, \cdots, w_{d0})$, and $\boldsymbol{\beta} = (\beta_1, \beta_2, \cdots, \beta_d)$.

Network training to estimate the parameters is typically done using the least squares criterion to minimize

$$E = \sum_{k=1}^{n} (y_k - \hat{y}_k)^2.$$

An instantaneous version of the above criterion, $E = (y_k - \hat{y}_k)^2$, is used in a stochastic gradient algorithm and leads to parameters being updated as follows:

$$\Delta\beta_j = \beta_j^{t+1} - \beta_j^t = -\rho\frac{\partial E}{\partial \beta_j} = -\rho(y_k - \hat{y}_k)b_j,$$

$$\Delta w_{ji} = w_{ji}^{t+1} - w_{ji}^t = -\rho\frac{\partial E}{\partial w_{ji}} = -\rho\frac{\partial E}{\partial b_j}\frac{\partial b_j}{\partial v_j}\frac{\partial v_j}{\partial w_{ji}},$$

where β_j^t and w_{ji}^t are the parameter values at the tth iteration, ρ is a learning rate, and

$$\frac{\partial v_j}{\partial w_{ji}} = x_i,$$

$$\frac{\partial E}{\partial b_j} = \frac{\partial}{\partial b_j}\{\frac{1}{2}(y_k - \hat{y}_k)^2\} = -(y_k - \hat{y}_k)\frac{\partial \hat{y}_k}{\partial b_j} = (y_k - \hat{y}_k)\beta_j,$$

$$\frac{\partial b_j}{\partial v_j} = b(v_j)[1 - b_j(v_j)] \text{ for } b_j(v_j) = \frac{1}{1 + e^{-\lambda v_j}},$$

$$\frac{\partial b_j}{\partial v_j} = 1 - b^2(v_j) \text{ for } b_j(v_j) = \tanh(v_j).$$

Therefore, weights between the input and hidden layers are updated as follows:

$$\Delta w_{ji} = \rho(y_k - \hat{y}_k)\beta_j\frac{\partial b_j}{\partial v_j}x_i.$$

The above parameter updating is known as *back propagation learning* (Rumelhart et al. (1986)). There are numerous other training algorithms based on gradient optimization that are more efficient (in the sense of faster convergence) than the above algorithm. Interested readers should consult the considerable literature on neural networks (Haykin (1998), Hassoun (1995), and Bishop (1995)). The steps for training MLP networks can be summarized as follows.

1. **Normalization:** Normalize the data set such that $u_i = \frac{(x_i - \overline{x}_i)}{s_i}$, where $\overline{x}_i$ and s_i are the sample mean and standard deviation of x_i, respectively. This normalization helps the training process to ensure that the inputs to the hidden units are comparable to one another and to avoid saturation (i.e., the sum products of the weights and the inputs resulting in values which are too large or too small for a logistic function) of the activation functions.

2. **Select network architecture:**

   - Select the number of hidden layers (a single hidden layer is sufficient; see Theorem 12).

- Select the proper number of units in the hidden layer. Too few or too many units will create under- or over-fitting, respectively. Over-fitting causes poor capability of predicting untried points.

- Select activation functions. *Sigmoid* or *tanh* activation functions are the most popular choices for the units in the hidden layer while the linear activation function for the output unit is appropriate for regression problems.

3. **Select learning rate** ρ**:** Too large a value of learning rate may cause the learning process to fail to converge while too small a learning rate may cause slow convergence. Many variations of learning algorithms to speed up the convergence rate employ different strategies of learning rate (e.g., Darken and Moody (1992), Le Cun, Simard and Pearlmutter (1993)).

4. **Initialization:** Initialize the weights, $\mathbf{W}$ and $\boldsymbol{\beta}$, randomly within small ranges. Because the network training employs gradient search, the training process may be trapped to a local minima. Therefore, several initial starting values of $\mathbf{W}$ and $\boldsymbol{\beta}$ may be tried to get the best result.

5. **Training:** Train the network with the algorithm of choice (e.g., back propagation) until sufficient fitting error is achieved for the training data set. Many practices in neural networks suggest splitting the data sets into training and testing sets. The former is used for network training while the latter is used to stop the training when prediction error on the testing data set achieves a minimum. When the size of the data set is small, however, this approach may be unjustified. Other practices include *early stopping* of training. That is, the training iteration process is stopped after a small number of iterations before the fitting errors are too small to avoid overfit. This heuristics, however, is very ad hoc as it is usually difficult to determine when to stop. Some authors have suggested using a penalty function in addition to the least square criterion function to eliminate some of the units in the hidden layer or to prune some of the weights. In the example below we shall employ a post-training penalty approach to eliminate some units in the hidden layer to improve network prediction capability.

5.6.2 A case study

An exhaust manifold is one of the engine components that is expected to endure harsh and rapid thermal cycling conditions ranging from sub-zero to, in some cases, more than 1000°C. The increasingly reduced product development cycle has forced engineers to rely on computer-aided engineering tools to predict design performance before any actual physical testing. Hazime, Dropps, Anderson and Ali (2003) presented a transient non-linear finite element method to simulate inelastic deformation used to predict the thermomechanical fatigue life of cast exhaust manifolds. The method incorporates elastic-plastic and creep material models and transient heat transfer analysis to simulate the manifold behavior under transient thermal loads. The exhaust manifold assembly includes the exhaust manifold component, fastener, gasket, and a portion of the cylinder head (see Figure 5.14). In addition to predicting fatigue life, the model also predicts dynamic sealing pressure on the gasket to identify potential exhaust gas leaks. The transient effects of temperature and the effect of creep change the behavior of the manifold, in terms of its internal stress distribution and contact pressure at the gasket, in turn affecting the sealing integrity from cycle to cycle.

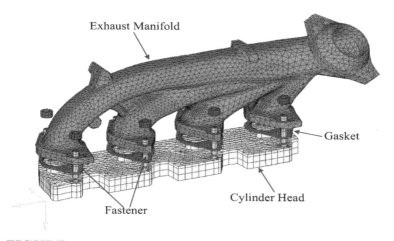

FIGURE 5.14
Exhaust manifold finite element model.

To optimize the gasket design to prevent *leaks*, a symmetric Latin hypercube (SLH) design with 17 runs and 5 factors is applied to the simulation model (see Section 2.3 for the definition of SLH). The SLH design matrix is optimized with respect to maximin distance (see Section 2.4.3) as shown in Table 5.11.

Three-layer MLP networks are used as the metamodel. Sigmoidal activation functions are employed for the hidden layer while the linear activation function is used for the output unit. Several networks with different numbers of hidden units and training processes are considered:

1. A network with 2 hidden units which is expected to under-fit the training data.

2. A network with 15 hidden units which is expected to over-fit the training data. For the latter network, several training strategies are evaluated:

TABLE 5.11

Training Data Set for the Exhaust Manifold Sealing
Performance

Run	x_1	x_2	x_3	x_4	x_5	y
1	0.00	0.38	0.13	0.63	0.06	14.30
2	0.75	1.00	0.63	0.06	0.44	11.08
3	0.81	0.06	0.06	0.31	0.63	14.36
4	0.13	0.19	0.31	0.25	0.69	10.01
5	0.38	0.56	0.75	0.19	1.00	11.19
6	0.94	0.25	0.56	0.00	0.25	2.70
7	0.69	0.69	0.19	0.88	0.88	19.55
8	0.44	0.88	0.00	0.44	0.19	19.68
9	0.25	0.31	0.88	1.00	0.50	9.65
10	0.56	0.13	1.00	0.56	0.81	9.63
11	0.31	0.50	0.81	0.13	0.13	7.94
12	0.06	0.75	0.44	0.50	0.75	16.98
13	0.63	0.44	0.25	0.81	0.00	19.51
14	0.88	0.81	0.69	0.75	0.31	22.88
15	0.19	0.94	0.94	0.69	0.38	18.01
16	0.50	0.00	0.38	0.94	0.56	10.20
17	1.00	0.63	0.50	0.38	0.94	17.68

- Perform training until the fitting errors are very small (i.e., 10,000 training iterations);

- Perform early stopping with only 200 iterations in the training process;

- Perform post-training process by applying the penalized least square method to the connection between the hidden and output layers.

To evaluate the performance of the networks, the trained networks are used to predict the testing data set in Table 5.12. The results for these networks in terms of the root mean square errors for the testing dataset are summarized in Table 5.13.

Note that the two unit hidden layers do not provide sufficient fit to the training data and also provide poor prediction to the testing data set. The decision for the early stopping strategy of 200 iterations is based on the best prediction (i.e., minimum RMSE) to the testing data set. In this respect, the good performance for this strategy is achieved unfairly because the testing data set is used during the training to make the early stopping decision while the other approaches did not use the testing data at all during the training. The RMSE for the MLP with 15 hidden units without early stopping exhibits over-fit problems for the data as indicated by a very small fitting RMSE but a larger prediction RMSE.

TABLE 5.12

Testing Data Set for the Exhaust Manifold Sealing Performance

Run	x_1	x_2	x_3	x_4	x_5	y
1	0.00	0.26	0.47	0.21	0.42	8.61
2	0.63	0.42	0.05	0.26	0.11	15.48
3	0.84	0.63	0.53	0.42	0.00	20.72
4	0.68	0.00	0.37	0.68	0.26	14.93
5	0.11	0.79	0.26	0.89	0.16	17.59
6	0.47	0.95	0.21	0.11	0.37	12.54
7	0.58	0.89	0.42	0.95	0.84	19.15
8	0.16	0.11	0.16	0.47	0.89	12.47
9	0.32	0.68	0.00	0.58	0.53	19.32
10	0.21	0.47	0.74	0.00	0.21	3.57
11	0.37	0.16	0.84	0.84	0.05	10.57
12	0.53	0.05	1.00	0.32	0.58	6.19
13	1.00	0.53	0.11	0.79	0.63	21.30
14	0.05	0.58	0.89	0.74	0.74	13.05
15	0.89	0.21	0.32	0.05	0.79	6.30
16	0.95	0.84	0.95	0.53	0.32	21.19
17	0.74	0.74	0.79	0.16	0.95	12.04
18	0.79	0.32	0.68	1.00	0.47	12.53
19	0.26	1.00	0.63	0.37	0.68	18.08
20	0.42	0.37	0.58	0.63	1.00	14.76

TABLE 5.13

Performance Comparison in Terms of RMSE (fitting training data and predicting testing data) of Different MLP Networks

Networks	RMSE (fit)	RMSE (prediction)
2 unit hidden layer	1.4882	3.1905
15 unit hidden layer with early stopping (200 iterations)	0.0102	2.1853
15 unit hidden layer without early stopping (10000 iterations)	0.0000	2.3258
15 unit hidden layer without early stopping and penalized least square post-training (3 out of 15 hidden units are maintained)	0.9047	2.2575

To alleviate the problem of unknown early stopping of training and to achieve better prediction capability (i.e., to minimize bias minimization to the training data and to minimize prediction variance), a simple post-training hidden unit selection is performed. Note that once the network is trained so that the weights between input and hidden layers, $\mathcal{W}$, are fixed, the network mapping can be considered a linear model in terms of the weight connection between hidden and output units, $\boldsymbol{\beta}$, with hidden units as sigmoidal basis functions. In this respect, a penalized least square method in (5.10) can be applied to select the sigmoidal basis functions,

$$\frac{1}{2}\|\mathbf{y} - \beta_0 - \mathbf{B}\boldsymbol{\beta}\|^2 + \sum_{j=1}^{s} p_\lambda(|\beta_j|),$$

where $\mathbf{B} = (b_1, b_2, \cdots, b_d)$. Here SCAD penalty function can be used, similar to the polynomial basis function selection. Note that the selection is conducted after a large size of networks (i.e., 15 hidden units) is fully trained. For fixed $\mathcal{W}$, the hidden unit selection process can be performed quickly and easily by using the standard linear least squares technique to estimate $\boldsymbol{\beta}$. For this example only 3 out of 15 hidden units are selected by the SCAD penalty function. The comparison between the prediction of the 3 out of 15 hidden unit MLP and the true values is shown in Figure 5.15.

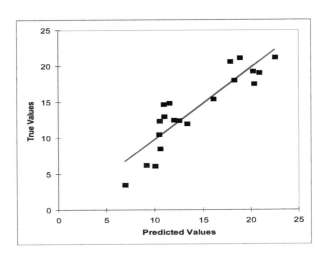

FIGURE 5.15
Predicted values Vs. True values from MLP network with 3 hidden units selected by the SCAD penalized least squares to the 15 hidden unit networks.

5.6.3 Radial basis functions

Radial basis function (RBF) methods are techniques for exact interpolation of data in multi-dimensional space (Powell (1987)). The RBF maps the inputs to the outputs using a linear combination of the basis functions

$$y(\mathbf{x}) = \mu + \sum_{j=1}^{n} \beta_j b_j(\|\mathbf{x} - \mathbf{x}_j\|),$$

where μ is the mean of y. In a matrix form, the interpolation for the data set can be written as

$$\mathbf{y} = \mu + \mathbf{B}\beta,$$

where $\mathbf{y} = (y_i, \cdots, y_n)'$, and the square matrix, $\mathbf{B} = \{b_{ij}(\|\mathbf{x} - \mathbf{x}_j\|)\}, i, j = 1, 2, \cdots, n$. Once the basis functions are set, the parameter β can be easily calculated using the least squares method

$$\beta = (\mathbf{B}'\mathbf{B})^{-1}\mathbf{B}(\mathbf{y} - \mu).$$

There are several popular choices of basis functions such as

Linear	$b_j(\|\mathbf{x} - \mathbf{x}_j\|) = \|\mathbf{x} - \mathbf{x}_j\|$
Cubic	$b_j(\|\mathbf{x} - \mathbf{x}_j\|) = \|\mathbf{x} - \mathbf{x}_j\|^3$
Thin-plate spline	$b_j(\|\mathbf{x} - \mathbf{x}_j\|) = \|\mathbf{x} - \mathbf{x}_j\|^2 \log \|\mathbf{x} - \mathbf{x}_j\|$
Gaussian	$b_j(\|\mathbf{x} - \mathbf{x}_j\|) = \exp(-\theta_j \|\mathbf{x} - \mathbf{x}_j\|^2)$

A number of modifications to the exact interpolation have been proposed (Moody and Darken (1989), Bishop (1995)) such as

- The number of basis functions need not be equal to the number of data points and typically are less than n;

- The basis functions are not constrained to be at the data points, but they can be selected as part of the network training process (i.e., parameter estimation);

- The *width* parameters θ_j can be varied for each basis function and determined during the training process.

Here, we will use the *Gaussian basis function*. Note that θ_j is the reciprocal of the variance in a normal distribution; thus, it can be interpreted as the (inverse) width of the Gaussian kernel. These parameters θ_j in RBF are often heuristically chosen, such as by setting all θ_j equal ($\theta_j = \theta$) to some multiple average distance between the basis function centers to ensure some overlap so that the interpolation is relatively smooth. Too large or too small θ produces an under- or over-smoothing, respectively. When Gaussian basis functions are chosen, the RBF model is closely related to the Gaussian Kriging model (see Li and Sudjianto (2005)); therefore, maximum likelihood estimation or its penalized function as discussed in Section 5.4 can be applied to determine the parameter θ as follows:

$$Q(\mu, \theta) = -\frac{1}{2} \log |\mathbf{B}(\theta)|^{-\frac{1}{2}} - \frac{1}{2}(\mathbf{y} - \mathbf{1}_N \mu)^T \mathbf{B}^{-1}(\theta)(\mathbf{y} - \mathbf{1}_N \mu) - \sum_{j=1}^{s} p_\lambda(\theta),$$

where $p_\lambda(.)$ is a given non-negative penalty function with a regularization parameter λ, and $\mathbf{B}(\theta)$ is an $n \times n$ matrix with the (i,j)-element $b(x_i, x_j)$. As mentioned in Sections 5.1 and 5.4, the SCAD penalty function can be employed as the penalty terms to regularize the estimation of θ, i.e.,

$$p'_\lambda(\theta) = \lambda \{ I(\theta \le \lambda) + \frac{(a\lambda - \theta)_+}{(a-1)\lambda} I(\theta > \lambda) \}.$$

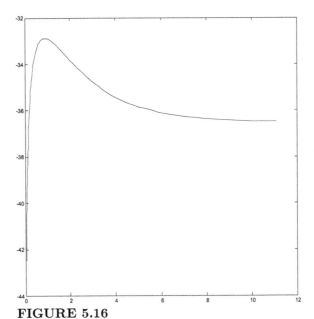

FIGURE 5.16
SCAD penalized likelihood function for the exhaust manifold data.

The parameter λ can be set equal to $\alpha \{ ln(n)/n \}^{\frac{1}{2}}$. From our experience (Li and Sudjianto (2005)), the optimal value for θ is insensitive with respect to the change of λ for $\alpha = 0.5 \sim 1.25$. Maximizing the penalized likelihood yields a penalized likelihood estimate $\hat{\mu}$ and $\hat{\theta}$ for μ and θ, respectively. When μ is set equal to the mean of y, a one-dimensional line search (see, for example, Press, Teukolsky, Vetterling and Flannery (1992)) can be employed to find the optimal θ that maximizes the penalized likelihood criterion. The steps to build RBF networks are as follows:

1. Normalize the data set such that $u_i = (x_i - \bar{x}_i)/s_i$, where $\bar{x}_i$ and s_i are the mean and standard deviation of x_i, respectively. This normalization helps the training process to make the input scales comparable to each other; thus, equal θ for all inputs can be applied.

2. Apply a line search algorithm to find θ that maximizes the penalized likelihood criterion.

3. Apply least squares (or penalized least squares as discussed below) to estimate β.

To illustrate the process, the exhaust manifold sealing example described in Section 5.6 is used. The penalized likelihood function using a SCAD penalty function for the data set from Table 5.11 is shown in Figure 5.16. The maximum likelihood value is attained for $\theta = 0.8744$.

Once the parameter is determined, the second stage of parameter estimation for β can be easily done using the least squares method. If desired, penalized least squares for a linear model, as presented in Section 5.1, may be employed to reduce the number of basis functions for the purpose of improving the prediction capability, i.e.,

$$\frac{1}{2}\|\mathbf{y} - \mu - \mathbf{B}\beta\|^2 + \sum_{j=1}^{s} p_\lambda(|\beta_j|).$$

The data set in Table 5.11 is used to illustrate the selection of parameters for the network. The performance of the networks in predicting the test data set (Table 5.12) in terms of root mean squares of errors for different sets of parameters is summarized in Table 5.14. The RMSE of arbitrarily selected $\theta = 2$ is compared to that of θ selected using penalized likelihood as well as βs from the least squares and (SCAD) penalized least squares.

TABLE 5.14
Root Mean Squares of Prediction Errors for RBFs with Different
Parameter Estimation Techniques

Estimation Technique	RMSE
$\theta = 2$; least squares	7.1466
$\theta = 2$; penalized least squares	6.4510
Penalized likelihood $\theta = 0.8744$; least squares	2.2504
Penalized likelihood $\theta = 0.8744$; penalized least squares	2.2421

5.7   Local Polynomial Regression

Local polynomial regression is a popular modeling technique for nonparametric regression. Fan and Gijbels (1996) present a systematic account of theoretical properties of local polynomial regression. Many applications and illustrations of local polynomial regression can be found in the statistical literature. However, local polynomial regression has not been a popular approach in modeling computer experiments, as it is more powerful for lower-dimensional data sets. Høst (1999) discussed how to use the local polynomial regression techniques with Kriging for modeling computer experiments. Tu (2003) and Tu and Jones (2003) gave some case studies on how to use local polynomial regression to fit computer experiment data. These works demonstrate the potential of local polynomial regression to model computer experiments. In this section, the motivation for local polynomial regression is introduced. We will further establish the connections between local polynomial regression and other computer experiment modeling techniques introduced in the previous sections.

5.7.1   Motivation of local polynomial regression

Let us begin with a typical one-dimensional nonparametric regression model. Suppose that (x_i, y_i), $i = 1, \cdots, n$ is a random sample from the following model:

$$Y = m(X) + \varepsilon, \tag{5.35}$$

where ε is random error with $E(\varepsilon|X = x) = 0$, and $\text{Var}(\varepsilon|X = x) = \sigma^2(x)$. Here $m(x) = E(Y|X = x)$ is called a *regression function*. In the context of nonparametric regression, it is assumed only that $m(x)$ is a smooth function of x. We want to estimate $m(x)$ without imposing a parametric form on the regression function.

For a given data set, one may try to fit the data by using a linear regression model. If a nonlinear pattern appears in the scatter plot of Y against X, one may employ polynomial regression to reduce the modeling bias of linear regression. As demonstrated in Example 18, polynomial regression fitting may nonetheless have substantial biases because the degree of polynomial regression cannot be controlled continuously, and individual observations can have a great influence on remote parts of the curve in polynomial regression models. Local polynomial regression can repair the drawbacks of polynomial fitting.

Suppose that the regression function $m(\cdot)$ is smooth. Applying the Taylor expansion for $m(\cdot)$ in a neighborhood of x, we have

$$m(u) \approx \sum_{j=0}^{p} \frac{m^{(j)}(x)}{j!} (u - x)^j \overset{\text{def}}{=} \sum_{j=0}^{p} \beta_j (u - x)^j. \tag{5.36}$$

Thus, for x_i close enough to x,

$$m(x_i) \approx \sum_{j=0}^{p} \beta_j (x_i - x)^j \overset{\text{def}}{=} \mathbf{x}_i^T \boldsymbol{\beta},$$

where $\mathbf{x}_i = (1, (x_i - x), \cdots, (x_i - x)^p)'$ and $\boldsymbol{\beta} = (\beta_0, \beta_1, \cdots, \beta_p)'$. Let $K(x)$ be a function satisfying $\int K(x)\,dx = 1$, called a *kernel function*, and let h be a positive number called a *bandwidth* or a *smoothing parameter*. Intuitively, data points close to x carry more information about $m(x)$. This suggests using a locally weighted polynomial regression

$$\sum_{i=1}^{n} (y_i - \mathbf{x}_i' \boldsymbol{\beta})^2 K_h(x_i - x), \qquad (5.37)$$

where $K_h(x_i - x)$ is a weight. Denote by $\hat{\beta}_j$ $(j = 0, \cdots, p)$ the minimizer of (5.37). Then an estimator for the regression function $m(x)$ is

$$\hat{m}(x) = \hat{\beta}_0(x). \qquad (5.38)$$

Furthermore, an estimator for the ν-th order derivative of $m(x)$ at x is

$$\hat{m}_\nu(x) = \nu! \hat{\beta}_\nu(x).$$

It is well known that the estimate $\hat{m}(x)$ is not very sensitive to the choice of K, scaled in a canonical form as discussed by Marron and Nolan (1988). Thus, it is common to take the kernel function to be a symmetric probability density function. The most commonly used kernel function is the Gaussian density function given by

$$K(x) = \frac{1}{\sqrt{2\pi}} \exp(-x^2/2).$$

Other popular kernel functions include the uniform kernel

$$K(x) = 0.5, \quad \text{for} \quad -1 \le x \le 1,$$

and the Epanechikov kernel

$$K(x) = 0.75(1 - x^2), \quad \text{for} \quad -1 \le x \le 1.$$

The smoothing parameter h controls the smoothness of the regression function. The choice of the bandwidth is of crucial importance. If h is chosen too large, then the resulting estimate misses fine features of the data, while if h is selected too small, then spurious sharp structures become visible. There are many reported results on bandwidth selection in the literature. Jones et al. (1996a,b) gave a systematic review of this topic. In practice, one may use a data-driven method to choose the bandwidth or simply select a bandwidth by visually inspecting the resulting estimated regression function.

When $p = 0$, local polynomial fitting becomes kernel regression. The kernel regression estimator is given by

$$\hat{m}_h(x) = \frac{\sum_{i=1}^{n} K_h(x_i - x)y_i}{\sum_{i=1}^{n} K_h(x_i - x)}, \tag{5.39}$$

which is the solution of a local constant fitting. This estimator is also referred to as an NW-kernel estimator because it was proposed by Nadaraya (1964) and Watson (1963) independently.

When $p = 1$, the local polynomial fitting is referred to as local linear fitting. The local linear fitting not only estimates the regression curve, but also its first derivative. This is one advantage of local linear regression over kernel regression.

While local polynomial regression was originally proposed for nonparametric regression in which error is assumed to present, it can be applied directly for modeling computer experiments in which there is no random error, or equivalently, in which $\mathrm{Var}(Y|X = x) = 0$. We now illustrate the local polynomial techniques via application to the data in Example 18.

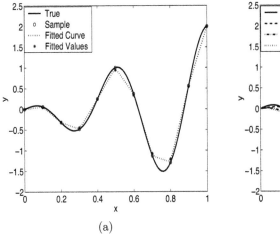
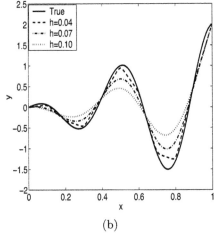

(a)   (b)

FIGURE 5.17
Plots for local linear fit for data in Example 18. (a) Plot of the true model, scatter plot of collected sample, fitted curve, and fitted values. (b) Plot of local polynomial fits with different bandwidth.

Example 18 (Continued)

We refit the data in Example 18 using local linear regression. The local polynomial fit using Gaussian kernel with $h = 0.04$ is depicted in Figure 5.17 (a). We next demonstrate the importance of selecting a proper bandwidth. Figure 5.17(b) presents the local linear fit using Gaussian kernel with three different bandwidths, and illustrates that a larger bandwidth yields a fit with a larger bias.

5.7.2 Metamodeling via local polynomial regression

Due to the need for metamodeling in most experiments, the idea of local polynomial regression in the previous section can be generalized to the multivariate case. Høst (1999) considered the following model

$$y(\mathbf{x}) = m(\mathbf{x}) + z(\mathbf{x}),$$

where $m(\mathbf{x})$ is a smoothing function of $\mathbf{x}$, and $z(\mathbf{x})$ is residual. He further proposed using local polynomial regression to model the overall trend $m(\mathbf{x})$ and then using the Kriging approach to interpolate residuals $z(\mathbf{x})$, because local polynomial regression usually does not yield an exact interpolation over the observed sample.

For simplicity of presentation, we limit ourselves to local linear regression. Applying the Taylor expansion for $m(\cdot)$ in a neighborhood of $\mathbf{x}$, it follows that

$$m(\mathbf{u}) \approx \beta_0 + \sum_{j=1}^{s} (\frac{\partial m(\mathbf{x})}{\partial u_j})(u_j - x_j) \overset{\text{def}}{=} \beta_0 + \sum_{j=1}^{s} \beta_j(u_j - x_j). \qquad (5.40)$$

Local linear regression minimizes the following weighted least squares

$$\sum_{i=1}^{n} \left\{ y_i - \beta_0 - \sum_{j=1}^{s} \beta_j(x_{ij} - x_j) \right\}^2 K_{\mathbf{h}}(\mathbf{x}_i - \mathbf{x}),$$

where

$$K_{\mathbf{h}}(\mathbf{x}_i - \mathbf{x}) = \frac{1}{h_1 \cdots h_s} K\left(\frac{x_{11} - x_1}{h_1}, \cdots, \frac{x_{1s} - x_s}{h_s} \right)$$

and $K(u_1, \cdots, u_s)$ is an s-dimensional kernel function satisfying $\int K(\mathbf{u}) \, d\mathbf{u} = 1$.

Denote $W = \text{diag}\{K_{\mathbf{h}}(\mathbf{x}_1 - \mathbf{x}), \cdots, K_{\mathbf{h}}(\mathbf{x}_1 - \mathbf{x})\}$, $\boldsymbol{\beta} = (\beta_0, \cdots, \beta_s)'$, and

$$\mathbf{X} = \begin{pmatrix} 1 & x_{11} - x_1 & \cdots & x_{1s} - x_s \\ \vdots & \vdots & \vdots & \vdots \\ 1 & x_{n1} - x_1 & \cdots & x_{ns} - x_s \end{pmatrix}.$$

Then, we have

$$\hat{\boldsymbol{\beta}} = (\mathbf{X}'W\mathbf{X})^{-1}\mathbf{X}'W\mathbf{y}$$

and

$$\hat{m}(\mathbf{x}) = \mathbf{e}_1' \hat{\boldsymbol{\beta}} = \mathbf{e}_1' (\mathbf{X}'W\mathbf{X})^{-1}\mathbf{X}'W\mathbf{y}, \qquad (5.41)$$

where $\mathbf{e}_1$ is an $s+1$-dimensional column vectors whose first element equals 1 and whose others equal 0.

5.8 Some Recommendations

We conclude this chapter by giving some recommendations for metamodeling. Since Kriging models, splines approach, neural network methods and local polynomial regression all are smoothing methods, we first present some connections among these modeling techniques.

5.8.1 Connections

Silverman (1984) has shown that smoothing splines are equivalent to variable kernel regression. In this section, we discuss how Kriging, radial basis functions, and local polynomial regression are related to one another.

From (5.39) and (5.41), we can re-express the kernel regression and local polynomial regression estimator as

$$\hat{m}(\mathbf{x}) = \sum_{i=1}^{n} \tilde{w}_i(\mathbf{x}) K_{\mathbf{h}}(\mathbf{x}_i - \mathbf{x}), \qquad (5.42)$$

where $\tilde{w}_j(\mathbf{x})$ is viewed as a weight of the j-th observation at location $\mathbf{x}$. Furthermore, $\tilde{w}_j(\mathbf{x})$ is a linear combination of y_i, $i = 1, \cdots, n$. Although the estimator $\hat{m}(\mathbf{x})$ is derived from nonparametric regression models, it can be applied directly to modeling computer experiments.

Gaussian Kriging approach linearly interpolates the residual $\mathbf{y} - \mathbf{B}\hat{\boldsymbol{\beta}}$ in the following way:

$$\mathbf{r}'(\mathbf{x})R^{-1}(\mathbf{y} - \mathbf{B}\hat{\boldsymbol{\beta}}),$$

which can be rewritten as

$$\sum_{i=1}^{n} a_i(\mathbf{x}) r(\mathbf{x}_i), \qquad (5.43)$$

where $\mathbf{r}(\mathbf{x}) = (r(\boldsymbol{\theta}, \mathbf{x}, \mathbf{x}_1), \cdots, r(\boldsymbol{\theta}; \mathbf{x}, \mathbf{x}_n))'$, and $\mathbf{a}(\mathbf{x}) = (a_1(\mathbf{x}), \cdots, a_n(\mathbf{x}))' = R^{-1}(\mathbf{y} - \mathbf{B}\hat{\boldsymbol{\beta}})$, which is a linear combination of $\mathbf{y}$ since $\hat{\boldsymbol{\beta}}$ is a linear estimator of $\boldsymbol{\beta}$. In particular, if a correlation function is taken as

$$r(\boldsymbol{\theta}, \mathbf{s}, \mathbf{t}) = \exp\{-\sum_{k=1}^{s} \theta_k |s_k - t_k|^2\},$$

then (5.42) with a Gaussian kernel and (5.43) have the same form because the normalized constant in the kernel function does not matter. The difference between these two methods is in how they assign weights to each observation. The Gaussian Kriging method employs the property of multivariate normal distributions and yields a BLUP, while local polynomial regression assigns the weights using the idea of local modeling, and it results in a minimax estimate in a certain sense (Fan (1992)). In the Gaussian Kriging model, the θ_ks are regarded as unknown parameters which can be estimated by maximum likelihood. From the point of view of local modeling, the θ_ks are considered smoothing parameters. Selection of smoothing parameters has been extensively studied in the statistical literature. The cross validation procedure is an easily understand approach for selecting smoothing parameters, although it is computationally intensive.

The radial basis function approach can also be viewed as a version of local modeling. In the radial basis function approach, one constructs a basis as follows:

$$\left\{ K\{\|(\mathbf{x} - \mathbf{x}_i)/\theta\|^2\} : i = 1, \cdots, n \right\}. \tag{5.44}$$

It is not difficult to verify that for certain choices of $K(\cdot)$, such as $K(t) = \exp(-\frac{1}{2}t^2)$, the set of functions is linear independent. Therefore, $y(\mathbf{x})$ can be interpolated by

$$\sum_{i=1}^{n} \beta_i K\{\|(\mathbf{x} - \mathbf{x}_i)/\theta\|^2\}.$$

The least squares estimator for the β_is can be obtained and is a linear combination of the y_is. Thus, the predictor based on the radial basis function is given by

$$\hat{y}(\mathbf{x}) = \sum_{i=1}^{n} \hat{\beta}_i K\{\|(\mathbf{x} - \mathbf{x}_i)/\theta\|^2\}. \tag{5.45}$$

This has the same form as that in (5.43) regarding $K\{\|(\mathbf{x} - \mathbf{x}_i)/\theta\|^2\}$ as a kernel function. In the Gaussian Kriging model, to avoid the optimization problem for $\boldsymbol{\theta}$ over a high-dimensional domain, it is common to set $\theta_j = \theta\sigma_j$, where σ_j is the sample standard deviation of the j-component of $\mathbf{x}$. This is equivalent to using the radial basis function to interpolate the residuals.

5.8.2 Recommendations

Simpson, Peplinski, Koch and Allen (2001) presented a survey and conducted some empirical comparisons of metamodeling techniques, including mainly polynomial models, neural network methods, and Kriging. They recommended the following:

1. Polynomial models are primarily intended for regression with random error; however, they have been used widely and successfully in many

applications and case studies of modeling computer experiments. Polynomial modeling is the best established metamodeling technique, and is probably the easiest to implement. Although polynomial models have some limitations (see Section 5.2), they are recommended for exploration in deterministic applications with a few fairly well-behaved factors.

2. Kriging may be the best choice in the situation in which the underlying function to be modeled is deterministic and highly nonlinear in a moderate number of factors (less than 50, say).

3. Multi-layer perceptron networks may be the best choice (despite their tendency to be computationally expensive to create) in the presence of many factors to be modeled in a deterministic application.

The regression splines method introduced in Section 5.3 can be used to improve polynomial models and is quite easy to implement. As described in Section 5.3, the number of parameters in the regression spline model will dramatically increase as the number of factors increase. This implies that the regression splines method is recommended only in the presence of a small number of well-behaved factors.

Since local polynomial regression is relatively new in the literature, Simpson et al. (2001) did not include this method. From the connections among the radial basis function approach, Kriging models, and local polynomial regression, we would expect the performance of all these methods to be similar. However, more numerical comparisons and further research are needed. All examples in this chapter were conducted using MATLAB code. The code is free and available through the authors and is posted at www.stat.psu.edu/~rli/DMCE. The website is updated frequently.

6

Model Interpretation

This chapter focuses on how to give reasonable interpretations for models in computer experiments, focusing on the concept of sensitivity analysis. Global sensitivity analyses, including Sobol's indices and their extensions, are introduced. The Fourier amplitude sensitivity test (FAST) is also discussed.

6.1   Introduction

Computer models are often used in science and engineering fields to describe complicated physical phenomena that are governed by a set of equations, including linear, nonlinear, ordinary, and partial differential equations. Such complicated models are usually difficult to interpret. We may, however, be able to give a good interpretation using a metamodel built by some techniques introduced in the previous chapter. For example, we want to rank the importance of the input variables and their interactions. These questions are particularly important when we are interested in gaining insight into the relationships and effects of input variables to the response variable. While it may be easy to interpret low-order polynomial models using traditional ANOVA or by simply inspecting the regression coefficients, it would be difficult to directly understand metamodels with sophisticated basis functions such as Kriging or neural networks. Therefore, a general approach, extending the idea of traditional ANOVA decomposition, for model interpretation is needed. The so-called *sensitivity analysis* (SA) was motivated by this. Sensitivity analysis of the output aims to quantify how much a model depends on its input variables. Section 1.7 stated that the investigator may be concerned with

• how well the metamodel resembles the system or the true model, i.e., model validation;

• which input variables and which interactions contribute most to the output variability;

• whether we can remove insignificant input factors and improve the efficiency of the metamodel;

- which level-combination of the input variables can reach the maximum or minimum of the output y.

Methods in SA can be grouped into three classes:

- *Factor screening:* Factor screening is used in dealing with models in computer experiments that are computationally expensive to evaluate and have a large number of input factors. The aim of factor screening is to identify which input factors among the many potentially important factors are really important.

- *Local SA:* The local SA concentrates on the local impact of the input factors on the model and involves computing partial derivatives of the metamodel with respect to the input variables. For the computation of the derivatives numerically, the input parameters are allowed to vary within a small interval.

- *Global SA*: Global SA focuses on apportioning the output uncertainty to the uncertainty in the input variables. It typically takes a sampling approach to distributions for each input variable.

A rich literature exists on sensitivity analysis. A comprehensive review and detailed introduction to sensitivity analysis are found in Saltelli et al. (2000). This chapter focuses only on global SA. To easily understand the SA approach, in Section 6.2 we review how to implement the traditional ANOVA decomposition techniques to rank importance of marginal effects. Though this technique is borrowed from linear regression, the approach can be extended and applied to more sophisticated models. We introduce global sensitivity analysis in Section 6.3 based on functional ANOVA decomposition. Global sensitivity indices can be applied to the metamodels discussed in the previous chapter to rank the importance of the input factors.

6.2   Sensitivity Analysis Based on Regression Analysis

Suppose that a sample of the input variables is taken by some sampling strategy and the corresponding sequence of output values is computed using the model under analysis. Let us fit data $\{(\mathbf{x}_i, y_i) = (x_{i1}, \cdots, x_{is}, y_i), i = 1, \cdots, n\}$ using a linear model

$$y_i = \alpha + \beta_1 x_{i1} + \cdots + \beta_s x_{is} + \varepsilon_i, \qquad (6.1)$$

where ε_i is random error, and $\alpha, \beta_1, \cdots, \beta_s$ are unknown regression coefficients. Their least squares estimators are denoted by $a, b_1, \cdots, b_s$.

6.2.1   Criteria

There are many criteria for measuring the importance of factors in the model as well as each input variable. The following are some of them.

A. Standardized regression coefficients: One approach to assess the importance of the input variables is to use their regression coefficients. More reasonable procedure concerns standardizing all the variables:

$$\frac{y_i - \bar{y}}{s_y}, \frac{x_{ij} - \bar{x}_j}{s_j}, j = 1, \cdots, s, \tag{6.2}$$

where $\bar{y}$ and s_y are the sample mean and sample standard deviation of y, and $\bar{x}_j$ and s_j are the sample mean and sample standard deviation of x_{ij}, respectively. Model (6.1) fits the standardized data; the corresponding regression coefficients are called *standardized regression coefficients* (SRCs) and are given by $b_j s_j / s_y, j = 1, \cdots, s$. Because all of the variables have been transferred into the same scale, the output is most sensitive to those inputs whose SRCs are largest in absolute value.

B. Coefficient of multiple determination: Denote the total sum of squares, error sum of squares, and regression sum of squares of model (6.1) by

$$\text{SSTO} = \sum_{i=1}^{n}(y_i - \bar{y})^2, \quad \text{SSE} = \sum_{i=1}^{n}(y_i - \hat{y}_i)^2, \quad \text{SSR} = \sum_{i=1}^{n}\hat{y}_i^2,$$

respectively. We have $\text{SSTO} = \text{SSR} + \text{SSE}$. The *coefficient of multiple determination* or *model coefficient of determination* is defined as the positive square root of

$$R^2 = 1 - \frac{\text{SSE}}{\text{SSTO}} = \frac{\text{SSR}}{\text{SSTO}}.$$

R^2 provides a measure of model fit.

C. Partial sum of squares: The partitioning process above can be expressed in the following more general way. Let $\mathbf{x}_{(1)}$ be a sub-vector of $\mathbf{x}$. Let $\text{SSR}(\mathbf{x}_{(1)})$ and $\text{SSE}(\mathbf{x}_{(1)})$ be the regression sum of squares and the error sum of squares, respectively, when only $\mathbf{x}_{(1)}$ is included in the regression model. For example, take $\mathbf{x}_{(1)} = (x_1, x_2)$; then $\text{SSR}(x_1, x_2)$ is the regression sum of squares when only x_1 and x_2 are included in the regression model. The difference $(\text{SSR}(x_1, x_2) - \text{SSR}(x_1))$ reflects the increase of regression sum of squares associated with adding x_2 to the regression model when x_1 is already in the model. Thus, we define an *extra sum of squares*, written as $\text{SSR}(x_2|x_1)$:

$$\text{SSR}(x_2|x_1) = \text{SSR}(x_1, x_2) - \text{SSR}(x_1).$$

Note that

$$\text{SSTO} = \text{SSR}(x_1, x_2) + \text{SSE}(x_1, x_2) = \text{SSR}(x_1) + \text{SSE}(x_1).$$

Thus,

$$\text{SSR}(x_2|x_1) = \text{SSE}(x_1) - \text{SSE}(x_1, x_2).$$

This implies that $SSR(x_2|x_1)$ reflects the additional reduction in the error sum of squares associated with x_2, given that x_1 is already included in the model. Similarly, we have the following notation:

$$SSR(\mathbf{x}_{(2)}|\mathbf{x}_{(1)}) = SSR(\mathbf{x}_{(1)}, \mathbf{x}_{(2)}) - SSR(\mathbf{x}_{(1)}), \qquad (6.3)$$

where $\mathbf{x}_{(1)}$ and $\mathbf{x}_{(2)}$ are two non-overlaid sub-vectors of $\mathbf{x}$. It is clear that

$$SSR(\mathbf{x}_{(2)}|\mathbf{x}_{(1)}) = SSE(\mathbf{x}_{(1)}) - SSE(\mathbf{x}_{(1)}, \mathbf{x}_{(2)}).$$

Therefore, the extra sum of squares $SSR(\mathbf{x}_{(2)}|\mathbf{x}_{(1)})$ measures the marginal reduction in the error sum of squares when $\mathbf{x}_{(2)}$ is added to the regression model, given that $\mathbf{x}_{(1)}$ is already in the model. In particular, taking $\mathbf{x}_{(2)} = x_i$ and $\mathbf{x}_{(1)} = x_{-i} = \{x_1, \cdots, x_{i-1}, x_{i+1}, \cdots, x_s\}$,

$$p_i = SSR(x_i|x_{-i}), i = 1, \cdots, s,$$

measures the importance of input variable x_i to model (6.1), and p_i is called the *partial sum of squares* of variable x_i. Forward stepwise regression chooses, at each step, the most important variable under the current model; this determination can be based on the partial sum of squares.

D. Predictive error sum of squares: Consider model (6.1) that is constructed on $n-1$ data points except the kth data point. For the removed y_k, let $\hat{y}_{-k}$ be estimator of y_k by the model. The difference $(y_k - \hat{y}_{-k})$ gives predictive error at $\mathbf{x}_k$. The *predictive error of sum of squares* (PRESS) is defined by $PRESS = \sum_{k=1}^{n}(y_k - \hat{y}_{-k})^2$. The PRESS based on the regression model with $s-1$ variables except x_i is denoted by $PRESS_{-i}$. The larger the $PRESS_{-i}$, the more important the variable x_i.

E. Partial correlation coefficients: When one wants to estimate correlation between y and input variable x_i, *partial correlation coefficients* (PCCs) are useful. Consider the following two regression models:

$$\hat{y}_k = \mathbf{b}_0 + \sum_{l \neq i} b_l x_{kl}, \quad \hat{x}_{ki} = c_0 + \sum_{l \neq i} c_l x_{kl}, \quad k = 1, \cdots, n. \qquad (6.4)$$

The partial correlation coefficient between y and x_i is defined as the correlation coefficient between $y_k - \hat{y}_k$ and $x_{ki} - \hat{x}_{ki}$. The PCC provides the strength of the correlation between y and x_i after adjustment for any effect due to the other input variables $x_l, l \neq i$.

F. Coefficient of partial determination: From (6.3), we have

$$SSR(\mathbf{x}_{(1)}, \mathbf{x}_{(2)}) = SSR(\mathbf{x}_{(1)}) + SSR(\mathbf{x}_{(2)}|\mathbf{x}_{(1)}).$$

This can be viewed as decomposition of the regression sum of squares.

Similar to the coefficient of multiple determination, define the *coefficient of partial determination* as

$$r^2_{\mathbf{x}_{(2)}|\mathbf{x}_{(1)}} = \frac{\text{SSR}(\mathbf{x}_{(2)}|\mathbf{x}_{(1)})}{\text{SSE}(\mathbf{x}_{(1)})},$$

and call the positive square root of $r^2_{\mathbf{x}_{(2)}|\mathbf{x}_{(1)}}$ the partial correlation coefficient. This gives another way to motivate this criterion.

G. Criteria based on rank transformation: The rank transformation has been widely used in the construction of testing statistics due to its robustness. Rank transformed statistics provide a useful solution in the presence of long-tailed input and output distributions. The rank transformation replaces the data with their corresponding ranks. More exactly, sort $y_1, \cdots, y_n$ from the smallest to the largest and denote the sorted data by $y_{(1)} \le y_{(2)} \le \cdots \le y_{(n)}$. The original outputs $y_{(k)}$ are replaced by k. A similar procedure applies to each input variable $x_{1i}, \cdots, x_{ni}$, for $i = 1, \cdots, s$. Based on the ranked data, we find standardized regression coefficients and partial correlation coefficients that are called the *standardized rank regression coefficients* (SRRC) and *partial rank correlation coefficients* (PRCC), respectively.

6.2.2 An example

Example 2 (Continued) Consider a quadratic polynomial metamodel:

$$g(\mathbf{x}) = \alpha x_0 + \sum_{j=1}^{s} \beta_j x_j + \sum_{j=1}^{s} \gamma_j x_j^2 + \sum_{1 \le k < l \le s} \tau_{kl} x_k x_l,$$

where $s = 8$ and $x_0 \equiv 1$. Firstly, we have $\text{SSR}(x_0) = 128.7075$. Due to the hierarchical structure of the quadratic polynomial model, we begin with the linear effects. The extra sums of squares for the linear effects $\text{SSR}(\cdot|x_0)$ are listed in the first row of Table 6.1, from which we can see that $\text{SSR}(x_6|x_0)$ has the greatest value among all the linear effects. Thus x_6 is the most important linear effect. We next compute $\text{SSR}(x_k|x_0, x_6)$, which is given in the second row of Table 6.1. Thus, x_1 is the most important linear effect given that x_6 is already in the model. The extra sums of squares $\text{SSR}(x_k|x_0, x_1, x_6)$, listed in the third row of Table 6.1, are small.

We now consider the quadratic effects. The extra sums of squares $\text{SSR}(x_k^2|x_0, x_1, x_6)$ are listed in the diagonal position of Table 6.2, and $\text{SSR}(x_k x_l|x_0, x_1, x_6)$s are shown in the off-diagonal position of Table 6.2. From these values, we can find that the interaction between x_1 and x_6 is the most important among all of the second-order terms (i.e., the quadratic terms and interaction terms).

TABLE 6.1

Extra Sum of Squares

SS	x_1	x_2	x_3	x_4	x_5	x_6	x_7	x_8
SSR$(\cdot\|x_0)$	4.1280	0.1250	0.2200	0.2267	0.0392	**4.7432**	0.0288	0.8978
SSR$(\cdot\|x_0, x_6)$	**4.6487**	0.0543	0.2200	0.3576	0.0392	—	0.0848	0.8978
SSR$(\cdot\|x_0, x_1, x_6)$	—	0.0512	0.2200	0.3657	0.0392	—	0.0887	0.8978

TABLE 6.2

Extra Sum of Squares SSR$(x_i x_j \| x_0, x_1, x_6)$

SS	x_1	x_2	x_3	x_4	x_5	x_6	x_7	x_8
x_1	1.2986							
x_2	0.0192	0.0571						
x_3	0.1244	0.1658	0.2270					
x_4	0.1618	0.5679	0.2417	0.5450				
x_5	0.1765	0.3003	0.0785	0.9177	0.0778			
x_6	**5.8905**	0.0026	0.2349	0.6278	0.1830	0.0075		
x_7	0.0533	0.0892	0.6373	0.4280	0.2485	0.1032	0.0473	
x_8	0.3805	0.1486	0.1230	0.0117	0.2831	0.6937	0.3969	0.9899

We further compute SSR$(\cdot\|x_0, x_1, x_6, x_1 x_6)$ for the variables that have not been included in the model. In this case, all of them are small. Thus, we consider the following metamodel:

$$g(\mathbf{x}) = \alpha x_0 + \beta_1 x_1 + \beta_6 x_6 + \tau_{16} x_1 x_6.$$

From the above analysis, this model was obtained by starting with x_0 and then adding $x_6, x_1, x_1 x_6$ in order. From the first row of Table 6.1, SSR$(x\|x_0) = $ 4.1280, 4.7432 for $x = x_1, x_6$, respectively, and we can find SSR$(x\|x_1 x_6) = $ 0.3737. However, we can calculate SSR$(x\|x_0, x_6) = 4.6487$, 8.2624 for $x = $ $x_1, x_1 x_6$. Thus, when x_0 and x_6 are already in the model, the marginal effect of the interaction of $x_1 x_6$ is more significant than that of x_1. This is consistent with the model found in Section 5.2. Finally, we have SSR$(x_1\|x_0, x_6, x_1 x_6) = $ 2.2768. Hence, when x_0, x_6 and $x_1 x_6$ are already in the model, the contribution of x_1 to reducing variation is not as significant as those of x_6 and $x_1 x_6$. This implies that penalized least squares with the SCAD penalty discussed in Section 5.2 yields a very good model.

Campolongo et al. (see Chapter 2 in Saltelli et al. (2000)) applied the criteria introduced in this section to many models. Their examples are very helpful for understanding the concepts introduced in this section.

6.3 Sensitivity Analysis Based on Variation Decomposition

The decomposition of sum of squares is the key point in analysis of variance (ANOVA). The latter has played an essential role in experimental design and regression analysis. The method can be extended to decomposition of a metamodel by *functional analysis of variance*.

6.3.1 Functional ANOVA representation

Suppose that the metamodel $g(\mathbf{x})$ is an integrable function defined in the experimental domain $T \subset R^s$. For simplicity of presentation, it is assumed, in this section, that T is the s-dimensional unit cube $C^s = [0,1]^s$. It can then be shown (see, for instance, Sobol' (2001, 2003)) that $g(\mathbf{x})$ can be represented in the following *functional ANOVA* form:

$$g(\mathbf{x}) = g_0 + \sum_i g_i(x_i) + \sum_{i<j} g_{ij}(x_i, x_j) + \cdots + g_{1\cdots s}(x_1, \cdots, x_s), \qquad (6.5)$$

where

$$\int_0^1 g_{i_1, \cdots, i_t}(x_{i_1}, \cdots, x_{i_t}) dx_k = 0, \quad k = i_1, \cdots, i_t. \qquad (6.6)$$

Integrating (6.5) over C^s we have

$$\int g(\mathbf{x})\, d\mathbf{x} = g_0,$$

$$\int g(\mathbf{x}) \prod_{k \neq i} dx_k = g_0 + g_i(x_i),$$

$$\int g(\mathbf{x}) \prod_{k \neq i,j} dx_k = g_0 + g_i(x_i) + g_j(x_j) + g_{ij}(x_i, x_j),$$

and so on. It follows from (6.6) that all summands in (6.5) are orthogonal in the sense that

$$\int g_{i_1, \cdots, i_u} g_{j_1, \cdots, j_v}\, d\mathbf{x} = 0, \qquad (6.7)$$

whenever $(i_1, \cdots, i_u) \neq (j_1, \cdots, j_v)$.

The term ANOVA is used here because the representation in (6.5) provides us with the same interpretation as that of a classical ANOVA model. For example, $g_i(x_i)$ can be viewed as the main effects, while $g_{ij}(x_i, x_j)$ may be regarded as first-order interaction effects, and so on.

Assume further that $g(\mathbf{x})$ is square integrable. Then all the $g_{i_1 \cdots i_t}$ in the decomposition (6.5) are square integrable, and due to orthogonality of the decomposition in (6.7), we have

$$\int g^2\,d\mathbf{x} = g_0^2 + \sum_i \int g_i^2(x_i)\,dx_i + \sum_{i<j} \int g_{ij}^2(x_i, x_j)\,dx_i dx_j$$

$$+ \cdots + \int g_{12\cdots s}^2(x_1, \cdots, x_s)\,dx_1 \cdots dx_s.$$

Define the total variance and partial variances by

$$D = \int g^2(\mathbf{x})\,d\mathbf{x} - g_0^2 \quad \text{and} \quad D_{i_1\cdots i_s} = \int g_{i_1\cdots i_s}^2(x_{i_1}, \cdots, x_{i_s})\,dx_{i_1} \cdots dx_{i_s},$$

respectively. Note that

$$D = \sum_{k=1}^{s} \sum_{i_1 < \cdots < i_k} D_{i_1\cdots i_k},$$

which is similar to the decomposition in the traditional ANOVA.

Definition 9 The ratios

$$S_{i_1\cdots i_k} = \frac{D_{i_1\cdots i_k}}{D} \tag{6.8}$$

are called global sensitivity indices.

The integer k is often called the order or the dimension of the index (6.8). All $S_{i_1\cdots i_k}$s are non-negative and their sum is

$$\sum_{k=1}^{s} \sum_{i_1 < \cdots i_k}^{s} S_{i_1\cdots i_k} = 1.$$

The indices $S_{i_1\cdots i_k}$ are sometimes called *Sobol' indices*, as they were proposed in Sobol' (2001). For a (piecewise) continuous function $g(\mathbf{x})$, the equality $S_{i_1\cdots i_k} = 0$ implies that $g_{i_1\cdots i_k} \equiv 0$. Thus, the functional structure of $g(\mathbf{x})$ can be studied by estimating the indices.

Global sensitivity indices can be used to rank the importances of input variables. In the context of modeling computer experiments, when the interest is in understanding the main effects and the first-order interactions, it may be enough to just compute D_i and D_{ij}, for $i = 1, \cdots, s$ and $j = i+1, \cdots, s$, and rank the importance of input variables by sorting the D_is for the main effects and D_{ij}s for the first-order interactions. When one is interested in understanding the total effect of each variable including its interactions with other variables, the *total sensitivity index* as proposed by Homma and Saltelli (1996) can be employed by partitioning the input vector, $\mathbf{x}$, into two complementary sets $\mathbf{x}_{-i}$, and x_i, where $\mathbf{x}_{-i}$ is the set excluding the ith variable and x_i is the variable of interest. Specifically,

$$ST_i = S_i + S_{(i,-i)} = 1 - S_{-i},$$

where S_{-i} is the sum of all $S_{i_1 \cdots i_k}$ terms which do not include the ith variable, i.e., the total fraction of variance complementary to variable x_i. This index is adapted from the "freezing inessential variables" approach proposed by Sobol' (1993).

6.3.2 Computational issues

It is challenging to compute the integrals presented in the definition of $D_{i_1 \cdots i_k}$. In practice, the simplest way to estimate $D_{i_1 \cdots i_k}$ is by using a Monte Carlo method: generate random vectors $\mathbf{x}_1, \cdots, \mathbf{x}_N$ from a uniform distribution over the unit cube in R^s, the experimental domain. Then $D_{i_1 \cdots i_k}$ can be estimated by

$$\frac{1}{N} \sum_{j=1}^{N} g_{i_1 \cdots, i_k}(\mathbf{x}_j).$$

It can be shown that this estimate is consistent and possesses a $\sqrt{n}$ convergence rate.

Methods of approximating integrals have been widely studied by mathematicians in the fields of number theory and computational mathematics. See, for example, Hua and Wang (1981) and Niederreiter (1992). Quasi-Monte Carlo methods can be used to improve the rate of convergence of the Monte Carlo approach. In fact, the theory of uniform design itself originated from quasi-Monte Carlo methods. Using quasi-Monte Carlo methods, we can generate a set of points $\{\mathbf{x}_k, k = 1, \cdots, N\}$ which is space filling over the unit cube in R^s. With properly chosen $\mathbf{x}_j$s, the quantity

$$\hat{D}_{i_1 \cdots i_k} = \frac{1}{N} \sum_{j=1}^{N} g_{i_1 \cdots, i_k}(\mathbf{x}_j)$$

converges to $D_{i_1 \cdots i_k}$ much faster than randomly generated $\mathbf{x}_j$s. Hence, estimates of the global sensitivity calculations are as follows. It is clear that

$$\hat{g}_0 = \frac{1}{N} \sum_{j=1}^{N} g(\mathbf{x}_j), \quad \text{and} \quad \hat{D} = \frac{1}{N} \sum_{j=1}^{N} g^2(\mathbf{x}_j) - \hat{g}_0^2.$$

Estimation of D_i is a little more complicated. For ease of presentation, let us begin with $i = 1$. By its definition,

$$g_1(x_1) = \int g(\mathbf{x}) \prod_{k \neq 1} dx_k - g_0.$$

Therefore,

$$D_1 = \int g_1^2(x_1) \, dx_1,$$

which equals

$$\int \{\int g(\mathbf{x}) \prod_{k \neq 1} dx_k\}^2 \, dx_1 - g_0^2$$

$$= \int \int \int g((x_1, \mathbf{u}')')g((x_1, \mathbf{v}')') \, d\mathbf{u} d\mathbf{v} dx_1 - g_0^2,$$

where $\mathbf{u}$ and $\mathbf{v}$ are $(s-1)$-dimensional vectors. Therefore, estimating D_1 involves $(2s-1)$-dimensional integration. To implement the quasi-Monte Carlo method, we generate a set of points $\{\mathbf{t}_k, k = 1, \cdots, N\} \subset C^{2s-1}$ over the experiment domain. Let $x_{k1} = t_{k1}$, $\mathbf{u}_k = (t_{k2}, \cdots, t_{ks})'$, and $\mathbf{v}_k = (t_{k(s+1)}, \cdots, t_{k(2s-1)})'$. Then

$$\hat{D}_1 = \frac{1}{N} \sum_{j=1}^{N} g((x_{k1}, \mathbf{u}_k')')g((x_{k1}, \mathbf{v}_k')') - \hat{g}_0^2.$$

Similarly, we can compute $\hat{D}_i$ for $i = 2, \cdots, s$. To compute D_{-1}, we generate a set of points $\{\mathbf{t}_k, k = 1, \cdots, N\} \subset C^{(s+1)}$. Let $\mathbf{x}_k = (t_{k1}, \cdots, t_{k(s-1)})'$, $u_k = t_{ks}$, and $v_k = t_{k(s+1)}$. Then

$$\hat{D}_{-1} = \frac{1}{N} \sum_{j=1}^{N} g((u_k, \mathbf{x}_k')')g((v_k, \mathbf{x}_k')') - \hat{g}_0^2,$$

$$\widehat{ST}_1 = 1 - \frac{\hat{D}_{-1}}{\hat{D}}.$$

Similarly, we may compute $\hat{D}_{-i}$ and $\widehat{ST}_{-i}$ for $i = 2, \cdots, s$.

Example 20 As an illustration, we consider a simple metamodel

$$g(\mathbf{x}) = (x_1 - 0.5)(x_2 - 0.5) + 0.5e^{x_3 - 0.5},$$

where $x_i = [0, 1]$ for $i = 1, 2, 3$. In this example, $g_0 = 0.5 \int_0^1 e^{x_3 - 1/2} \, dx_3 = 0.5211$ since $\int_0^1 (x_1 - 0.5) \, dx_1 = \int_{-1}^1 (x_2 - 0.5) \, dx_2 = 0$. Moreover,

$$g_1(x_1) = g_2(x_2) = 0,$$

and

$$g_3(x_3) = 0.5e^{x_3 - 0.5} - g_0.$$

Furthermore,

$$g_{12}(x_1, x_2) = (x_1 - 0.5)(x_2 - 0.5), \quad g_{13}(x_1, x_3) = 0, \quad g_{23}(x_2, x_3) = 0.$$

We can further calculate the Sobol' indices for the main and total effects. Table 6.3 summarizes these indices.

The indices identify that x_1 and x_2 have no main effect, but in terms of total effects all variables are equally important.

TABLE 6.3
Sobol' indices for main and total effects

Variable	x_1	x_2	x_3
Main Effects ($\hat{S}_i$)	0.0000	0.0000	0.7622
Total Effects ($\widehat{ST}_i$)	0.2378	0.2378	0.7622

Both Monte Carlo methods and quasi-Monte Carlo methods can deal with any square integrable functions. However, the metamodels for computer experiments often have a particular form, which allows us to derive a closed expression for the constants $D_{i_1 \cdots i_k}$. Jin, Chen and Sudjianto (2004) studied this issue in detail. In what follows, we summarize their procedure.

Note that the metamodels introduced in the previous chapter have the following form

$$g(\mathbf{x}) = \beta_0 + \sum_j \beta_j B_j(\mathbf{x}), \tag{6.9}$$

where β_js are some constants which may depend on the data but not $\mathbf{x}$, and $B_j(\mathbf{x})$ can be represented as

$$B_j(\mathbf{x}) = \prod_{k=1}^{s} h_{jk}(x_k)$$

for some functions $h_{jk}(\cdots)$. For example, the Kriging model with a covariance function

$$\mathrm{Cov}\{y(\mathbf{s}), y(\mathbf{t})\} = \sigma^2 \exp(-\sum_{k=1}^{s} \theta_k |s_k - t_k|^2)$$

can be represented by

$$g(\mathbf{x}) = \mu + \sum_{i=1}^{n} \beta_i r_i(\mathbf{x}), \tag{6.10}$$

where

$$r_i(\mathbf{x}) = \mathrm{Corr}\{y(\mathbf{x}), y(\mathbf{x}_i)\} = \prod_{k=1}^{s} \exp(-\theta_k |x_k - x_{ik}|^2),$$

where $\mathbf{x} = (x_1, \cdots, x_s)'$, and $\{\mathbf{x}_k, \ k = 1, \cdots, n\}$ is a set of design points in which the response $y(\mathbf{x})$s are collected.

Note that the square of $g(\mathbf{x})$ in (6.9) has the same form as $g(\mathbf{x})$. Let us illustrate the procedure using the Kriging model in (6.10). The way to compute $\int g_{i_1 \cdots i_k}^2 \, dx_{i_1} \cdots x_{i_k}$s is similar. Thus, let us begin with $\int g(\mathbf{x}) \, d\mathbf{x}$. It is clear that

$$\int g(\mathbf{x}) \, d\mathbf{x} = \beta_0 + \sum_j \beta_j \int B_j(\mathbf{x}) \, d\mathbf{x}.$$

It is easy to see that

$$\int B_j(\mathbf{x})\,d\mathbf{x} = \prod_{k=1}^{s} \int h_{jk}(x_k)\,dx_k,$$

which converts the s-fold integral to a one-dimensional integral.

Here we only illustrate how to compute $\int g^2(\mathbf{x})\,d\mathbf{x}$. In order to get a closed form for $\int g^2(\mathbf{x})\,d\mathbf{x}$, it suffices to get a closed form for $\int r_i(\mathbf{x})\,d\mathbf{x}$ and $\int r_i(\mathbf{x})r_j(\mathbf{x})\,d\mathbf{x}$.

$$a_i \stackrel{\text{def}}{=} \int_{[0,1]^s} r_i(\mathbf{x})\,d\mathbf{x} = \prod_{k=1}^{s} \int_0^1 \exp\{-\theta_k(x_k - x_{ik})^2\}\,dx_k$$

$$= \prod_{k=1}^{s} \sqrt{2\pi/\theta_k}[\Phi\{(1 - x_{ik})/\theta_k\} - \Phi(-x_{ik}/\theta_k)],$$

where $\Phi(\cdot)$ is the cumulative distribution function of the standard normal distribution $N(0,1)$, which can be easily computed by many software packages. Similarly, we have

$$a_{ij} \stackrel{\text{def}}{=} \int_{[0,1]^s} r_i(\mathbf{x})r_j(\mathbf{x})\,d\mathbf{x}$$

$$= \prod_{k=1}^{s} \int_0^1 \exp\{-\theta_k\{(x_k - x_{ik})^2 + (x_k - x_{jk})^2\}\}\,dx_k$$

$$= \prod_{k=1}^{s} \int_0^1 \exp(-2\theta_k[\{x_k - (x_{ik} + x_{jk})/2\}^2 + (x_{ik} - x_{jk})^2/4])\,dx_k$$

$$= \exp\{-\sum_{k=1}^{s} \theta_k(x_{ik} - x_{jk})^2/2\}$$

$$\times \prod_{k=1}^{s} \sqrt{\pi/\theta_k}[\Phi\{(1 - (x_{ik} + x_{jk}))/2\theta_k\} - \Phi(-(x_{ik} + x_{jk})/2\theta_k)].$$

Therefore,

$$\int_{[0,1]^s} g(\mathbf{x})\,d\mathbf{x} = \beta_0^2 + 2\boldsymbol{\beta}'\mathbf{a} + \boldsymbol{\beta}'\mathbf{A}\boldsymbol{\beta},$$

where $\boldsymbol{\beta} = (\beta_1, \cdots, \beta_n)^T$, $\mathbf{a} = (a_1, \cdots, a_n)$, and $\mathbf{A}$ is an $n \times n$ matrix with (i,j)-elements a_{ij}. We can further compute the lower-dimensional integrals.

6.3.3 Example of Sobol' global sensitivity

We now conduct a global sensitivity analysis for the case study introduced in Section 5.4.3.

The main effects $g_i(x_i)$ are depicted in Figure 6.1, from which we can see that clearance and pin offset have the strongest main effects. We computed $D = 2.8328$, S_i and S_{ij}, tabulated in Table 6.4, where the diagonal elements correspond to S_i and the off-diagonal elements to S_{ij}. Results in Table 6.4 tell us that the clearance, pin offset, and their interaction are most important variables, as the sum of their global sensitivity indices $(0.4628+0.2588+0.2102)$ is about 93% of the total variation. Table 6.4 also indicates that skirt length may also be an important variable, as the global sensitivity index for its main effect is about 3% of the total variation and its interaction with the clearance is about 4% of the total variation. Note that, in this application, clearance variation due to manufacturing variability is a variable that engineers choose not to control (in robust design terminology, this variable is a "noise" variable). The strong interaction between clearance and pin offset implies that we should adjust the pin offset in such a way that it minimizes the the effect of clearance on the output variable.

TABLE 6.4
Indices S_i and S_{ij}

Factor	Clearance	Press	Length	Profile	Ovality	Pin Offset
Clearance	0.4628					
Press	0.0000	0.0000				
Length	0.0391	0.0001	0.0276			
Profile	0.0000	0.0000	0.0001	0.0000		
Ovality	0.0000	0.0000	0.0001	0.0000	0.0000	
Pin Offset	0.2588	0.0000	0.0012	0.0000	0.0000	0.2102

6.3.4 Correlation ratios and extension of Sobol' indices

Many authors regard the input variables **x** as random variables, and hence, outputs of computer experiments are viewed as realizations of a random variable (see, for instance, Chan, Saltelli and Tarantola (1997)). Denote by Y the response variable and X_i the i-th component of **x**. Consider the following variance decomposition from McKay (1995)

$$\text{Var}(Y) = \text{Var}\{E(Y|X_i)\} + E\{\text{Var}(Y|X_i)\},$$

where $\text{Var}\{E(Y|X_i)\}$ is called *variance conditional expectation* (VCE). Motivated by this decomposition, we define

$$\text{CR}_i = \frac{\text{Var}\{E(Y|X_i)\}}{\text{Var}(Y)}$$

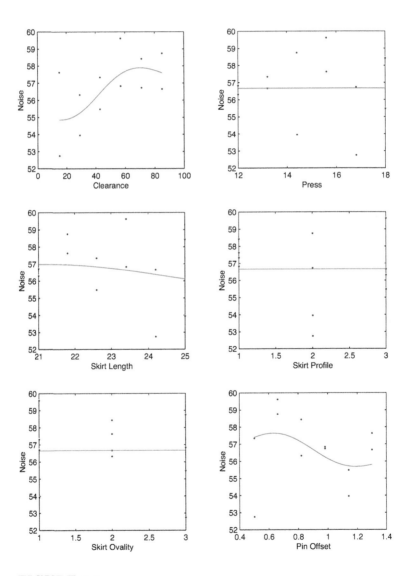

FIGURE 6.1
Main effects plots for piston noise. The solid line stands for $g_0 + g_i(x_i)$, where $g_0 = 56.6844$, and the dots stand for the individual observations.

for $i = 1, \cdots, s$. The CR_i is called the *correlation ratio* in McKay (1995). It is clear that the CR_i reflects the the proportion of variation of outputs explained by X_i. The correlation ratios play a similar role in the nonlinear regression setting to that of the usual correlation coefficient ρ for linear relationships between the output and the input variables (Kendall and Stuart (1979) and Krzykacz (1990)). Hora and Iman (1989) advocated using the square root of the VCE as a measure of the importance of X_i. Note that the square root of VCE is not scale-invariant. Iman and Hora (1990) further suggested the use of

$$\mathrm{LCR}_i = \frac{\mathrm{Var}[E\{\log(Y)|X_i\}]}{\mathrm{Var}\{\log(Y)\}}$$

as another measure of the importance of X_i. Saltelli et al. (1993) discussed a modified version of the Hora and Iman (1989)'s approach which relates to Krzykacz (1990). The idea of the correlation ratio can be extended to the partial correlation ratio, paralleling the partial correlation coefficient in linear models (see, Section 6.2).

Note that when the joint distribution of input variables is the uniform distribution over the unit cube C^s, the correlation ratios CR_i are equal to the Sobol' indices S_i defined in Section 6.3.1. The Sobol' indices can be extended to the case in which $\mathbf{x}$ has a density function with supports C^s. See, for example, Jin et al. (2004).

Denote $p(\mathbf{x})$ to be the density function of $\mathbf{x}$, and $p(x_{i_1}, \cdots, x_{i_k})$ to be the marginal density function of $(x_{i_1}, \cdots, x_{i_k})$. Similar to the functional ANOVA decomposition in (6.5), we have

$$\int g(\mathbf{x})p(\mathbf{x}) \, d\mathbf{x} = g_0,$$

$$\int g(\mathbf{x})p(\mathbf{x}_{-i}) \prod_{k \neq i} dx_k = g_0 + g_i(x_i),$$

$$\int g(\mathbf{x})p(\mathbf{x}_{-(ij)}) \prod_{k \neq i,j} dx_k = g_0 + g_i(x_i) + g_j(x_j) + g_{ij}(x_i, x_j),$$

and so on.

Based on the orthogonality feature of the above decomposition, it can be shown that the variance of g can be expressed as the summation of the effect variances of $g_{i_1 \cdots i_k}(x_{i_1}, \cdots, x_{i_k})$. That is,

$$\mathrm{Var}\{g(\mathbf{x})\} = \sum_{k=1}^{s} \sum_{i_1 < \cdots < i_k} \mathrm{Var}\{g_{i_1 \cdots i_k}(x_{i_1}, \cdots, x_{i_k})\},$$

where

$$\mathrm{Var}\{g(\mathbf{x})\} = \int g^2(\mathbf{x})p(\mathbf{x}) \, d\mathbf{x}$$

and

$$\text{Var}\{g_{i_1 \cdots i_k}(x_{i_1}, \cdots, x_{i_k})\} = \int g_{i_1 \cdots i_k}^2(x_{i_1}, \cdots, x_{i_k}) p(x_{i_1}, \cdots, x_{i_k}) \prod_{l=1}^{k} dx_{i_l}$$

as $E\{g_{i_1 \cdots i_k}(x_{i_1}, \cdots, x_{i_k})\} = 0$.

We can further define the *generalized Sobol' indices* as

$$GS_{i_1 \cdots i_k} = \frac{\text{Var}\{g_{i_1 \cdots i_k}(x_{i_1}, \cdots, x_{i_k})\}}{\text{Var}\{g(\mathbf{x})\}}.$$

It is clear that when the joint distribution of $\mathbf{x}$ is the uniform distribution over the unit cube, the generalized Sobol' indices are exactly the same as the Sobol' indices defined in Section 6.3.1. It is common to assume the input variables are independent. In other words,

$$p(\mathbf{x}) = \prod_{i=1}^{s} p_i(x_i),$$

where $p_i(x_i)$ stands for the marginal density function of x_i. Under the assumption of independence, the generalized Sobol' indices GS_i, $i = 1, \cdots, s$ coincide with the correlation ratios CR_i. This provides insight into why the correlation ratio CR_i can be used as a measure of importance of X_i.

6.3.5 Fourier amplitude sensitivity test

Another popular index to rank the importance of input variables is the *Fourier amplitude sensitivity test* (FAST), proposed by Cukier, Levine and Schuler (1978). Reedijk (2000) provided a detailed description of FAST indices, computational issues, and applications. The FAST indices provide a way to estimate the expected value and variance of the output and the contribution of individual input variables to this variance. The greatest advantage of the FAST is that the evaluation of sensitivity estimates can be carried out independently for each variable using just a single set of runs. In what follows, we give a brief introduction of the FAST. Again, for simplicity of presentation, it is assumed that the domain of input variables is the s-dimensional unit cube C^s. The main idea of FAST is to convert the s-dimensional integral of variance computation into a one-dimensional integral. This can be done by applying the ergodic theorem (Weyl (1938)) representing C^s along a curve, defined by a set of parametric equations

$$x_i(u) = G_i(\sin(w_i u)), \quad i = 1, \cdots, s,$$

where u is a scalar variable with range R^1, and $\{w_i\}$ is a set of integer angular frequencies.

With proper choice of w_i and G_i, it can be shown (Weyl (1938)) that

$$g_0 \overset{\text{def}}{=} \int_{C^s} g(x_1, \cdots, x_s) \, dx_1 \cdots dx_s$$

$$= \frac{1}{2\pi} \int_{-\pi}^{\pi} g(G_1(\sin(w_1 u)), \cdots, G_s(\sin(w_s u))) \, du.$$

This representation allows us to apply Fourier analysis for $g(\mathbf{x})$. Using Parseval's theorem, we have the following decomposition:

$$D = \int_{C^s} g(\mathbf{x})^2 \, dx - g_0^2$$

$$\approx 2 \sum_{w=1}^{\infty} (A_w^2 + B_w^2), \tag{6.11}$$

where A_w and B_w are the Fourier coefficients, defined by

$$A_w = \frac{1}{2\pi} \int_{-\pi}^{\pi} g(\mathbf{x}(u)) \cos(wu) \, du$$

and

$$B_w = \frac{1}{2\pi} \int_{-\pi}^{\pi} g(\mathbf{x}(u)) \sin(wu) \, du,$$

and $\mathbf{x}(u) = (G_1(\sin(w_1 u)), \cdots, G_s(\sin(w_s u)))'$. Further define

$$D_{w_i} = 2 \sum_{p=1}^{\infty} (A_{pw_i}^2 + B_{pw_i}^2).$$

The summation over p in the definition of D_{w_i} is meant to include all the harmonics related to the frequency associated with the input variable x_i. The coefficient D_{w_i} can be interpreted as a "main effect" in the model. Therefore, the D_{w_i} are referred to as the first-order indices or classical FAST. Unfortunately, the fraction of the total variance due to interactions cannot be computed by this technique at present. Following Cukier et al. (1978), the part of variance due to the input variable x_i is D_{w_i}, the frequency assigned to x_i so that the quantity

$$S_i = \frac{D_{w_i}}{D}$$

is the fraction of the variance of g due to the input variable x_i. Therefore, it can be used to rank the importance of input variables.

To compute S_i, one must choose G_i and w_i. Cukier et al. (1978) proposed a systematic approach for determining the search curve $G_i(u)$ by solving a partial differential equation with boundary condition $G_i(0) = 0$. For the analysis of computer experiments, we may simply take

$$G_i(t) = \frac{1}{2} + \frac{1}{\pi} \arcsin(t)$$

for all $i = 1, \cdots, s$. That is,

$$x_i(u) = \frac{1}{2} + \frac{1}{\pi} \arcsin(\sin(w_i u)).$$

The search curve drives arbitrarily close to any point $\mathbf{x} \in C^s$ if and only if the set of frequencies w_i is incommensurate. A set of frequencies is said to be incommensurate if none of them can be obtained by a linear combination of the other frequencies with integer coefficients. The easiest way to obtain an *incommensurate set of frequencies* is by using irrational numbers. See Reedijk (2000) for details. In practice, due to limited computer precision, one may use integer frequencies, which is actually an approximation. When the frequencies are incommensurate the generated curve is space-filling. As an example, a two-dimensional case generated from curves with $w_1 = 11$ and $w_2 = 21$ as shown in Figure 6.2.

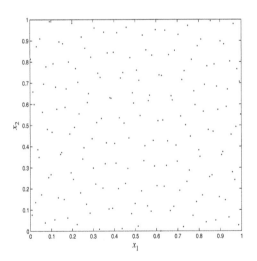

FIGURE 6.2

Scatter plot of sampling points for two dimensional problem with $\{w_1, w_2\} = \{11, 21\}$.

The summations in the definition of D and D_{w_i} are taken from 1 to ∞. Since the Fourier amplitudes decrease as p increases, D_{w_i} can be approximated by

$$2 \sum_{p=1}^{M} (A_{pw_i}^2 + B_{pw_i}^2),$$

where M is the maximum harmonic, which is usually taken to be 4 or 6 in practice.

Saltelli et al. (2000) proposed *extended FAST* (EFAST) to not only calculate the main effects but also total effects to allow full quantification of the importance of each variable. Consider the frequencies that do not belong to the set pw_i. These frequencies contain information about interactions among factors at all orders not accounted for by the main effect indices. To calculate the total effects, we assign a high frequency w_i for the ith variable and a different low frequency value w_{-i} to the remaining variables. By evaluating the spectrum at w_{-i} frequency and its harmonic, we can calculate the partial variance D_{-i}. This is the total effect of any order that does not include the ith variable and is complementary to the variance of the ith variable. For example, consider a function with three variables, x_1, x_2, and x_3. By assigning $w_1 = 21$ and $w_2 = w_3 = 1$, we can calculate the the total effect of x_1. Similar to Sobol' indices, the total variance due to the ith variable is $D_{Ti} = D - D_{-i}$ and the total effect is

$$ST_i = S_i + S_{(i,-i)} = 1 - S_{-i},$$

where S_{-i} is the sum of all $S_{i_1 \cdots i_k}$ terms which do not include the term i. Note that while the total effect indices do not provide full characterization of the system of $2^s - 1$ interactions, they allow a full quantification of the importance of each variable.

Example 21 As a demonstration, we consider a simple quadratic polynomial metamodel:

$$g(\mathbf{x}) = x_1 x_2 + x_3^2$$

where $x_i = [-1, 1]$ for $i = 1, 2, 3$. The calculated FAST indices for the main and total effects are summarized in Table 6.5. The indices identify that x_1 and x_2 have no main effect, but in terms of total effects all variables are equally important.

TABLE 6.5
FAST Indices for Main and Total Effects

Variable	x_1	x_2	x_3
Main Effects ($\hat{S}_i$)	0.00	0.00	0.44
Total Effects ($\widehat{ST}_i$)	0.56	0.56	0.44

6.3.6   Example of FAST application

Consider again the borehole example presented in Section 1.5. Instead of a polynomial model as described in Section 1.5, we use a Gaussian Kriging metamodel to fit the data. Thereafter, the FAST method is applied to the metamodel with the set of frequencies $\{19, 59, 91, 113, 133, 143, 149, 157\}$ corresponding to the eight variables $\{r_w, r, T_u, T_l, H_u, H_l, L, K_w\}$. The FAST main effects for these variables are $\{0.8257, 0.0000, 0.0001, 0.0000, 0.0414, 0.0411, 0.0390, 0.0093\}$ for which r_w has the largest main effect followed by H_u and H_l. This result is in agreement with that of traditional ANOVA decomposition from polynomial models as discussed in Section 1.5. The computation for the total effects is conducted using $w_i = 21$ and $w_{\sim i} = 1$. To avoid potential overlap, only the first $w_{max}/2$ components are used in calculating $D_{\sim i}$.

$$D_{\sim i} = 2 \sum_{k=1}^{w_{max}/2} (A_k^2 + B_k^2).$$

The FAST total effects for the variables are $\{0.8798, 0.0270, 0.0269, 0.0268, 0.0866, 0.0878, 0.0850, 0.0451\}$. In this case, the orders of importance suggested by the main and total effects are about the same. This result indicates that the effects are dominated by the main effects.

7

Functional Response

This chapter concerns computer experiments with functional response. Three classes of models, spatial temporal models, functional linear models, and partially functional models, along with related estimation procedures, will be introduced. Some case studies are given for illustration.

7.1 Computer Experiments with Functional Response

Functional data are data collected over an interval of some index with an intensive sampling rate. For example, suppose that data for each experiment unit were collected, each 20 seconds over a period of an hour, and we want to know how the factors affect the shape of the resulting curves. Functional data can be in the form of one-dimensional data such as a curve or higher dimensional data such as a two- or three-dimensional image. With the advent of computer model and visualization, outputs of computer models are often presented in a higher dimensional image to gain better physical insight. Figure 7.1 shows an output of a computational fluid dynamics mode for cylinder head port flow simulation in an internal combustion engine.

With the advent of modern technology and devices for collecting data, people can easily collect and store functional data, and functional data analysis is becoming popular in various research fields. Ramsay and Silverman (1997) present many interesting examples of functional data and introduce various statistical models to fit the functional data. Similarly, computer experiments with functional responses have increased in complexity leading up to today's sophisticated three-dimensional computer models. However, there is little literature on modeling of computer experiments with functional response, and modeling computer experiments is typically limited to point outputs. Here, we provide a more sophisticated approach to naturally analyzing functional responses which may suggest more insightful conclusions that may not be apparent otherwise. Let us introduce two typical examples of functional response in computer experiments.

Example 22 below is a typical example of functional response with sparse sampling rate. The output was collected over different rotations per minute (RPMs), and a total of 6 outputs were collected for each design condition. This

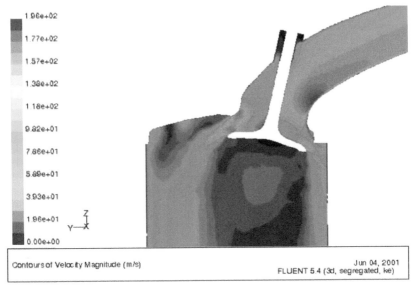

1.96e+02	
1.77e+02	
1.57e+02	
1.38e+02	
1.18e+02	
9.82e+01	
7.86e+01	
5.89e+01	
3.93e+01	
1.96e+01	
0.00e+00	

Contours of Velocity Magnitude (m/s)   Jun 04, 2001
                                       FLUENT 5.4 (3d, segregated, ke)

FIGURE 7.1
Cylinder head port flow generated by a computational fluid dynamic model.

response is similar to that of growth curve models in the statistical literature
(see, for example, Pan and Fang (2002)). Although one may view such output
as multi-response, we refer to it instead as functional response because the
data could conceivably have been collected over the whole interval of RPM,
and because the response is several outputs of one variable rather than outputs
of several variables. Example 23 is a typical example of functional response
with intensive sampling rate. The response was recorded for every degree of
crank angle over [0, 500].

Example 22 (Engine Noise, Vibration and Harshness)
    Computer models are commonly employed in the design of engine structure
to minimize radiated noise. The model includes multi-body dynamic simulation to estimate acting mechanical forces during engine operations from
the cranktrain and valvetrain, as well as combustion forces. Together with
the finite element structural model of the engine component or system (see
Figure 7.2 below), this model is used to evaluate structural response during
operating conditions as a function of engine RPM and frequency. The vibro-
acoustic relationship between the engine vibrations and the acoustic pressure
field may then be applied to calculate the radiated engine noise (see, for example, Gérard, Tournour, El Masri, Cremers, Felice and Selmane (2001)).

    To optimize the design of the cylinder block, 17 design variables (e.g., bulk-

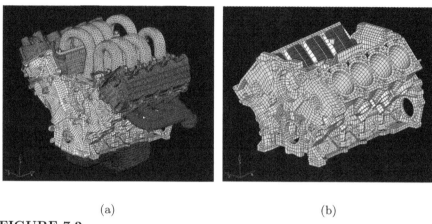

(a)   (b)

FIGURE 7.2

Engine models for Example 22: (a) engine system finite element model, and (b) engine block finite element model.

head thickness, oil pan rail thickness, valley wall ribbing, etc.; a detailed list is given in Table 7.1) were chosen. A uniform design with 30 runs was used for the experiment and is presented in Table 7.2. In this example, the selected response variable is the structural response of the oil pan flange in terms of acceleration at various engine RPMs (e.g., 1000, 2000,..., 6000 RPM), which directly influences the radiated noise through the vibro-acoustic relationship. The outputs are presented in Table 7.3. This experiment can be treated either as multi-response (i.e., response is a vector of numbers) or functional response (i.e., response is a sample of points from a function) since for each run six responses can be fitted by a curve (cf. Figure 7.5).

TABLE 7.1

Name of Design Variables

Var.	Name	Var.	Name
x_1	Bulkhead thickness	x_{10}	Side wall ribbing
x_2	Oil pan rail thickness	x_{11}	RFOB ribbing
x_3	Skirt wall thickness	x_{12}	FFOB ribbing
x_4	Main cap thickness	x_{13}	Valley wall ribbing
x_5	Liner thickness	x_{14}	Bearing beam
x_6	Valley wall thickness	x_{15}	Dam with value 1
x_7	Bore wall thickness	x_{16}	Dam with value 2
x_8	Valley plate thickness	x_{17}	Den
x_9	Skirt ribbing	x_{18}	Young modulus

Example 23 (*Valvetrain*) The valvetrain system is one of the most important elements of internal combustion engines for the delivery of desired engine performance in terms of horsepower/torque, noise and vibration, fuel economy, and emissions. Design optimization is performed to maximize performance capability while maintaining the durability of the system. During the optimization process, engineers synthesize the design of components in the valvetrain system (e.g., cam profile, spring, etc.) to meet the intended performance target. To achieve such goals, various computer-aided engineering models are simultaneously employed to achieve the best balance of performance and durability attributes. Multi-body dynamic computer models are commonly employed to optimize the dynamic behavior of a valvetrain, especially at high engine speeds, where the the spring behavior becomes highly nonlinear (see Figure 7.3 below) in an internal combustion engine (Philips, Schamel and Meyer (1989)). The model can be used to study typical valvetrain system characteristics in terms of durability, noise, and valve sealing. Detailed three-dimensional geometric information as well as other physical properties such as valve seat, spring (spring surge and coil clash), camshaft torsional and bending, cam-follower hydrodynamic, and lash adjuster dynamics are included in the models. The model is very crucial for guiding valvetrain performance optimization while maintaining a stable valvetrain dynamic behavior. One indicator of a stable valvetrain is that the valve movement must follow a prescribed motion determined by the camshaft profile. At high speed valvetrain operations, however, this may not be the case, as the effect of inertia on the dynamic behavior becomes prominent. This criterion is especially crucial during valve closing to minimize valve bounce. Figure 7.4 shows the motion errors of the valve compared to the prescribed motion by the camshaft for the first four designs listed in Table 7.4. A perfectly stable valve train should have a straight line or zero error throughout the crank angles. From Figure 7.4, we can see that the variability of the motion errors varies for various design configurations.

In this situation, engineers attempt to minimize motion errors by finding the best level-combination of the factors, such as cylinder head stiffness, rocker arm stiffness, hydraulic lash adjuster, spring, cam profile, etc. To achieve this goal, one is interested in understanding the effects of each design variable on the motion errors. In this case, the response variable (i.e., amount of motion error) is defined over a crank angle range, instead of a single value; thus, it is considered a functional response.

Table 7.4 depicts an experimental matrix which was applied to the computer model to minimize the motion error, particularly to minimize valve bounce during valve closing. In practice, it is reasonable to model the data in Example 22 using linear models for repeated measurements. For Example 23, nonparametric and semiparametric modeling techniques may be used to construct a good predictor for the functional response. A general framework for modeling is necessary.

TABLE 7.2
Design of Engine Noise Vibration and Harshness

Run #	x_1	x_2	x_3	x_4	x_5	x_6	x_7	x_8	x_9	x_{10}	x_{11}	x_{12}	x_{13}	x_{14}	$\{x_{15}, x_{16}\}$	x_{17}	x_{18}
1	25.5	9	7.5	22.5	5.5	9	5.5	8.5	1	0	1	0	1	0	3	2.91	77.49
2	25.5	9	6.5	25.5	3.5	13	7	13	0	1	1	1	1	1	2	2.77	70.11
3	18.5	21	7.5	18.5	3.5	11	7	13	1	1	0	1	0	0	1	2.91	77.49
4	22.5	13	8.5	25.5	2	13	5.5	8.5	0	0	0	0	0	1	3	2.77	77.49
5	22.5	17	5	20.5	2	6	9	3.5	0	1	1	1	1	1	1	2.77	77.49
6	24.5	9	6.5	22.5	3.5	4	7	13	0	0	0	1	0	1	1	2.63	73.8
7	22.5	13	5	25.5	5.5	13	5.5	6	1	1	1	0	0	1	2	2.77	77.49
8	24.5	17	8.5	20.5	3.5	13	9	6	0	1	0	0	1	0	3	2.63	70.11
9	18.5	17	5	20.5	5.5	9	5.5	3.5	1	1	1	1	1	0	3	2.63	73.8
10	20.5	9	9	18.5	3.5	11	5.5	3.5	1	0	0	0	0	1	2	2.77	77.49
11	18.5	9	9	24.5	3.5	11	9	3.5	1	0	0	0	0	0	3	2.91	73.8
12	20.5	13	6.5	24.5	2	6	9	8.5	1	1	0	0	0	0	1	2.91	73.8
13	25.5	13	7.5	22.5	5.5	13	9	8.5	1	1	1	1	0	0	1	2.77	77.49
14	25.5	13	9	25.5	2	4	7	13	0	1	1	0	1	0	2	2.77	77.49
15	24.5	9	8.5	18.5	2	6	9	10	1	0	0	1	1	0	2	2.63	73.8
16	18.5	21	5	20.5	5.5	9	5.5	8.5	0	0	0	0	1	1	2	2.91	70.11
17	22.5	21	6.5	25.5	2	11	5.5	10	1	0	1	1	1	0	2	2.63	77.49
18	24.5	23	9	22.5	3.5	11	9	10	0	0	0	0	1	0	1	2.91	73.8
19	25.5	21	7.5	24.5	2	4	7	3.5	0	1	0	1	0	1	3	2.77	70.11
20	20.5	17	9	22.5	5.5	9	9	13	1	0	0	1	0	0	2	2.63	70.11
21	18.5	23	7.5	24.5	5.5	11	9	13	0	1	0	0	0	0	2	2.63	73.8
22	24.5	23	8.5	20.5	2	13	7	3.5	0	0	1	0	1	1	2	2.91	73.8
23	20.5	23	6.5	25.5	2	6	5.5	10	0	1	1	1	0	1	3	2.77	73.8
24	22.5	23	8.5	24.5	3.5	6	7	6	1	1	1	0	1	1	3	2.63	70.11
25	24.5	21	5	20.5	2	9	7	6	1	0	1	0	0	0	1	2.91	70.11
26	22.5	17	6.5	18.5	3.5	4	9	8.5	0	1	0	1	1	1	1	2.91	70.11
27	20.5	13	9	22.5	5.5	4	7	10	1	1	0	1	1	1	3	2.77	77.49
28	20.5	21	8.5	24.5	5.5	6	5.5	6	0	0	1	0	0	0	1	2.63	70.11
29	25.5	17	5	18.5	3.5	4	5.5	6	0	1	1	1	0	0	3	2.91	73.8
30	18.5	23	7.5	18.5	5.5	9	7	10	1	0	1	0	1	1	1	2.63	70.11

TABLE 7.3

Response of Engine Noise Vibration and Harshness

Run #	1000	2000	3000	4000	5000	6000
1	0.09	0.39	1.01	2.84	8.51	14.73
2	0.11	0.46	1.23	3.46	9.39	18.32
3	0.09	0.38	1.03	3.15	8.76	14.89
4	0.10	0.42	1.11	3.35	9.09	14.76
5	0.12	0.49	1.27	3.95	10.76	18.16
6	0.11	0.46	1.20	3.72	11.08	17.17
7	0.10	0.42	1.14	3.28	8.95	14.73
8	0.09	0.41	1.07	3.16	9.02	15.92
9	0.09	0.39	1.01	3.09	8.42	13.20
10	0.10	0.42	1.12	3.52	9.76	16.02
11	0.10	0.41	1.10	3.41	7.92	14.89
12	0.11	0.44	1.19	3.81	9.60	17.33
13	0.10	0.41	1.13	3.10	8.62	13.83
14	0.10	0.43	1.17	3.46	9.76	17.37
15	0.10	0.45	1.15	3.48	9.39	16.54
16	0.10	0.43	1.14	3.67	8.17	14.63
17	0.09	0.38	1.01	3.17	8.31	12.32
18	0.09	0.40	1.05	3.09	8.95	14.77
19	0.10	0.42	1.10	3.70	9.06	18.14
20	0.10	0.42	1.17	3.19	9.46	17.35
21	0.09	0.40	1.06	3.13	8.52	14.09
22	0.10	0.44	1.16	3.41	8.92	16.99
23	0.09	0.39	1.02	3.16	8.77	14.81
24	0.09	0.39	1.04	3.24	8.54	15.47
25	0.11	0.44	1.20	4.04	10.30	19.03
26	0.11	0.48	1.27	4.33	9.72	18.48
27	0.09	0.39	1.07	3.15	8.74	14.71
28	0.10	0.46	1.23	3.46	9.62	19.06
29	0.10	0.40	1.13	3.63	9.79	18.72
30	0.10	0.41	1.14	3.50	10.01	17.42

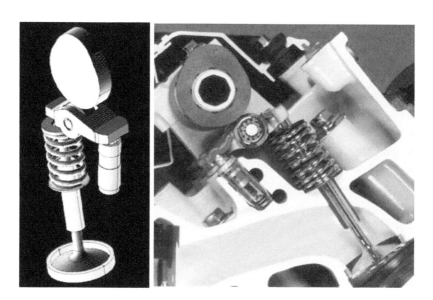

FIGURE 7.3
Roller finger follower valvetrain system.

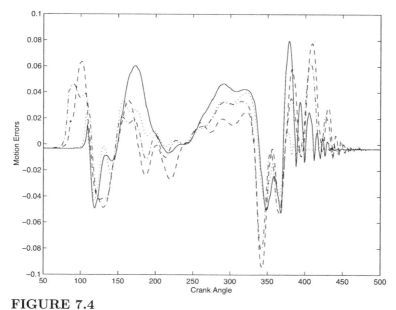

FIGURE 7.4
Valve motion errors of the first four valvetrain designs listed in
Table 7.4.

TABLE 7.4

Design for Valvetrain Experiment

Head Stiffness	RA Stiffness	Lash Adjuster	Cam Phasing	Clearance	Spring Height	Ramp
1	1	1	1	1	1	1
2	1	1	1	1	2	3
1	2	1	1	2	1	3
2	2	1	1	2	2	1
1	1	2	1	2	2	2
2	1	2	1	2	1	2
1	2	2	1	1	2	2
2	2	2	1	1	1	2
1	1	1	2	2	2	2
2	1	1	2	2	1	2
1	2	1	2	1	2	2
2	2	1	2	1	1	2
1	1	2	2	1	1	3
2	1	2	2	1	2	1
1	2	2	2	2	1	1
2	2	2	2	2	2	3

A general framework for functional response

Various statistical models have been proposed for functional data in the statistical literature. In this chapter, we introduce the spatial temporal model, the functional linear model, and the partially functional linear model. All these three models can be written in a unified form:

$$y(t, \mathbf{x}) = \mu(t, \mathbf{x}) + z(t, \mathbf{x}), \tag{7.1}$$

where $\mu(t, \mathbf{x})$ is the mean function of $y(t, \mathbf{x})$, $z(t, \mathbf{x})$ is a random term with mean zero, $\mathbf{x}$ are the input variables, and t stands for some index. In the classic case of growth curve modeling, t is time, but, in general, there are many possibilities, e.g., in Example 23, t stands for the crank angle.

Different models propose different structures for the mean function; therefore, one needs different ways to estimate the parameters involved in the model. Section 7.2 introduces the spatial temporal model and a related estimation algorithm. Section 7.4 presents functional linear models and their estimation procedures. We introduce partially functional linear models and illustrate their applications in Section 7.5. Section 7.3 gives an overview of the penalized regression splines method which is used to estimate functional linear models and partially functional linear models.

7.2 Spatial Temporal Models

The spatial temporal model regards the mean function $\mu(t, \mathbf{x})$ in (7.1) as an overall trend and assumes that $z(t, \mathbf{x})$ is a Gaussian random field indexed by t and $\mathbf{x}$. Thus, it views the output of a computer experiment as a realization of a Gaussian random field. The spatial temporal model is a generalization of the Gaussian Kriging model and has been widely used in the literature of spatial statistics (see, for example, Carroll, Chen, Li, Newton, Schmiediche, Wang and George (1997)).

The *spatial temporal model* extends the Gaussian Kriging model by incorporating index-direction correlation by considering

$$y(t, \mathbf{x}) = \mu(t, \mathbf{x}) + z(t, \mathbf{x}), \tag{7.2}$$

where $\mu(t, \mathbf{x})$ is the mean function of $y(t, \mathbf{x})$, and $z(t, \mathbf{x})$ is a *Gaussian random field* (GRF) with mean 0. We can extend the estimation procedure and prediction procedure of the Gaussian Kriging model in Section 5.4 for the spatial temporal model.

7.2.1 Functional response with sparse sampling rate

Consider a spatial temporal model for functional response with sparse sample rate. Thus, the response is a vector of outputs of one response variable over different values of t, as in Example 22. Let J denote the number of outputs for each run. Let us use matrix notation for succinctness of presentation. Denote $\mathbf{Y}$ and $\mathbf{M}$ to be $n \times J$ matrices with (i, j)-element $y(t_j, \mathbf{x}_i)$ and $\mu(t_j, \mathbf{x}_i)$, respectively. Then model (7.2) can be represented as

$$\mathbf{Y} = \mathbf{M} + \mathbf{R}^{-1/2} \mathbf{Z} \Psi^{-1/2}, \tag{7.3}$$

where Ψ is a $J \times J$ positive definite matrix, $\mathbf{Z}$ is an $n \times J$ random matrix with all elements being independent variates on standard normal distribution $N(0, 1)$, and $\mathbf{R}$ is a correlation matrix whose (u, v)-element is

$$r(\boldsymbol{\theta}_t; \mathbf{x}_u, \mathbf{x}_v) = \text{Corr}\{y(t, \mathbf{x}_u), y(t, \mathbf{x}_v)\},$$

where $\boldsymbol{\theta}_t$ represents the unknown parameter in the correlation function r. Denote $\boldsymbol{\theta} = (\boldsymbol{\theta}_1', \cdots, \boldsymbol{\theta}_J')'$ and $\mathbf{R} = \mathbf{R}(\boldsymbol{\theta})$. When all $\boldsymbol{\theta}_j$s are the same, then $\boldsymbol{\theta}$ can be simply set to be $\boldsymbol{\theta}_1$. One may further assume that the correlation function has the correlation structure introduced in Section 5.4.

Let $\mathbf{m}(\mathbf{x}) = (\mu(t_1, \mathbf{x}), \cdots, \mu(t_J, \mathbf{x}))$. In practice, we assumed that $\mathbf{m}$ can be represented or approximated by

$$\mathbf{m}(\mathbf{x}) = \mathbf{b}'(\mathbf{t}, \mathbf{x})\boldsymbol{\beta},$$

where $\mathbf{b}(\mathbf{t}, \mathbf{x}) = (b_1(\mathbf{t}, \mathbf{x}), \cdots, b_L(\mathbf{t}, \mathbf{x}))'$ with $\mathbf{t} = (t_1, \cdots, t_J)$ and $\boldsymbol{\beta}$ is an $L \times J$ unknown parameter matrix. Note that the ith-row of $\mathbf{M}$ corresponds to $\mathbf{m}(\mathbf{x}_i)$. Therefore the mean matrix has the following form

$$\mathbf{M} = \mathbf{B}\boldsymbol{\beta},$$

where $\mathbf{B} = (\mathbf{m}(\mathbf{x}_1)', \cdots, \mathbf{m}(\mathbf{x}_n)')'$ is an $n \times L$ known matrix. Thus, model (7.3) can be written in a form of linear model:

$$\mathbf{Y} = \mathbf{B}\boldsymbol{\beta} + \mathbf{R}^{-1/2}(\boldsymbol{\theta})\mathbf{Z}\boldsymbol{\Psi}^{-1/2}. \tag{7.4}$$

Using the theory of linear models, we can have an explicit form for the maximum likelihood estimate of $\boldsymbol{\beta}$ and $\boldsymbol{\Psi}$ if $\mathbf{R}$ is known. Specifically, for a given $\mathbf{R}$,

$$\hat{\boldsymbol{\beta}} = (\mathbf{B}'\mathbf{R}^{-1}(\boldsymbol{\theta})\mathbf{B})^{-1}\mathbf{B}'\mathbf{R}^{-1}(\boldsymbol{\theta})\mathbf{Y}, \tag{7.5}$$

and

$$\hat{\boldsymbol{\Psi}} = \frac{1}{n}(\mathbf{Y} - \mathbf{B}\hat{\boldsymbol{\beta}})'\mathbf{R}^{-1}(\boldsymbol{\theta})(\mathbf{Y} - \mathbf{B}\hat{\boldsymbol{\beta}}). \tag{7.6}$$

The maximum likelihood estimator for $\boldsymbol{\theta}_t$ does not have a closed form. Numerical algorithms such as the Newton-Raphson or Fisher scoring algorithms may be used to search for the solution. In summary, we have the following algorithm.

Algorithm 7.1:

Step 1: Set the initial value of $\boldsymbol{\beta}$ to be $(\mathbf{B}'\mathbf{B})^{-1}\mathbf{B}\mathbf{Y}$, the least squares estimator of $\boldsymbol{\beta}$;

Step 2: For a given $\boldsymbol{\beta}$, we update $\boldsymbol{\Psi}$ using (7.6) and update $\boldsymbol{\theta}$ by solving the following equations

$$\partial \ell(\boldsymbol{\theta})/\partial \boldsymbol{\theta} = 0;$$

where

$$\ell(\boldsymbol{\theta}) = \log(|\mathbf{R}(\boldsymbol{\theta})|) + \text{tr}\{\mathbf{R}^{-1}(\boldsymbol{\theta})(\mathbf{Y} - \mathbf{B}\hat{\boldsymbol{\beta}})\hat{\boldsymbol{\Psi}}^{-1}(\mathbf{Y} - \mathbf{B}\hat{\boldsymbol{\beta}})'\}.$$

In this step, we need numerical iteration algorithms, such as the Newton-Raphson algorithm and the Fisher scoring algorithm, to solve the score equation.

Step 3: For a given $\boldsymbol{\theta}$, we update $\boldsymbol{\beta}$ using (7.4);

Step 4: Iterate Step 2 and Step 3 until it converges.

After estimating the unknown parameters, we can predict the response at the unobserved site $\mathbf{x}$. Let $\mathbf{y}(\mathbf{x}) = (y(t_1, \mathbf{x}), \cdots, y(t_J, \mathbf{x}))$. Then the prediction of $\mathbf{y}(\mathbf{x})$ is

$$\hat{\mathbf{y}}(\mathbf{x}) = \mathbf{b}'(\mathbf{t}, \mathbf{x})\hat{\boldsymbol{\beta}} + \mathbf{r}(\hat{\boldsymbol{\theta}}; \mathbf{x})\mathbf{R}^{-1}(\hat{\boldsymbol{\theta}})(\mathbf{Y} - \mathbf{B}\hat{\boldsymbol{\beta}}),$$

where
$$\mathbf{r}(\boldsymbol{\theta};\mathbf{x}) = (r(\boldsymbol{\theta}_1;\mathbf{x}_1,\mathbf{x}),\cdots,r(\boldsymbol{\theta}_n;\mathbf{x}_n,\mathbf{x})).$$

Furthermore, the covariance matrix of $\hat{\mathbf{y}}(\mathbf{x})$ may be estimated by

$$\widehat{\mathrm{Cov}}\{\hat{\mathbf{y}}(\mathbf{x})\} = \left[1 - (\mathbf{b}'(\mathbf{x}),\mathbf{r}'(\mathbf{x}))\begin{pmatrix}\mathbf{0} & \mathbf{B}'\\ \mathbf{B} & \mathbf{R}(\boldsymbol{\theta})^{-1}\end{pmatrix}\begin{pmatrix}\mathbf{b}(\mathbf{x})\\ \mathbf{r}(\mathbf{x})\end{pmatrix}\right]\boldsymbol{\Psi}.$$

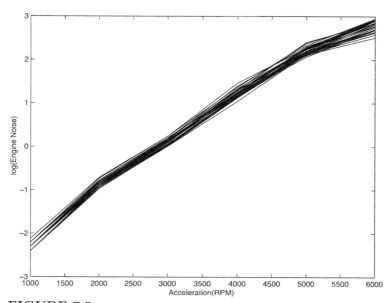

FIGURE 7.5

Plot of log(engine noise) versus engine acceleration.

Example 22 (Continued) We are interested in studying the impact of the design variables on the ratio of responses (engine noise, EN for short) at different speeds. Figure 7.5 depicts the plot of the logarithm of EN versus the acceleration (RPM), and indicates that the EN increases approximately exponentially as the acceleration increases. The following response variables are defined to analyze the rate of increase of EN as the engine speed increases. Define

$y_1 = \log(\text{EN at 2000RPM}) - \log(\text{EN at 1000RPM});$
$y_2 = \log(\text{EN at 3000RPM}) - \log(\text{EN at 2000RPM});$
$y_3 = \log(\text{EN at 4000RPM}) - \log(\text{EN at 3000RPM});$
$y_4 = \log(\text{EN at 5000RPM}) - \log(\text{EN at 4000RPM});$
$y_5 = \log(\text{EN at 6000RPM}) - \log(\text{EN at 5000RPM}).$

Let $\mathbf{y} = (y_1, \cdots, y_5)$. Consider model (7.3) with $\mathbf{M} = \mathbf{1}_n \times \boldsymbol{\mu}$ and $\boldsymbol{\mu} = (\mu_1, \cdots, \mu_5)$. The elements of $\mathbf{R}(\boldsymbol{\theta})$ are assumed to have the following form:

$$r(\boldsymbol{\theta}; \mathbf{u}, \mathbf{v}) = \exp(-\sum_{k=1}^{s} \theta_k |u_k - v_k|^2).$$

The above estimation procedure is applied for the model. The resulting estimate is

$$\hat{\boldsymbol{\mu}} = (1.4464, 0.9799, 1.1120, 0.9898, 0.5582),$$

$$\hat{\Psi} = \frac{1}{1000} \begin{pmatrix} 1.5470 & -0.5139 & -0.5969 & -0.0640 & 0.4787 \\ -0.5139 & 0.6829 & -0.1786 & 0.2182 & 0.2455 \\ -0.5969 & -0.1786 & 3.1579 & -1.6537 & -0.1561 \\ -0.0640 & 0.2182 & -1.6537 & 4.8234 & -2.1072 \\ 0.4787 & 0.2455 & -0.1561 & -2.1072 & 5.4517 \end{pmatrix},$$

and

$$(\hat{\theta}_1, \hat{\theta}_2, \hat{\theta}_5, \hat{\theta}_6, \hat{\theta}_8, \hat{\theta}_9, \hat{\theta}_{13}, \hat{\theta}_{17}, \hat{\theta}_{18}) =$$
$$\frac{1}{100}(8.4729, 9.3003, 3.2912, 2.7130, 14.5859, 4.2431, 17.3692, 79.1475, 63.7816),$$

and all other $\hat{\theta}$s are less than 10^{-10}. We can obtain a predictor for untried sites using the resulting spatial temporal model. Furthermore, we can conduct sensitivity analysis for each component of the response vector using the methods described in Chapter 6.

7.2.2   Functional response with intensive sampling rate

In practice, one may estimate the mean function $\mu(t, \mathbf{x})$ using nonparametric smoothing methods as described in Sections 7.4 and 7.5, and then apply the spatial temporal model for the resulting residuals. As usual, the mean function of the residuals is typically close to zero. Thus, one may consider a simple case in which $\mu(t, \mathbf{x}) = 0$. In this section, we assume $\mu(t, \mathbf{x})$ has the following structure:

$$\mu(t, \mathbf{x}) = \sum_{l=0}^{L} B_l(\mathbf{x})\beta_l(t) \stackrel{\text{def}}{=} \mathbf{b}(\mathbf{x})'\boldsymbol{\beta}_t.$$

This yields a spatial temporal model

$$y(t, \mathbf{x}) = \mathbf{b}(\mathbf{x})'\boldsymbol{\beta}_t + z_t(\mathbf{x}), \qquad (7.7)$$

where $z_t(\mathbf{x}) = z(t, \mathbf{x})$ is a Gaussian random field with mean 0, and we assume its covariance function has the following form:

$$\text{Cov}\{z_t(\mathbf{u}), z_t(\mathbf{v})\} = \sigma_t^2 r(\boldsymbol{\theta}_t; |\mathbf{u} - \mathbf{v}|),$$

where the parameter $\boldsymbol{\theta}_t$ depends on t. Thus, for a given t, model (7.7) can be viewed as a universal Gaussian Kriging model (5.16).

Suppose that $y(t_j, \mathbf{x}_i)$ is the output associated with the input $(t_j, \mathbf{x}_i)$, $i = 1, \cdots, n$ and $j = 1, \cdots, J$. For $t = t_j, j = 1, \cdots, J$, therefore, we can estimate $\boldsymbol{\beta}_t, \sigma^2(t)$, and $\boldsymbol{\theta}_t$ using the the estimation procedure described in Section 5.4. We also obtain the BLUP for $y(t, \mathbf{x})$ at $t = t_j$ and the unobserved site $\mathbf{x}$. We further predict $y(t, \mathbf{x})$ at unobserved t and $\mathbf{x}$ by using a linear interpolation of $y(t_1, \mathbf{x}), \cdots, y(t_J, \mathbf{x})$. Let us summarize the above procedure as an algorithm for calculating a prediction of $y(t, \mathbf{x})$ at the unobserved site $\mathbf{x}$ and over the whole interval $[t_1, t_J]$.

Algorithm 7.2:

Step 1: For $j = 1, \cdots, J$, apply the estimation procedure described in Section 5.4 for model (7.7) with data $\{\mathbf{x}_i, y(t_j, \mathbf{x}_i)\}, i = 1, \cdots, n$, and obtain estimates $\hat{\boldsymbol{\beta}}_t, \hat{\sigma}_t^2, \hat{\boldsymbol{\theta}}_t$ for $t = t_j$.

Step 2: Calculate prediction $y(t_j, \mathbf{x})$ using (5.19) based on data $\{\mathbf{x}_i, y(t_j, \mathbf{x}_i)\}$, $i = 1, \cdots, n$ and estimates $\hat{\mu}_t, \hat{\sigma}_t^2, \hat{\boldsymbol{\theta}}_t, t = t_j$.

Step 3: Obtain the prediction $y(t, \mathbf{x})$ for $t \in [t_1, t_J]$ using linear interpolation of $\hat{y}(t_1, \mathbf{x}), \cdots, \hat{y}(t_J, \mathbf{x})$.

The above algorithm is easily implemented for those who have used Gaussian Kriging models for modeling computer experiments. When the sampling rate is intensive, i.e., the gaps between two consecutive index values are small, this procedure is expected to work well. We will illustrate this estimation procedure in Sections 7.4 and 7.5.

7.3   Penalized Regression Splines

Nonparametric regression methods (mainly splines smoothing and local polynomial regression) can be employed for estimating the nonparametric smoothing coefficients in the functional linear models introduced in Section 7.4 and the semiparametric regression models discussed in Section 7.5. In this chapter, we will systematically use the penalized splines method due to its ease of implementation and lower computational cost. The penalized splines method was proposed for nonparametric regression which is assumed to present random error by Eilers and Marx (1996) and Ruppert and Carroll (2000). Parts of the material in this section and next section are extracted from Li, Sudjianto and Zhang (2005).

Let us start with a simple nonparametric regression model. Suppose that $\{(t_i, y_i), i = 1, \cdots, n\}$ is a random sample from

$$y_i = \mu(t_i) + \varepsilon_i, \tag{7.8}$$

where ε_i is a random error with zero mean, and $\mu(t)$ is an unspecified smooth mean function. To estimate $\mu(t)$, we expand $\mu(t)$ using a set of power truncated splines basis with a given set of knots $\{\kappa_1, \cdots, \kappa_K\}$:

$$1, t, t^2, \cdots, t^p, (t - \kappa_1)_+^p, \cdots, (t - \kappa_K)_+^p$$

where $a_+ = (a + |a|)/2$, the positive part of a (see (1.25)). In other words, model (7.8) is approximated by

$$y_i \approx \beta_0 + \sum_{j=1}^p \beta_j t_i^j + \sum_{k=1}^K \beta_{k+p}(t_i - \kappa_k)_+^p + \varepsilon_i. \tag{7.9}$$

The ordinary least squares method can be used to estimate the coefficient β_js. However, the approximation to $\mu(t)$ in (7.9) may be over-parameterized in order to avoid large modeling bias. Over-parameterization causes some problems, the most severe of which is that the resulting least squares estimate of $\mu(t)$ may have large variance and yield a very poor prediction, mainly because of collinearity. To solve these problems, one selects significant terms in (7.9) using some variable selection procedure to linear regression model. This method is termed the regression splines approach (Stone (1985)). In the presence of a large number of basis terms, variable selection for model (7.9) is very challenging. Therefore, some authors (Eilers and Marx (1996), Ruppert and Carroll (2000)) advocate instead the use of penalized splines methods, which is described below.

The penalized splines approach is to estimate β_j by minimizing the following penalized least squares function:

$$\sum_{i=1}^n \left[y_i - \left\{ \beta_0 + \sum_{j=1}^p \beta_j t_i^j + \sum_{k=1}^K \beta_{k+p}(t_i - \kappa_k)_+^p \right\} \right]^2 + \lambda \sum_{k=1}^K \beta_{k+p}^2,$$

where λ is a tuning parameter, which can be selected by data driven methods, such as cross validation or generalized cross validation. Define

$$\mathbf{b}(t) = [1, t, \cdots, t^p, (t - \kappa_1)_+^p, \cdots, (t - \kappa_K)_+^p]'$$

which is a $(K + p + 1)$-dimensional column vector, and denote

$$B = [\mathbf{b}(t_1), \cdots, \mathbf{b}(t_n)]'$$

which is an $n \times (K + p + 1)$ matrix. Thus, the penalized least squares estimate of $\boldsymbol{\beta} = [\beta_0, \cdots, \beta_{K+p+1}]'$ is given by

$$\hat{\boldsymbol{\beta}} = (B'B + \lambda D)^{-1} B' \mathbf{y},$$

where $\mathbf{y} = (y_1, \cdots, y_n)'$ and D is a diagonal matrix with the first $p + 1$ diagonal element being 0 and other diagonal elements being 1. That is, $D = \text{diag}\{0, \cdots, 0, 1, \cdots, 1\}$. Thus, $\mu(t)$ can be estimated by

$$\hat{\mu}(t) = \mathbf{b}(t)' \hat{\boldsymbol{\beta}}.$$

A crucial question for penalized splines is how to select the tuning parameter. Note that

$$\hat{\mu}(t) = \mathbf{b}(t)'(B'B + \lambda D)^{-1}B'\mathbf{y},$$

which is the linear combination of y_i. Thus, this kind of estimator is termed a linear smoother. For linear smoothers, the generalized cross validation (GCV, Craven and Wahba (1979)) method can be used to select the tuning parameter. Note that

$$\hat{\mathbf{y}} = B(B'B + \lambda D)^{-1}B'\mathbf{y}.$$

Thus, $P_\lambda = B(B'B + \lambda D)^{-1}B'$ can be viewed as a projection matrix and

$$e(\lambda) = \text{trace}(P_\lambda)$$

as a degree of freedom (or effective number of parameters) in the model fitting. Thus, the GCV score is defined as

$$\text{GCV}(\lambda) = \frac{1}{n} \frac{\|\mathbf{y} - B\hat{\boldsymbol{\beta}}\|^2}{\{1 - (e(\lambda)/n)\}^2}$$

and we may select λ to be

$$\hat{\lambda} = \min_\lambda \text{GCV}(\lambda).$$

The minimization is carried out by computing the GCV score over a set of a grid point for λ.

In practical implementation, two natural questions arise here: how to determine the order p, and how to choose the knots. Empirical studies show that the resulting penalized splines estimate is not very sensitive to the choice of knots or the order p. In practice, one usually sets $p = 3$, which corresponds to a cubic spline. Cubic spline has continuous second-order derivatives, but piecewise continuous third-order derivatives. Throughout this chapter, p is taken to be 3.

As to the choice of knots, if t_i is evenly distributed over an interval, then the knots can be simply taken to be evenly distributed over the interval. Alternatively, the knot κ_k is taken to be $100k/(K+1)$-th sample percentile of t_is for unevenly distributed t_i.

Example 24 As an illustrative example, we generate a random sample of size 50 from the model

$$y = \sin(2\pi t) + \varepsilon,$$

where $t \sim U(0, 1)$ and $\varepsilon \sim N(0, 1)$. In Sections 7.3 and 7.4, our focus is on estimating the mean function $\mu(t, \mathbf{x})$. Thus, for this example, we simulate a random error in order to view the error as the portion of the model that cannot be explained by the mean function. Figure 7.6 depicts the resulting estimates of penalized spline with tuning parameter selected by GCV. To demonstrate

the impact of choice of knots, we compare the resulting estimate with $K = 10$
and 30. From Figure 7.6, we can see that the resulting estimate is not sensitive
to choice of knots. Indeed, the GCV selects a larger λ (i.e., a stronger penalty
on the coefficients) for $K = 30$ than for $K = 10$. As an illustration, Figure 7.6
also depicts the resulting spline estimate without penalty. The spline estimate
with $K = 10$ offers reasonable results, but it yields a very under-smoothed
estimate when $K = 30$. This implies that spline regression without penalty
may be sensitive to choice of knots.

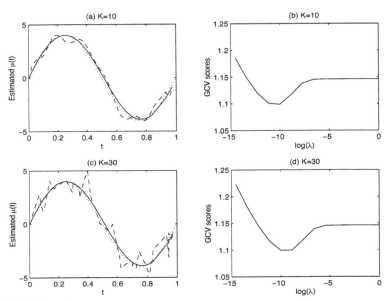

FIGURE 7.6
Spline and penalized spline estimates. In (a) and (c), solid, dashed, and
dotted lines are the penalized spline estimate, spline estimate, and true
function, respectively. (b) and (d) are the plot of GCV scores versus the
values of $\log(\lambda)$.

7.4   Functional Linear Models

Due to the curse of dimensionality (Bellman (1961)), it is very challenging
to estimate a multivariate surface. Furthermore, the **x** in our setting can be

discrete variables. Fully nonparametric models cannot directly be applied for estimating $\mu(t, \mathbf{x})$ in (7.1), so a certain structure should be imposed on $\mu(t, \mathbf{x})$. As a natural extension of linear regression model, we consider allowing the coefficient to vary over t

$$\mu(t, \mathbf{x}) = \beta_0(t) + \mathbf{x}'\boldsymbol{\beta}(t). \tag{7.10}$$

This model enables us to examine the possible index-varying effect of $\mathbf{x}$, and is referred to as a functional linear model because its coefficients are assumed to be smooth functions of t, and because it becomes a linear model for given t. Functional linear models have been employed to analyze functional data and longitudinal data in the statistical literature (see, for example, Chiang, Rice and Wu (2001), Fan and Zhang (2000), Faraway (1997), Hoover, Rice, Wu and Yang (1998), Huang, Wu and Zhou (2002), Nair, Taam and Ye (2002), Hastie and Tibshirani (1993), Wu, Chiang and Hoover (1998), and references therein).

7.4.1 A graphical tool

Model (7.1) with the mean function (7.10) becomes

$$y(t, \mathbf{x}) = \beta_0(t) + \mathbf{x}'\boldsymbol{\beta}(t) + z(t, \mathbf{x}). \tag{7.11}$$

For given t, (7.11) is a linear regression model. Thus, the least squares method can be used to estimate $\beta_0(t)$ and $\boldsymbol{\beta}(t)$. The plot of the resulting estimate can serve as a graphical tool for checking whether the coefficient functions really vary over the index t. Let us demonstrate the idea by a simulated data set.

Example 25 In this example, we generate a typical random sample from the following model:

$$y_i(t_j) = \beta_0(t_j) + \beta_1(t_j)x_{1i} + \beta_2(t_j)x_{2i} + \beta_3(t_j)x_{3i} + \varepsilon_i(t_j), \tag{7.12}$$

where $t_j = j/(J+1)$, $\mathbf{x}_i = (x_{1i}, x_{2i}, x_{3i})' \sim N_3(\mathbf{0}, I_3)$, and $\varepsilon_i(t_j) \sim N(0,1)$, for $i = 1, \cdots, n$ and $j = 1, \cdots, J$ with $n = 50$, and $J = 80$. Furthermore, $\mathbf{x}_i$ and $\varepsilon_i(t_j)$ are independent. Here we consider

$$\beta_0(t) = \beta_1(t) = 1, \beta_2(t) = \sin(2\pi t), \text{ and } \beta_3(t) = 4(t - 0.5)^2.$$

The least squares estimate for the coefficient functions is depicted in Figure 7.7.

Figure 7.7(a) and (b) show that the estimates for both $\beta_0(t)$ and $\beta_1(t)$ hover around 1. This indicates that the true coefficients likely are constant. From Figure 7.7(c) and (d), we can see that the estimates of both $\beta_2(t)$ and $\beta_3(t)$ vary over the index t. This implies that the true coefficient functions are index-dependent. Figure 7.7 shows that the mean function in this example is semiparametric, namely, some coefficients are index-invariant, while the others are index-variant. Note that the varying-coefficient model (7.11) is an example of a semiparametric regression model; these models will be systematically introduced in Section 7.5. This example demonstrates the usefulness of the graphical tool.

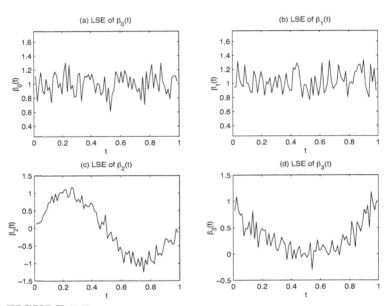

FIGURE 7.7
Least squares estimates of the coefficient functions.

7.4.2 Efficient estimation procedure

The least squares estimates of regression coefficients $\beta_0(t), \beta_1(t), \cdots$ only use the data collected at t, and therefore, their efficiency can be dramatically improved by smoothing over t. The smoothing method can efficiently estimate the coefficients using data collected not only at t but also at other places. Thus, we propose a two-step estimation procedure to $\beta_0(t)$ and $\boldsymbol{\beta}(t)$ as follows:

Step 1 For each t at which data were collected, calculate the least squares estimates of $\beta_0(t)$ and $\boldsymbol{\beta}(t)$, denoted by $\tilde{\beta}_0(t)$ and $\tilde{\boldsymbol{\beta}}(t) = (\tilde{\beta}_1(t), ..., \tilde{\beta}_d(t))$, where d is the dimension of $\boldsymbol{\beta}(t)$.

Step 2 Use penalized spline to smooth $\tilde{\beta}_0(t), \tilde{\beta}_1(t), \cdots, \tilde{\beta}_d(t)$ over t component-wise. Denote the smoothed estimates as $\hat{\beta}_0(t)$ and $\hat{\boldsymbol{\beta}}(t) = (\hat{\beta}_1(t), \cdots, \hat{\beta}_d(t))$.

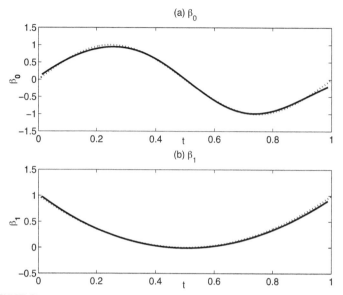

FIGURE 7.8

Penalized spline estimates for the coefficient functions. Solid line is the smoothed estimate. Dotted line is the true value.

This estimation procedure can be easily implemented. Furthermore, it allows different coefficient functions using different tuning parameters. The two-step estimation procedure was proposed for longitudinal data by Fan and Zhang (2000), although they use local polynomial regression in the second step. Before we demonstrate our application, let us give a simulation example to illustrate the estimation procedure.

Example 26 In this example, we generate a typical random sample from the following model:

$$y_i(t_j) = \beta_0(t_j) + \beta_1(t_j)x_i + \varepsilon_i(t_j), \tag{7.13}$$

where $x_i \sim U(0,1)$, $t_j \sim U(0,1)$, and $\varepsilon_i(t_j) \sim N(0,1)$, for $i = 1, \cdots, n$ and $j = 1, \cdots, J$ with $n = 50$, $J = 80$. Moreover, x_i, t_i and $\varepsilon_i(t_j)$ are independent. Furthermore, we take

$$\beta_0(t) = \sin(2\pi t), \text{ and } \beta_1(t) = 4(t - 0.5)^2$$

which are depicted in the dotted line in Figure 7.4.1, in which the smoothed estimates are also displayed as the solid line. From Figure 7.4.1, we can see that the estimate and the true curve are very close. This demonstrates that the proposed estimation method performs very well.

It is interesting to compare the smoothed estimate $\hat{\beta}$ obtained in *Step 2* with the simple least squares estimate (SLES) $\tilde{\beta}$ obtained in *Step 1*. In other words, it is interesting to investigate why *Step 2* is necessary. Let us introduce some notation. Define the mean squared errors (MSE) for an estimate $\hat{\beta}$ of β to be

$$\text{MSE}(\hat{\beta}) = \frac{1}{J} \sum_{j=1}^{J} (\beta(t_j) - \hat{\beta}(t_j))^2.$$

In order to make a direct comparison between $\hat{\beta}$ and $\tilde{\beta}$, we consider the ratio of MSE (RMSE) which is defined as

$$\text{RMSE} = \frac{\text{MSE}(\tilde{\beta})}{\text{MSE}(\hat{\beta})}.$$

TABLE 7.5
Summary of Mean Squared Errors

	β_0	β_1
	Mean (sd)	Mean (sd)
SLSE($\tilde{\beta}$)	0.0204 (0.0036)	0.0214 (0.0056)
Smoothed Est. ($\hat{\beta}$)	0.0019 (0.0011)	0.0014 (0.0011)

We now generate $N = 5000$ simulation data sets from model (7.13), and for each simulation data set, we calculate the MSE and RMSE. The sample mean and standard deviation of the 5000 MSE for each coefficient are depicted in Table 7.5, from which we can see that the smoothed estimate performs much better than the simple least squares estimate. Overall, the smoothed estimate reduces MSE by 90% relative to the simple least squares estimate. This is easily explained since the simple least squares estimate uses only data at t_j to estimate $\beta(t_j)$, while the smoothed estimate uses the full data set to estimate $\beta(t_j)$. Table 7.5 presents a global comparison of $\hat{\beta}$ and $\tilde{\beta}$. To demonstrate the comparison for each simulation, we present the box-plot of $\log_{10}(\text{RMSE})$ in Figure 7.9, from which we can see that even in the worst case, the RMSE is about $10^{0.5} = 3.16$, which means that the smoothing in *Step 2* reduced MSE by two-thirds in the worst case.

7.4.3 An illustration

In this section, we apply the proposed functional linear model and its estimation procedure for the valvetrain system discussed in the beginning of this

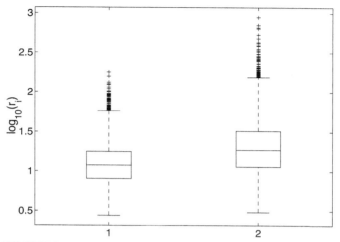

FIGURE 7.9

Boxplot of the log ratio MSEs.

chapter. In particular, we are interested in studying the dynamic behavior of the valve closing event (crank angle from 360° to 450°) at a high speed (i.e., 5500 rpm). In this study, engineers attempt to minimize motion error by adjusting the following design variables:

- cylinder head stiffness (HS)
- rocker arm stiffness (RS)
- hydraulic lash adjuster (LA)
- cam phasing (CP)
- clearance (Cl)
- spring stiffness (SS)
- ramp profile (Ra)

To do this, one must understand the effect of each design variable on motion errors. In this case, the response variable (i.e., motion errors), instead of being a single value, is observed during a closing event which is defined over a crank angle range. A computer experiment with 16 runs is depicted in Table 7.4.

The motion errors for the first four designs listed in Table 7.4 are shown in Figure 7.4, from which we can see that the variability of the motion errors varies for various design configurations. A perfectly stable valve train should have a straight line or zero error throughout the crank angles.

Denote HS by x_1, and so on. Since there are three levels for ramp, we use two dummy codes for this factor. Specifically, $x_7 = 1$ when the ramp is set at level 1, otherwise, $x_7 = 0$; similarly, $x_8 = 1$ when ramp is set at level 2, otherwise, $x_8 = 0$. In this example, we consider the following model:

$$y(t, \mathbf{x}) = \beta_0(t) + \beta_1(t)x_1 + \cdots + \beta_8(t)x_8 + z(t, \mathbf{x}),$$

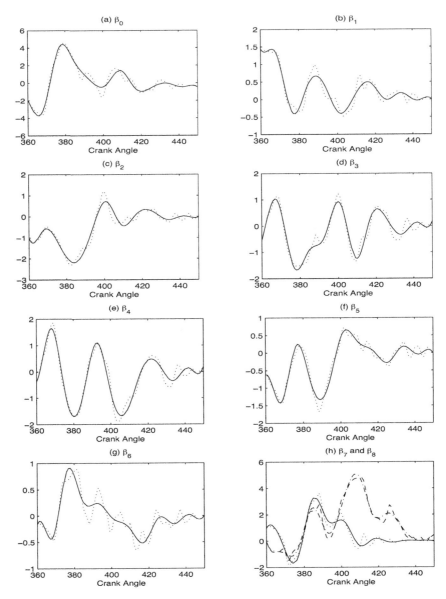

FIGURE 7.10

Estimated coefficient functions within the crank angle region $(360, 450)$ using the functional linear model for the valvetrain data. In (a)–(g), dotted and solid lines are the simple least square estimate and smoothed estimates, respectively. In (h), the dotted and solid lines are the simple least squares estimates and smoothed estimate for $\beta_7(t)$, respectively, and the dashed line and dash-dotted line are the simple least squares estimate and smoothed estimate for $\beta_8(t)$, respectively.

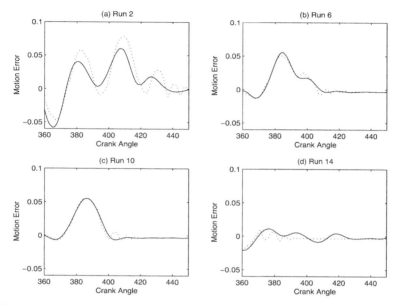

FIGURE 7.11
Valve motion errors compared to FLM estimates. The solid line is FLM estimates, and the dotted line is CAE simulations.

in which $y(t, \mathbf{x})$ is set to be 100 times the amount of motion error. The scale transformation is merely to avoid rounding error in numerical computation. The proposed estimated procedure is used to estimate the coefficient functions. The resulting estimates are depicted in Figure 7.10, in which the simple least squares estimates are displayed as dotted lines. From Figure 7.10, we can see that the smoothed estimates follow the trend of the simple least squares estimates very well. We can see from Figure 7.10 that all coefficient functions vary dramatically with angle.

After estimating the coefficient functions, we further compute the fitted values for all runs. Fitted values for four typical runs (Runs 2, 6, 10, and 14) are presented in Figure 7.11. The functional linear model fits the data quite well.

We apply the spatial temporal model proposed in Section 7.2.2 for the residuals yielded obtained by the simple least squares approach and the estimation procedure introduced in Section 7.4.2, respectively. The resulting estimate $\hat{\mu}(t)$ is depicted in Figure 7.12(a). From the figure, we can see that $\hat{\mu}(t) = 0$ for residuals obtained by simple least squares approach; this is because this approach does not involve smoothing and hence provides an unbiased estimate for $\mu(t)$. The resulting estimate $\hat{\sigma}^2$ is displayed in Figure 7.12(b), from which we can find that $\hat{\sigma}^2(t)$ are very close for these two different residuals. Figure 7.13 displays the resulting estimate $\hat{\gamma}(t)$. The estimate of $\gamma(t)$ based on the residuals of the FLM model hovers around that based on residuals of the simple least squares estimate.

7.5 Semiparametric Regression Models

In the previous section, we introduced functional linear models. Commonly, some coefficient functions in the functional linear models change across t, while others do not. This leads us to consider semiparametric regression models for the mean function. In Section 7.5.1, we introduce the partially linear model, the most frequently used semiparametric regression model in the statistical literature. We will further introduce two estimation procedures for the partially linear model. In Section 7.5.2, we consider partially functional linear models and introduce their estimation procedure. An illustration is given in Section 7.5.3.

7.5.1 Partially linear model

Let $\mathbf{x}_i$, $i = 1, \cdots, n$ be the input vector, and y_{ij} be the output collected at t_j, $j = 1, \cdots, J$, corresponding to the input $\mathbf{x}_i$. The partially linear model is defined as model (7.1) with

$$y_{ij} = \alpha(t_j) + \mathbf{x}_i'\boldsymbol{\beta} + z(t_j, \mathbf{x}_i). \tag{7.14}$$

This partially linear model naturally extends the linear regression model by allowing its intercept to depend on t_j. This model has been popular in the statistical literature (see, for example, Engle, Granger, Rice and Weiss (1986), Fan and Li (2004), Heckman (1986), Lin and Ying (2001), Speckman (1988), and references therein).

Direct approach. A direct approach to estimating $\alpha(t)$ and $\boldsymbol{\beta}$ is to approximate $\alpha(t)$ using a spline basis for given knots $\{\kappa_1, \cdots, \kappa_K\}$:

$$\alpha(t) \approx \alpha_0 + \sum_{k=1}^{p} \alpha_k t^k + \sum_{k=1}^{K} \alpha_{k+p}\alpha_k(t - \kappa_k)_+^p.$$

Thus,

$$y_{ij} \approx \alpha_0 + \sum_{j=1}^{p} \alpha_k t_j^k + \sum_{k=1}^{K} \alpha_{k+p}\alpha_k(t_j - \kappa_k)_+^p + \mathbf{x}_i'\boldsymbol{\beta} + z(t_j, \mathbf{x}_i). \tag{7.15}$$

Thus, the least squares method can be used to estimate α_j and $\boldsymbol{\beta}$.

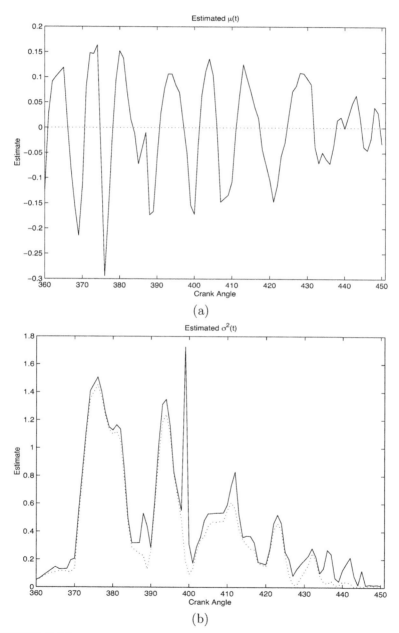

FIGURE 7.12

Estimate of $\mu(t)$ and $\sigma^2(t)$. The solid line is the estimate based on residuals of the FLM model, and the dotted line is the estimate based on residuals of the SLSE estimate.

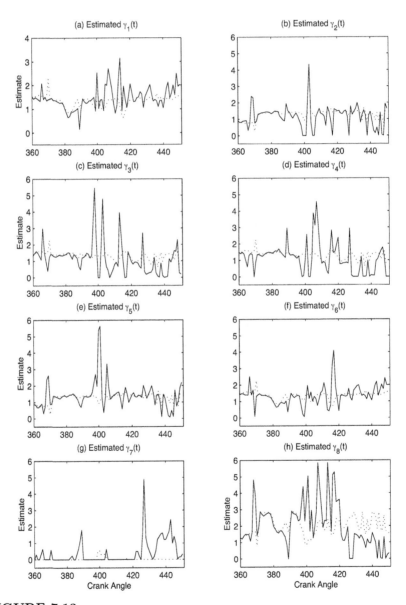

FIGURE 7.13

Estimate of $\gamma(t)$. The solid line is the estimate of $\gamma(t)$ based on the residuals of the FLM model, and the dotted line is the estimate of $\gamma(t)$ based on the residuals of the SLSE estimate.

Partial Residual Approach. Several other estimation procedures have been proposed for the partially linear model. The partial residual approach (Speckman (1988)) is intuitive and may be easily implemented. Denote

$$\bar{y}_{\cdot j} = \frac{1}{n}\sum_{i=1}^{n} y_{ij}, \quad \bar{\mathbf{x}} = \frac{1}{n}\sum_{i=1}^{n} \mathbf{x}_i, \quad \bar{z}_j = \frac{1}{n}\sum_{i=1}^{n} z(t_j, \mathbf{x}_i).$$

Then, from model (7.14),

$$\bar{y}_{\cdot j} = \alpha(t_j) + \bar{\mathbf{x}}'\boldsymbol{\beta} + \bar{z}(t_j).$$

Furthermore,

$$y_{ij} - \bar{y}_{\cdot j} = (\mathbf{x}_i - \bar{\mathbf{x}})'\boldsymbol{\beta} + \{z(t_j, \mathbf{x}_i) - \bar{z}(t_j)\},$$

and therefore,

$$\bar{y}_{i\cdot} - \bar{y} = (\mathbf{x}_i - \bar{\mathbf{x}})'\boldsymbol{\beta} + (\bar{z}_i - \bar{z}),$$

where

$$\bar{y}_{i\cdot} = \frac{1}{J}\sum_{i=1}^{J} y_{ij}, \quad \bar{y} = \frac{1}{J}\sum_{j=1}^{J} \bar{y}_{\cdot j}, \quad \text{and} \quad \bar{z}_i = \frac{1}{J}\sum_{j=1}^{J} z(t_j, \mathbf{x}_i).$$

Thus, a least squares estimate for β is

$$\hat{\boldsymbol{\beta}} = (\mathbf{X}_c'\mathbf{X}_c)^{-1}\mathbf{X}_c'\mathbf{y}_c$$

where $\mathbf{X}_c = (\mathbf{x}_1 - \bar{\mathbf{x}}, \cdots, \mathbf{x}_n - \bar{\mathbf{x}})'$ and $\mathbf{y}_c = (y_{1\cdot} - \bar{y}, \cdots, y_{n\cdot} - \bar{y})'$. After we obtain an estimate of β, we are able to calculate partial residuals $r_j = \bar{y}_{\cdot j} - \bar{\mathbf{x}}'\hat{\boldsymbol{\beta}}$. Denote $\mathbf{r} = (r_1, \cdots, r_J)$. Then we smooth r_j over t_j using penalized splines, introduced in Section 7.3, and thus we obtain an estimate of $\alpha(t)$. We next illustrate some empirical comparisons between the direction approach and the partial residual approach.

Example 27 We generated 5000 data sets, each consisting of $n = 50$ samples, from the following partially linear model,

$$y_{ij} = \alpha(t_j) + \mathbf{x}_i'\boldsymbol{\beta} + \varepsilon_{ij}, \tag{7.16}$$

where $t_j = j/(J+1)$ with $J = 80$, $\mathbf{x}_i$ is five-dimensional, and each component of $\mathbf{x}_i$ is independent and identically distributed from Bernoulli distribution with success probability 0.5. $\alpha(t) = \sin^2(2\pi t)$ and β is a 5-dimensional vector with all elements being 1. For each i, ε_{ij} is an AR(1) random error:

$$\varepsilon_{ij} = \rho\varepsilon_{ij-1} + e_j,$$

where $e_j \sim N(0,1)$. In our simulation, we take $\rho = 0.1$ and 0.5. Further ε_{ij}s are mutually independent for different i.

TABLE 7.6
Mean and Standard Deviation

ρ	β_1	β_2	β_3	β_4	β_5
0.1	0.999(0.037)	0.999(0.037)	1.001(0.037)	0.999(0.037)	1.001(0.037)
0.5	0.998(0.065)	0.998(0.066)	1.002(0.067)	0.998(0.066)	1.001(0.066)

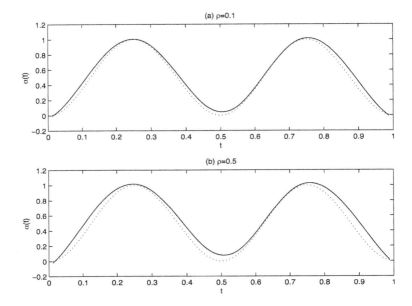

FIGURE 7.14
Estimated intercept function. Solid lines stand for the resulting estimate, and
the dotted lines are the true intercept function

The mean and standard error of the resulting estimates in the 5000 simulations are summarized in Table 7.6. We found that the results for the direct
approach and the partial residual approach are identical for this example.
However, in general, the partial residual approach may result in a more efficient estimate than the direct approach. Figure 7.14 gives plots of a typical
estimate of $\alpha(t)$.

7.5.2 Partially functional linear models

The partially functional linear model is a natural combination of functional
linear model and linear regression model. In fact, the partially linear model is
a special case of the partially functional linear model. Let $\{\mathbf{w}_i, \mathbf{x}_i\}$ consist of
input variables and y_{ij} be the output collected at t_j, corresponding to input

$\{\mathbf{w}_i, \mathbf{x}_i\}$. The partially functional linear model is defined as follows:

$$y_{ij} = \mathbf{w}_i'\boldsymbol{\alpha}(t_j) + \mathbf{x}_i'\boldsymbol{\beta} + z(t_j, \mathbf{w}_i, \mathbf{x}_i), \tag{7.17}$$

To include an intercept function, we may set the first component of $\mathbf{w}_i$ to be 1. Alternatively, we may set the first component of $\mathbf{x}_i$ to be 1 in order to introduce a constant intercept in the model.

Both the direction estimation procedure and the profile least squares approach for partially linear models can be extended to model (7.17). But their implementation may be a little complicated. Here we introduce a back-fitting algorithm. Compared with the direct estimation procedure, the back-fitting algorithm yields a more accurate estimate for β; and compared with the profile least squares approach (Fan and Huang (2005)) for model (7.17), the back-fitting algorithm is easier to implement in practice.

Suppose that $\{t_j, \mathbf{w}_i, \mathbf{x}_i, y_{ij}\}, i = 1, \cdots, n$ is a sample from a computer experiment, and that model (7.17) is used to fit the data. The back-fitting algorithm is intuitive. For a given $\boldsymbol{\beta}$, define $y_{ij}^* = y_{ij} - \mathbf{x}_i'\boldsymbol{\beta}$, then

$$y_{ij}^* = \mathbf{w}_i'\boldsymbol{\alpha}(t_j) + z(t_j, \mathbf{w}_i, \mathbf{x}_i),$$

which is a functional linear model, and the estimation procedure for such a model can be used to estimate $\boldsymbol{\alpha}(t)$. On the other hand, for a given $\boldsymbol{\alpha}(t)$, let $\tilde{y}_{ij} = y_{ij} - \mathbf{w}_i'\boldsymbol{\alpha}(t_j)$. We have

$$\tilde{y}_{ij} = \mathbf{x}_i'\boldsymbol{\beta} + z(t_j, \mathbf{w}_i, \mathbf{x}_i).$$

Thus, we can use the least squares approach to estimate $\boldsymbol{\beta}$.

The back-fitting algorithm iterates the estimation of $\boldsymbol{\alpha}(t)$ and the estimation of $\boldsymbol{\beta}$ until they converge. The proposed estimation procedure can be summarized as follows:

Step 1. (*Initial values*) Estimate $\boldsymbol{\alpha}(t)$ using the estimation procedure of the functional linear model proposed in Section 7.4, regarding the $\boldsymbol{\beta}$ as constant. Set the resulting estimate to be the initial values of $\boldsymbol{\alpha}(t)$.

Step 2. (*Estimation of $\boldsymbol{\beta}$*) Substitute $\boldsymbol{\alpha}(t)$ with its estimates in model (7.17), and estimate $\boldsymbol{\beta}$ using the least squares approach.

Step 3. (*Estimation of $\boldsymbol{\alpha}(t)$*) Substitute $\boldsymbol{\beta}$ with its estimate in model (7.17), and apply the estimation procedure for functional linear models to estimate $\boldsymbol{\alpha}(t)$.

Step 4 (*Iteration*) Iterate *Step 2* and *Step 3* until the estimates converge.

Since the initial estimate obtained in *Step 1* will be a very good estimate for $\boldsymbol{\alpha}(t)$, we may stop after a few iterations rather than wait until the iteration fully converges. The resulting estimate will be as efficient as that obtained by a full iteration (see Bickel, 1975 for other settings).

Example 25 (**Continued**) The true model from which data are generated in this example is a partially functional linear model. Let us apply the back-fitting algorithm described above for the data we analyzed before. The algorithm converges very quickly. The resulting model is

$$\hat{y}(t, \mathbf{x}) = 1.0015 + 1.0306 x_1 + \hat{\beta}_2(t) x_2 + \hat{\beta}_3(t) x_3,$$

where $\hat{\beta}_2(t)$ and $\hat{\beta}_3(t)$ are depicted in Figure 7.15. Compared with the least squares estimate displayed in Figure 7.7, the resulting estimate of the back-fitting algorithm is smooth and very close to the true coefficient function.

7.5.3   An illustration

In this section, we illustrate the partially functional linear model and its estimation procedure using the valvetrain example presented in Section 7.1. As a demonstration, we consider the data collected over crank angles from $90°$ to $360°$. To determine which effects are angle-varying effects and which effects are constant, we first apply the functional linear model in Section 7.4.3 for the data and use the simple least squares approach to estimate the coefficients. The resulting simple least squares estimates are depicted in Figure 7.16 from which we further consider the following partially functional linear models

$$y = \beta_0(t) + \beta_2 x_1 + \beta_3 x_2 + \beta_4 x_3 + \beta_5 x_4 + \beta_5 x_5 + \beta_6 x_6 + \beta_7(t) x_7 + \beta_8(t) x_8 + z(t, \mathbf{x}).$$

In other words, the intercept and the coefficients of the covariate *Ramp* are considered to vary over the crank angle and other coefficients are considered to be constant.

The back-fitting algorithm is applied to the partially functional linear model, and the resulting model is given by

$$\hat{y} = \hat{\beta}_0(t) - 0.2184 x_1 + 0.0392 x_2 - 0.5147 x_3 - 0.0006 x_4 + 0.2409 x_5$$
$$- 0.2857 x_6 + \hat{\beta}_7(t) x_7 + \hat{\beta}_8(t) x_8,$$

where $\hat{\beta}_0(t)$, $\hat{\beta}_7(t)$, and $\hat{\beta}_8(t)$ are shown in Figure 7.17.

The fitted values of the partially functional linear models are computed for all runs. Figure 7.18 depicts four typical ones (Runs 1, 5, 9, 13). From Figure 7.18, we can see the reasonably good fit to the data.

We apply the spatial temporal model in Section 7.2.2 for the residuals derived from the simple least squares estimate approach for the functional linear model and from the partially function linear model. The resulting estimate $\hat{\mu}(t)$, $\hat{\sigma}^2(t)$, and $\hat{\gamma}(t)$ are depicted in Figure 7.19 and 7.20, respectively.

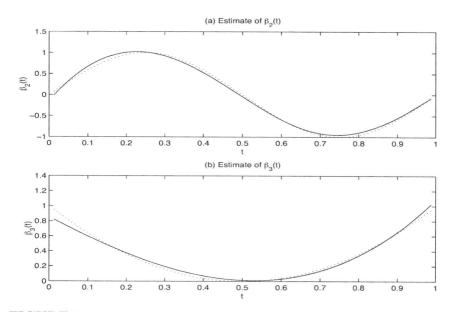

FIGURE 7.15

Estimates of $\beta_2(t)$ and $\beta_3(t)$. Solid lines stand for the estimated coefficient, and dotted lines for the true coefficient function.

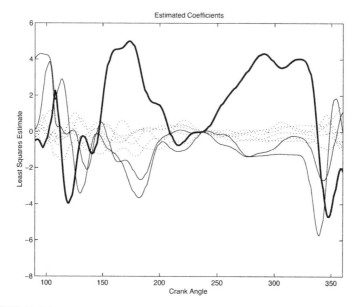

FIGURE 7.16

Estimated coefficients. The thick solid line stands for the intercept function $\hat{\beta}_0(t)$. The two thin solid lines are estimated coefficients of *Ramp*. Dotted lines are estimated coefficients of other factors.

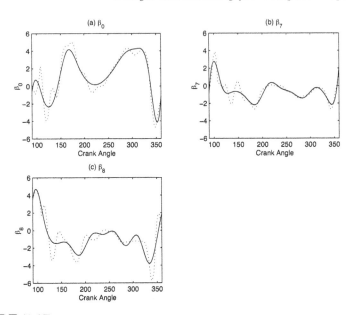

FIGURE 7.17

Estimated coefficient functions within crank angle region $(90, 360)$ using functional linear model for the valvetrain data. In (a)–(c), dotted and solid lines are the simple least square estimate and smoothed estimates, respectively.

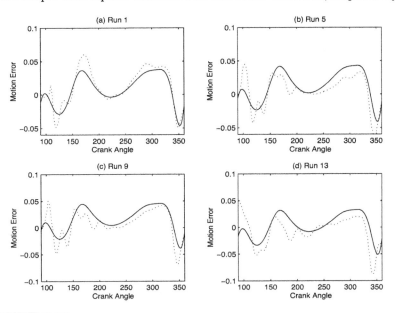

FIGURE 7.18

Valve motion errors compared to PFLM fit. The solid line is PFLM fit, and the dotted line is motion error.

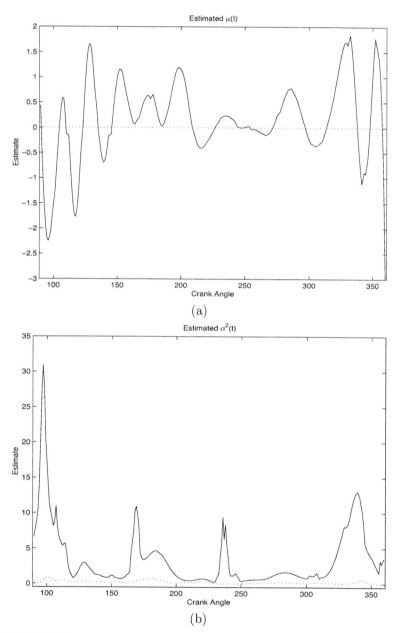

(a)

(b)

FIGURE 7.19
Estimate of $\mu(t)$ and $\sigma^2(t)$. The solid line is the estimate using the residuals
of the PFLM model, the dotted line is the estimate using the residuals of the
SLSE estimate.

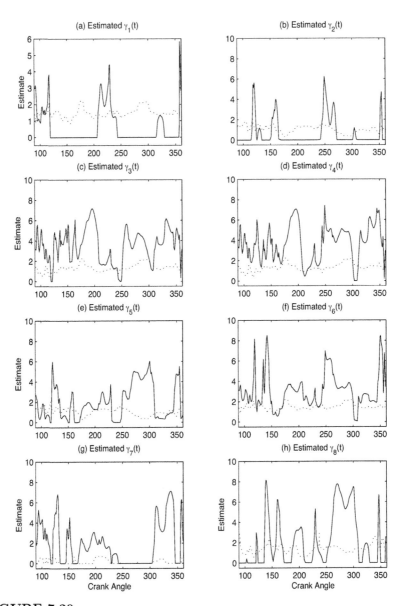

FIGURE 7.20

Estimate of $\gamma(t)$. The solid line is the estimate of $\gamma(t)$ based on the residuals of the PFLM model, and the dotted line is the estimate of $\gamma(t)$ based on the residuals of the SLSE estimate.

Appendix

To make this book more self-contained, in this appendix we introduce many basic concepts in matrix algebra, statistics, probability, linear regression analysis, and selection of variables in regression models. Due to limited space, we give only their definitions, simple properties, and illustration examples. A review of matrix algebra will be given in Section A.1. Some basic concepts in probability and statistics needed to understand in the book are reviewed in Section A.2. Linear regression analysis has played an important role in design and modeling for computer experiments. Section A.3 gives a brief introduction to basic concepts, theory, and method of linear regression analysis. Section A.4 reviews some variable selection criteria for linear regression models.

A.1   Some Basic Concepts in Matrix Algebra

This section reviews some matrix algebra concepts used in the book.

Matrix: An $n \times s$ *matrix* $\mathbf{A}$ is a rectangular array

$$\mathbf{A} = (a_{ij}) = \begin{bmatrix} a_{11} & \cdots & a_{1s} \\ \vdots & & \vdots \\ a_{n1} & \cdots & a_{ns} \end{bmatrix},$$

where a_{ij} is the element in the ith row and jth column. We always assume in the book that all the elements are *real numbers*.

Data sets can be expressed as a matrix of size $n \times s$, where n is the number of runs and s the number or variables (factors). For example, a data set of 6 runs and 4 variables is given by

$$D = \begin{bmatrix} 30 & 50 & 0.125 & 5.2 \\ 25 & 54 & 0.230 & 4.9 \\ 31 & 48 & 0.078 & 5.0 \\ 28 & 52 & 0.112 & 4.8 \\ 32 & 51 & 0.150 & 5.1 \\ 29 & 55 & 0.220 & 4.7 \end{bmatrix},$$

where rows represent runs and columns represent variables.

Transpose: The *transpose* of $\mathbf{A}$, denoted by $\mathbf{A}'$, is the $s \times n$ matrix with elements a_{ji}.

Row/column vector: If $n = 1$, $\mathbf{A}$ is a *row vector* or *s*-row vector; if $s = 1$, $\mathbf{A}$ is a *column vector*, or an *n*-column vector. We shall use a bold-face, lowercase letter, $\mathbf{a}$, to denote a column vector.

Zero matrix: A zero matrix is a matrix having 0's as its elements. In this case we write $\mathbf{A} = \mathbf{O}$.

Product: The *product of two matrices* $\mathbf{A} = (a_{ij}) : n \times m$ and $\mathbf{B} = (b_{kl}) :$ $m \times s$ is an $n \times s$ matrix given by

$$\mathbf{AB} = \left(\sum_{j=1}^{m} a_{ij} b_{jl} \right).$$

Square matrix: A matrix $\mathbf{A}$ is called a *square matrix* if $n = s$. A square matrix $\mathbf{A}$ is called *symmetric* if $\mathbf{A}' = \mathbf{A}$. A square matrix $\mathbf{A}$ is called a *diagonal matrix* if $a_{ij} = 0, i \neq j$, written $\mathbf{A} = \text{diag}(a_{11}, \cdots, a_{nn})$. If $\mathbf{A}$ is a diagonal matrix with $a_{ii} = 1, i = 1, \cdots, n$, then $\mathbf{A}$ is called the *identity matrix* or *unit matrix* of order n, written as $\mathbf{A} = \mathbf{I}_n$. A square matrix $\mathbf{A} = (a_{ij})$ is called an *upper triangle matrix* if $a_{ij} = 0$ for all $i > j$; and called a *lower triangle matrix* if $a_{ij} = 0$ for all $i < j$.

Trace: The sum of the diagonal elements of a square matrix $\mathbf{A}$ is called the *trace of the matrix* $\mathbf{A}$, i.e, $trace(\mathbf{A}) = \sum_{i=1}^{n} a_{ii}$.

Determinant: The *determinant of a square matrix* $\mathbf{A}$ is denoted by $|\mathbf{A}|$ or $det(\mathbf{A})$. $\mathbf{A}$ is called singular if $|\mathbf{A}| = 0$; otherwise $\mathbf{A}$ is non-singular. A square matrix with full rank is non-singular. The original definition of $|\mathbf{A}|$ is based on a complicated formula that we will not explain here due to space restrictions.

Rank: A *rank of a matrix* $\mathbf{A}$ of size $n \times s$ is said to have rank r, written as $rank(\mathbf{A}) = r$, if at least one of its r-square sub-matrices is non-singular while every $(r + 1)$-square sub-matrix is singular.
- $rank(\mathbf{A}) \leq \min(n, s)$;
- $rank(\mathbf{A}) = rank(\mathbf{A}')$;
- $rank(\mathbf{A}) = s < n$ is called the full column rank; $rank(\mathbf{A}) = n < s$ is called the full row rank;
- $rank(\mathbf{A}) = s = n$ is called the full rank.

Inverse of a matrix: Matrix $\mathbf{B}$ is said to be the *inverse of a square matrix* $\mathbf{A}$, if $\mathbf{AB} = \mathbf{BA} = \mathbf{I}$, where $\mathbf{I}$ is the identity matrix. A matrix has an inverse if and only if it is non-singular; in this case, the inverse is unique.

Eigenvalues and eigenvectors: The *eigenvalues of a square matrix* **A** of order n are solutions of the equation

$$|\mathbf{A} - \lambda\mathbf{I}| = 0.$$

There are n solutions, i.e., the matrix **A** has n eigenvalues, denoted by $\lambda_i, i = 1, \cdots, n$. The following properties are useful:
- there are n eigenvalues;
- the eigenvalues of **A** are real numbers if **A** is symmetric;
- the sum of the eigenvalues of **A** equals $trace(\mathbf{A})$;
- the product of the eigenvalues of **A** equals $|\mathbf{A}|$.

For each eigenvalue, λ_i, there exists a vector $\mathbf{l}_i$ such that

$$\mathbf{A}\mathbf{l}_i = \lambda_i\mathbf{l}_i.$$

The vector $\mathbf{l}_i$ is called the *eigenvector* of **A** associated to λ_i.

Non-negative definite/positive definite matrix: A symmetric matrix **A** of order n is called *non-negative definite*, denoted by $\mathbf{A} \geq 0$, if for any nonzero column vector $\mathbf{a} \in R^n$ we have $\mathbf{a}'\mathbf{A}\mathbf{a} \geq 0$; it is called *positive definite*, denoted by $\mathbf{A} > 0$, if $\mathbf{a}'\mathbf{A}\mathbf{a} > 0$. The following properties are useful:
- $\mathbf{X}'\mathbf{X} \geq 0$ for any $n \times s$ matrix **X**;
- the eigenvalues of **A** are non-negative, if $\mathbf{A} \geq 0$.
- the eigenvalues of **A** are positive, if $\mathbf{A} > 0$.

Projection matrix: A square matrix **A** satisfying $\mathbf{A}\mathbf{A} = \mathbf{A}$ is called *idempotent*; a symmetric and idempotent matrix is called a *projection matrix*.

Let **X** be an $n \times s$ matrix with rank $s \leq n$. Now $\mathbf{H} = \mathbf{X}(\mathbf{X}'\mathbf{X})^{-1}\mathbf{X}'$ is a projection matrix and is called a *hat matrix* in regression analysis.

Orthogonal matrix: A square matrix **A** is called an *orthogonal matrix* if $\mathbf{A}'\mathbf{A} = \mathbf{I}$. A square matrix **A** is called a *permutation matrix* if **A** is an orthogonal matrix with elements either 1 or 0.

Hadamard matrix: A *Hadamard matrix* of order n, **H**, is a square matrix of order n for which every entry equals either 1 or -1, and which satisfies that $\mathbf{H}\mathbf{H}' = n\mathbf{I}_n$. Many 2-level orthogonal designs are generated via Hadamard matrices by deleting the column of ones of **H** if **H** has such a column. For example,

$$\mathbf{H} = \begin{bmatrix} 1 & 1 & 1 & 1 \\ 1 & 1 & -1 & -1 \\ 1 & -1 & 1 & -1 \\ 1 & -1 & -1 & 1 \end{bmatrix}$$

is a Hadamard matrix of order 4. Deleting the first column we obtain $L_4(2^3)$ (cf. Section 1.2).

Hadamard product: The *Hadamard product*, also known as the *element-wise product*, of two same size matrices $\mathbf{A} = (a_{ij})$ and $\mathbf{B} = (b_{ij})$ is the matrix $\mathbf{C} = (c_{ij} = a_{ij}b_{ij})$ and denoted by $\mathbf{C} = \mathbf{A} \odot \mathbf{B}$.

Kronecker product: The *Kronecker product* (or *tensor product*) of $\mathbf{A} = (a_{ij})$ of size $n \times p$ and $\mathbf{B} = (b_{kl})$ of size $m \times q$ is an $nm \times pq$ matrix defined by

$$\mathbf{A} \otimes \mathbf{B} = (a_{ij}\mathbf{B}) = \begin{bmatrix} a_{11}\mathbf{B} & \cdots & a_{1p}\mathbf{B} \\ \vdots & & \vdots \\ a_{n1}\mathbf{B} & \cdots & a_{np}\mathbf{B} \end{bmatrix}. \tag{A.1}$$

Let

$$\mathbf{A} = \begin{bmatrix} 1 & 2 & 4 \\ 2 & 1 & 3 \\ 3 & 4 & 2 \\ 4 & 3 & 1 \end{bmatrix}, \quad \mathbf{B} = \begin{bmatrix} 1 & -1 & -1 \\ 1 & 1 & -1 \\ 1 & -1 & 1 \\ 1 & 1 & 1 \end{bmatrix} \quad \text{and} \quad \mathbf{C} = \begin{bmatrix} 1 & -1 \\ -2 & 3 \end{bmatrix}.$$

Then

$$\mathbf{A} \odot \mathbf{B} = \begin{bmatrix} 1 & -2 & -4 \\ 2 & 1 & -3 \\ 3 & -4 & 2 \\ 4 & 3 & 1 \end{bmatrix} \quad \text{and} \quad \mathbf{A} \otimes \mathbf{C} = \begin{bmatrix} 1 & -1 & 2 & -2 & 4 & -4 \\ -2 & 3 & -4 & 6 & -8 & 12 \\ 2 & -2 & 1 & -1 & 3 & -3 \\ -4 & 6 & -2 & 3 & -6 & 9 \\ 3 & -3 & 4 & -4 & 2 & -2 \\ -6 & 9 & -8 & 12 & -4 & 6 \\ 4 & -4 & 3 & -3 & 1 & -1 \\ -8 & 12 & -6 & 9 & -2 & 3 \end{bmatrix}.$$

A.2 Some Concepts in Probability and Statistics

A.2.1 Random variables and random vectors

In this subsection we review concepts of random variables/vectors and their distributions and moments.

Random variables: Suppose we collect piston-rings in a process. We find that values of the diameter of piston-rings are spread through a certain range. We might visualize piston-ring diameter as a *random variable*, because it takes on different values in the population according to some random mechanism. The number of telephone calls from 9:00 to 10:00 in a commercial center is another example of random variable. A random variable is called *continuous* if it is measured on a continuous scale and *discrete* if it is limited to a certain finite or countable set of possible values, such as the integers $0, 1, 2, \cdots$. The piston-ring diameter is a continuous random variable while the number of telephones is a discrete random variable.

Probability distribution function: A probability distribution function is a mathematical model that relates the value of a variable with the probability of occurrence of that value in the population. Let X be a random variable, the function, defined by $F(x) = P(X \leq x)$, and called the *cumulative distribution function* (cdf) or more simply, the *distribution function* of X. We write $X \sim F(x)$.

Discrete distribution: When the possible values of X are finite or countable values, $x_1, x_2, \cdots$ say, the probabilities $p_i = P(X = x_i), i = 1, 2, \cdots$ form a discrete distribution. The binomial distribution, the Poisson distribution, etc. are discrete distributions.

Continuous distribution: If there is a continuous function $p(x)$ such that $F(x) = \int_{-\infty}^{x} p(y)dy$ for any $-\infty < x < \infty$, we say that X follows a continuous distribution $F(x)$ with a probability density function (pdf) or density $p(x)$. The normal distribution and the exponential distribution are examples of continuous distribution.

Random Vector: A vector $\mathbf{x} = (X_1, \cdots, X_p)'$ is called a random vector if all the components $X_1, \cdots, X_p$ are random variables. The function $F(x_1, \cdots, x_p) = P(X_1 \leq x_1, \cdots, X_p \leq x_p)$ is called the *cumulative distribution function* (cdf) of $\mathbf{x}$. If there exists a non-negative and integrable function $p(x_1, \cdots, x_p)$ such that

$$F(x_1, \cdots, x_p) = \int_{-\infty}^{x_1} \cdots \int_{-\infty}^{x_p} p(y_1, \cdots, y_p)dy_1 \cdots dy_p,$$

the function p is called the *probability density function* (pdf) of $\mathbf{x}$.

The distribution of $(X_{i_1}, \cdots, X_{i_q})$, $q < p$, is called the *marginal distribution* of $(X_{i_1}, \cdots, X_{i_q})$. In this case $F(x_1, \cdots, x_p)$ is called the joint distribution of $X_1, \cdots, X_p$. The distribution of X_i given $X_j = x_j$ is called the *conditional distribution* and denoted by $F_i(x_i|X_j = x_j)$, where F_i is determined by $F(x_1, \cdots, x_p)$. As an extension, the conditional distribution $F_{i_1, \cdots, i_q}(x_{i_1}, \cdots, x_{i_q}|X_{j_i} = x_{j_1}, \cdots, X_{j_k} = x_{j_k})$ has a similar meaning to the previous one.

Independence: Let X and Y be two random variables. If their distribution functions satisfy

$$F(x, y) = F_x(x)F_y(y),$$

for any x and y in R, where $F(x, y)$ is the joint distribution function of (X, Y) and $F_x(\cdot)$ and $F_y(\cdot)$ are distributions of X and Y, respectively, we call X and Y statistically independent, or just independent.

Likewise, let $\mathbf{x} = (X_1, \cdots, X_p)$ and $\mathbf{y} = (Y_1, \cdots, Y_q)$ be two random vectors. If their distribution functions satisfy

$$F(\mathbf{x}_1, \cdots \mathbf{x}_p, \mathbf{y}_1, \cdots, \mathbf{y}_q) = F_x(\mathbf{x}_1, \cdots \mathbf{x}_p) F_y(\mathbf{y}_1, \cdots, \mathbf{y}_q),$$

for any $\mathbf{x} \in R^p$ and $\mathbf{y} \in R^q$, where F, F_x and F_y are the joint distribution function of $\mathbf{x}$ and $\mathbf{y}$, the distribution function of $\mathbf{x}$ and of $\mathbf{y}$, respectively, we call $\mathbf{x}$ and $\mathbf{y}$ statistically independent, or just independent.

Expected value: (a) The expected value of a discrete random variable X with $p_i = P(X = x_i)$, $i = 1, 2, \cdots$, is defined as

$$E(X) = \sum x_i p_i,$$

where the summation is taken over all possible values of X.

(b) The expected value of a continuous random variable X with density $p(x)$ is defined as

$$E(X) = \int x p(x)\, dx,$$

where the integral is taken over the range of X.

Covariance: The covariance of two random variables X and Y with a pdf $f(x, y)$ is defined by

$$\mathrm{Cov}(X, Y) = \int_{-\infty}^{\infty} \int_{-\infty}^{\infty} (x - E(X))(y - E(Y)) p(x, y) dx dy,$$

and we write $\mathrm{Cov}(X, Y) = E(X - E(X))(Y - E(Y))$. The covariance of X and Y has the following properties:

- $\mathrm{Cov}(X, Y) = \mathrm{Cov}(Y, X)$;
- $\mathrm{Cov}(X, X) = \mathrm{Var}(X)$;
- $\mathrm{Cov}(X, Y) = E(XY) - E(X)E(Y)$, where $E(XY) = \int \int xy p(x, y) dx dy$;
- Random variables X and Y are called *uncorrelated* if $\mathrm{Cov}(X, Y) = 0$.

Correlation coefficient: The correlation coefficient between two random variables X and Y is defined by

$$\mathrm{Corr}(X, Y) = \rho(X, Y) = \frac{\mathrm{Cov}(X, Y)}{\sqrt{\mathrm{Var}(X)\mathrm{Var}(Y)}}.$$

Obviously, $\mathrm{Corr}(X, X) = 1$.

Mean vector, covariance, and correlation matrices: Let a random vector $\mathbf{x} = (X_1, \cdots, X_p)'$ have a pdf $p(x_1, \cdots, x_p)$. Assume each X_i has a finite variance. The *mean vector* of $\mathbf{X}$ is defined by

$$E(\mathbf{x}) = \begin{bmatrix} E(X_1) \\ \vdots \\ E(X_p) \end{bmatrix}.$$

Let $\sigma_{ij} = \mathrm{Cov}(X_i, X_j)$ for $i, j = 1, \cdots p$. Obviously, $\sigma_{ii} = \mathrm{Cov}(X_i, X_i) = \mathrm{Var}(X_i)$. The matrix

$$\mathbf{\Sigma_x} = \mathrm{Cov}(\mathbf{x}) = \begin{bmatrix} \sigma_{11} & \cdots & \sigma_{1p} \\ \vdots & & \vdots \\ \sigma_{p1} & \cdots & \sigma_{pp} \end{bmatrix}$$

is called the *covariance matrix* of $\mathbf{x}$ and the matrix

$$\mathbf{R_x} = \mathrm{Corr}(\mathbf{x}) = \begin{bmatrix} \rho_{11} & \cdots & \rho_{1p} \\ \vdots & & \vdots \\ \rho_{p1} & \cdots & \rho_{pp} \end{bmatrix}$$

is called the *covariance matrix* of $\mathbf{x}$, where $\rho_{ij} = \mathrm{Corr}(X_i, X_j)$. Denote by σ_i the standard deviation of X_i. Then, $\mathrm{Cov}(X_i, X_j) = \sigma_{ij} = \sigma_i \sigma_j \rho_{ij}$. Let $\mathbf{S}$ be the diagonal matrix with diagonal elements $\sigma_1, \cdots, \sigma_p$. The relation between $\mathbf{\Sigma_x}$ and $\mathbf{R_x}$ is given by

$$\mathbf{\Sigma_x} = \mathbf{S} \mathbf{R_x} \mathbf{S}.$$

Divide $\mathbf{x}$ into two parts, $\mathbf{x} = (\mathbf{x}_1', \mathbf{x}_2')'$. If the covariance matrix of $\mathbf{x}$ has the form

$$\mathbf{\Sigma}(\mathbf{x}) = \begin{bmatrix} \mathbf{\Sigma}_{11} & \mathbf{0} \\ \mathbf{0} & \mathbf{\Sigma}_{22} \end{bmatrix},$$

where $\mathbf{\Sigma}_{11} = \mathrm{Cov}(\mathbf{x}_1)$ and $\mathbf{\Sigma}_{22} = \mathrm{Cov}(\mathbf{x}_2)$, we say that $\mathbf{x}_1$ and $\mathbf{x}_2$ are uncorrelated.

A.2.2 Some statistical distributions and Gaussian process

This subsection introduces the uniform distribution, univariate and multivariate normal distributions, and the Gaussian process.

Uniform distribution: The uniform distribution on a finite domain T has a constant density $1/\mathrm{Vol}(T)$, where $\mathrm{Vol}(T)$ is the volume of T. If a random vector $\mathbf{x}$ has the uniform distribution on T, then we denote $\mathbf{x} \sim U(T)$. In particular, the uniform distribution on $[0, 1]$, denoted by $U(0, 1)$, has a constant density 1 on [0,1], and the uniform distribution on the unit cube $C^s = [0, 1]^s$, denoted by $U(C^s)$, has a constant density 1 on C^s. A random variable X that follows $U(0, 1)$ is also called simply a *random number*. Such random numbers play an important role in simulation.

Normal distribution: For real number μ and positive number σ, if a random variable X has a density

$$p(x) = \frac{1}{\sqrt{2\pi}\sigma}\exp\left\{-\frac{1}{2}\left(\frac{x-\mu}{\sigma}\right)^2\right\},$$

then X is said to be a normal random variable with mean μ and variance σ^2, denoted by $X \sim N(\mu, \sigma^2)$. When $\mu = 0$ and $\sigma = 1$, the corresponding distribution is called the *standard normal distribution* and is denoted by $X \sim N(0,1)$. In this case, its density reduces to

$$p(x) = \frac{1}{\sqrt{2\pi}}e^{-\frac{1}{2}x^2}.$$

Multivariate normal distribution: If a p-dimensional random vector $\mathbf{x}$ has a pdf

$$p(\mathbf{x}) = (2\pi)^{-p/2}|\mathbf{\Sigma}|^{-1/2}\exp\left\{-\frac{1}{2}(\mathbf{x}-\boldsymbol{\mu})'\mathbf{\Sigma}^{-1}(\mathbf{x}-\boldsymbol{\mu})\right\},$$

then we say that $\mathbf{x}$ follows a multivariate normal distribution $N_p(\boldsymbol{\mu}, \mathbf{\Sigma})$. When $p = 1$, it reduces to a univariate normal distribution. The mean vector and covariance of $\mathbf{x} \sim N_p(\boldsymbol{\mu}, \mathbf{\Sigma})$ are

$$E(\mathbf{x}) = \boldsymbol{\mu}, \quad \text{Cov}(\mathbf{x}) = \mathbf{\Sigma},$$

respectively. If $\mathbf{x} \sim N_p(\boldsymbol{\mu}, \mathbf{\Sigma})$ and $\mathbf{A}$ is a $k \times p$ constant matrix with $k \leq p$, then $\mathbf{A}\mathbf{x}$ follows a multivariate normal distribution. In other words, any linear combination of a multivariate random vector is still a multivariate normal distribution. In particular, any marginal distribution of a multivariate distribution is a multivariate normal distribution. Partition $\mathbf{x}, \boldsymbol{\mu}$, and $\mathbf{\Sigma}$ into

$$\mathbf{x} = \begin{bmatrix} \mathbf{x}_1 \\ \mathbf{x}_2 \end{bmatrix}, \boldsymbol{\mu} = \begin{bmatrix} \boldsymbol{\mu}_1 \\ \boldsymbol{\mu}_2 \end{bmatrix}, \mathbf{x} = \begin{bmatrix} \mathbf{\Sigma}_{11} & \mathbf{\Sigma}_{12} \\ \mathbf{\Sigma}_{21} & \mathbf{\Sigma}_{22} \end{bmatrix}, \tag{A.2}$$

where $\mathbf{x}_1, \boldsymbol{\mu}_1 \in R^q$, $q < p$, and $\mathbf{\Sigma}_{11} : q \times q$. The marginal distributions of $\mathbf{x}_1$ and $\mathbf{x}_2$ are $\mathbf{x}_1 \sim N_q(\boldsymbol{\mu}_1, \mathbf{\Sigma}_{22})$ and $\mathbf{x}_2 \sim N_{p-q}(\boldsymbol{\mu}_2, \mathbf{\Sigma}_{11})$, respectively. The conditional distribution of $\mathbf{x}_1$ for given a $\mathbf{x}_2$ is $N_q(\boldsymbol{\mu}_{1.2}, \mathbf{\Sigma}_{11.2})$, where

$$E(\mathbf{x}_1|\mathbf{x}_2) = \boldsymbol{\mu}_{11.2} = \boldsymbol{\mu}_1 + \mathbf{\Sigma}_{12}\mathbf{\Sigma}_{22}^{-1}(\mathbf{x}_2 - \boldsymbol{\mu}_2), \tag{A.3}$$

$$\text{Cov}(\mathbf{x}_1|\mathbf{x}_2) = \mathbf{\Sigma}_{11.2} = \mathbf{\Sigma}_{11} - \mathbf{\Sigma}_{12}\mathbf{\Sigma}_{22}^{-1}\mathbf{\Sigma}_{21}. \tag{A.4}$$

Chi-squared distribution: If $X_1, \cdots X_p$ are independently identically variables following $N(0,1)$, then $Y = X_1^2 + \cdots + X_p^2$ is said to follow a *chi square distribution* with p degrees of freedom. The density of Y is given by

$$p(y) = \frac{e^{-y/2}y^{(p-2)/2}}{2^{p/2}\Gamma(p/2)}, y > 0,$$

and is denoted by $Y \sim \chi^2(p)$, where $\Gamma(\cdot)$ is the *gamma function* defined by

$$\Gamma(a) = \int_0^\infty e^{-t}t^{a-1}dt, \quad a > 0.$$

t-distribution: Let $X \sim N(0,1)$ and $Y \sim \chi^2(p)$ be independent. Then the distribution of random variable $t = X/\sqrt{Y/p}$ is called the *t-distribution* with p degrees of freedom and denoted by $t \sim t(p)$. Its density is given by

$$p(t) = \frac{\Gamma(\frac{p+1}{2})}{\sqrt{p\pi}\Gamma(\frac{p}{2})}\left(1 + \frac{t^2}{p}\right)^{-(p+1)/2}, \quad t \geq 0.$$

F-distribution: Let $X \sim \chi^2(p)$ and $y \sim \chi^2(q)$ be independent. Then the distribution of $F = \frac{X/p}{Y/q}$ is called the *F-distribution* with degrees of freedom p and q and denoted by $F \sim F(p,q)$. Its density is given by

$$p(x) = \frac{(\frac{p}{q})^{p/2}x^{(p-2)/2}}{B(\frac{p}{2}, \frac{q}{2})}\left(1 + \frac{p}{q}x\right)^{-(p+q)/2}, \quad x \geq 0,$$

where $B(\cdot, \cdot)$ is the *beta function* defined by $B(a,b) = \Gamma(a)\Gamma(b)/\Gamma(a+b)$ for any $a > 0$ and $b > 0$. If $X \sim t(p)$, then $X^2 \sim F(1,p)$. Therefore, many t-tests can be replaced by F-tests.

Gaussian processes: A stochastic process $\{X(\mathbf{t}), \mathbf{t} \in T\}$ indexed by $\mathbf{t}$ is said to be a *Gaussian process* if any of its finite dimensional marginal distribution is a normal distribution, i.e., if for any finite set $\{\mathbf{t}_1, \cdots, \mathbf{t}_p\}$, $(X(\mathbf{t}_1), \cdots, X(\mathbf{t}_p))$ has an p-dimensional normal distribution.

Denote the mean, variance, and covariance functions by

$$\mu(\mathbf{t}) = E\{Y(\mathbf{t})\}, \quad \sigma^2(\mathbf{t}) = \mathrm{Var}\{Y(\mathbf{t})\}, \quad \text{and} \quad \sigma(\mathbf{t}_1, \mathbf{t}_2) = \mathrm{Cov}(X(\mathbf{t}_1), X(\mathbf{t}_2)),$$

for any $\mathbf{t}_1, \mathbf{t}_2 \in T$.

A Gaussian process $\{X(\mathbf{t}), \mathbf{t} \in T\}$ indexed by $\mathbf{t}$ is said to be *stationary* if its $\mu(\mathbf{t})$ and $\sigma^2(\mathbf{t})$ are constant (independent of the index $\mathbf{t}$), and its $\mathrm{Cov}(X(\mathbf{t}_1), X(\mathbf{t}_2))$ depends only on $|t_{i1} - t_{i2}|, i = 1, \ldots, p$, where $\mathbf{t}_j = (t_{1j}, \ldots, t_{pj}), j = 1, 2$.

A.3   Linear Regression Analysis

There are many textbooks for linear regression analysis. This section gives a brief introduction to linear models, refreshing some of the basic and important concepts on linear regression analysis. Details on the linear models can be found in textbooks such as Weisberg (1985), Draper and Smith (1981), and Neter, Kutner, Nachtsheim and Wasserman (1996).

A.3.1   Linear models

A linear regression model is defined as

$$Y = \beta_0 + \beta_1 X_1 + \cdots + \beta_s X_s + \varepsilon. \tag{A.5}$$

The Y variable is called the response variable, or dependent variable, while the X variables are often called explanatory variables, or independent variables, or covariates, or regressors. The parameters βs are usually unknown and have to be estimated, while ε is termed random error. In general, the "random error" here refers to the part of the response variable that cannot be explained or predicted by the covariates.

The regression function, given the values of X variables, is the conditional expected value of the response variable, denoted by $m(x_1, \cdots, x_s)$, that is,

$$m(x_1, \cdots, x_s) = E(Y|X_1 = x_1, \cdots, X_s = x_s).$$

The term "linear model" refers to the fact that the regression function can be explicitly written as a linear form of βs, such as

$$m(x_1, \cdots, x_s) = \beta_0 + \beta_1 x_1 + \cdots + \beta_s x_s.$$

This assumption is not always valid and needs to be checked.

Example 28 Let $\{x_i, y_i\}, i = 1, \cdots, n$ be a bivariate random sample from the model

$$Y = m(X) + \varepsilon,$$

with $E(\varepsilon|X) = 0$ and $\text{var}(\varepsilon|X) = \sigma^2$. We may first try to fit a linear line through the data. In other words, we approximate the model by a *simple linear regression model*

$$Y = \beta_0 + \beta_1 X + \varepsilon.$$

In many situations, the simple linear regression model does not fit the data well. It is common to further consider the s-th degree *polynomial regression model*:

$$Y = \beta_0 + \beta_1 X + \cdots + \beta_s X^s + \varepsilon.$$

The polynomial regression, though nonlinear in x, is still called a linear model as it is linear in the parameters βs. This sometimes causes confusion, but by defining new variables

$$X_1 = X, X_2 = X^2, \cdots, X_s = X^s,$$

the polynomial model is now linear in the covariates $X_1, \cdots, X_s$. Note that the polynomial regression, being smooth everywhere, may not fit every data set well. Thus, other forms of regression functions, such as splines, may be used. Section 5.3 introduces spline regression method in details. In general, we consider linear models to be all models that are linear in unknown parameters.

A.3.2 Method of least squares

Suppose that $\{(x_{i1}, \cdots, x_{is}, y_i), i = 1, \cdots, n\}$ is a random sample from the linear model

$$y_i = \sum_{j=0}^{s} x_{ij}\beta_j + \varepsilon_i, \tag{A.6}$$

where we may set $x_{i0} \equiv 1$ to include the intercept in the model, and ε_i is random error. It is usually assumed that random errors $\{\varepsilon_i\}$ are homoscedastic, namely, they are uncorrelated random variables with mean 0 and common variance σ^2.

The method of least squares was advanced by Gauss in the early nineteenth century and has been pivotal in regression analysis. For data $\{(x_{i1}, \cdots, x_{ip}, y_i), i = 1, \cdots, n\}$, the *method of least squares* considers the deviation of each observation value y_i from its expectation and finds the coefficient βs by minimizing the sum of squared deviations:

$$S(\beta_0, \cdots, \beta_s) = \sum_{i=1}^{n}(y_i - \sum_{j=0}^{s} x_{ij}\beta_j)^2. \tag{A.7}$$

Matrix notation (cf. Section A.1) can give us a succinct expression of the least squares solution. Denote

$$\mathbf{y} = \begin{pmatrix} y_1 \\ \vdots \\ y_n \end{pmatrix}, \ \mathbf{X} = \begin{pmatrix} x_{10} & \cdots & x_{1s} \\ \vdots & \cdots & \vdots \\ x_{n0} & \cdots & x_{ns} \end{pmatrix}, \ \boldsymbol{\beta} = \begin{pmatrix} \beta_0 \\ \vdots \\ \beta_s \end{pmatrix}, \ \boldsymbol{\varepsilon} = \begin{pmatrix} \varepsilon_1 \\ \vdots \\ \varepsilon_n \end{pmatrix}.$$

Then model (A.6) can be written in the matrix form

$$\mathbf{y} = \mathbf{X}\boldsymbol{\beta} + \boldsymbol{\varepsilon}. \tag{A.8}$$

The matrix $\mathbf{X}$ is known as the design matrix and is of crucial importance in linear regression analysis. The $S(\boldsymbol{\beta})$ in (A.7) can be written as

$$S(\boldsymbol{\beta}) = (\mathbf{y} - \mathbf{X}\boldsymbol{\beta})'(\mathbf{y} - \mathbf{X}\boldsymbol{\beta}).$$

Differentiating $S(\boldsymbol{\beta})$ with respect to $\boldsymbol{\beta}$, we obtain the *normal equations*

$$\mathbf{X}'\mathbf{y} = \mathbf{X}'\mathbf{X}\boldsymbol{\beta}.$$

If $\mathbf{X}'\mathbf{X}$ is invertible, the normal equations yield the least squares estimator of $\boldsymbol{\beta}$

$$\hat{\boldsymbol{\beta}} = (\mathbf{X}'\mathbf{X})^{-1}\mathbf{X}'\mathbf{y}.$$

The properties of the least squares estimator have been well studied and are summarized as follows without proof.

PROPOSITION A.1
If the linear model (A.6) holds, then

(a) *Unbiasedness*: $E(\hat{\beta}|\mathbf{X}) = \beta$;

(b) *Covariance*: $\mathrm{Cov}(\hat{\beta}|\mathbf{X}) = \sigma^2(\mathbf{X}'\mathbf{X})^{-1}$;

(c) *Gauss-Markov theorem*: The least-squares estimator $\hat{\beta}$ is the best linear unbiased estimator (BLUE). That is, for any given vector $\mathbf{a}$, $\mathbf{a}'\hat{\beta}$ is a linear unbiased estimator of the parameter $\theta = \mathbf{a}'\beta$, and furthermore, for any linear unbiased estimator $\mathbf{b}'\mathbf{y}$ of θ, its variance is at least as large as that of $\mathbf{a}'\hat{\beta}$.

From Proposition A.1 (b), the covariance formula involves an unknown parameter σ^2. A natural estimator is based on the *residual sum of squares* (RSS), defined by

$$\mathrm{RSS} = \sum_{i=1}^{n}(y_i - \hat{y}_i)^2,$$

where $\hat{y}_i = \sum_{j=0}^{s} x_{ij}\hat{\beta}_j$ is the fitted value of the i-th datum point. An unbiased estimator of σ^2 is

$$\hat{\sigma}^2 = \mathrm{RSS}/(n - s - 1).$$

A.3.3 Analysis of variance

In practice, various hypothesis testing problems arise. For instance, one may ask whether some covariates have significant effects. This can be formulated as the following linear hypothesis problem:

$$H_0 : \mathbf{C}\beta = \mathbf{h} \quad \text{versus} \quad H_1 : \mathbf{C}\beta \neq \mathbf{h}, \tag{A.9}$$

where $\mathbf{C}$ is a $q \times p$ constant matrix with rank q and $\mathbf{h}$ is a $q \times 1$ constant vector. For example, if one wants to test $H_0 : \beta_1 = 0$ against $H_1 : \beta_1 \neq 0$, we can choose $\mathbf{C} = (0, 1, 0, \ldots, 0)$ and $\mathbf{h} = 0$, or if one wants to test $H_0 : \beta_1 = \beta_2$ against $H_1 : \beta_1 \neq \beta_2$, the corresponding $\mathbf{C} = (0, 1, -1, 0, \ldots, 0)$ and $\mathbf{h} = 0$.

Under H_0, the method of least squares is to minimize $S(\beta)$ with constraints $\mathbf{C}\beta = \mathbf{h}$. This leads to

$$\hat{\beta}_0 = \hat{\beta} - (\mathbf{X}'\mathbf{X})^{-1}\mathbf{C}'\{\mathbf{C}(\mathbf{X}'\mathbf{X})^{-1}\mathbf{C}'\}^{-1}(\mathbf{C}\hat{\beta} - \mathbf{h}). \tag{A.10}$$

The residual sum of squares, denoted by RSS_1 and RSS_0 under the full model and the null hypothesis, are

$$\mathrm{RSS}_1 = (\mathbf{y} - \mathbf{X}\hat{\beta})'(\mathbf{y} - \mathbf{X}\hat{\beta}),$$

and

$$\mathrm{RSS}_0 = (\mathbf{y} - \mathbf{X}\hat{\beta}_0)'(\mathbf{y} - \mathbf{X}\hat{\beta}_0),$$

respectively. The F-test for H_0 is based on the decomposition

$$\mathrm{RSS}_0 = (\mathrm{RSS}_0 - \mathrm{RSS}_1) + \mathrm{RSS}_1. \tag{A.11}$$

The sampling properties of these residuals sum of squares are as follows.

PROPOSITION A.2
If the linear model (A.6) holds and $\{\varepsilon_i\}$ are independently and identically distributed as $N(0, \sigma^2)$, then $RSS_1/\sigma^2 \sim \chi^2_{n-s-1}$. Further, under the null hypothesis H_0, $(RSS_0 - RSS_1)/\sigma^2 \sim \chi^2_q$ and is independent of RSS_1.

Proposition A.2 shows that under the hypothesis H_0 in (A.9)

$$F = \frac{(RSS_0 - RSS_1)/q}{RSS_1/(n - s - 1)}$$

has an $F_{q, n-s-1}$ distribution. The null hypothesis H_0 is rejected if the observed F-statistic is large. This test may be summarized in the analysis of variance as in Table A.1. The first column indicates the models or the variables. The second column is the sum of squares reduction (SSR), the third column shows the degree of freedom, and the fourth column is the mean squares, which are the sum of squares divided by its associated degree of freedom. One can include more information in the ANOVA table, by adding a column on F-statistic and a column on P-value. The first row shows the sum of squares reduction by using the full model, i.e., the one under the alternative hypothesis. This row is typically omitted. The total refers to the sum of the second through the fourth rows. The third row indicates the additional contribution of the variables that are in the full model but not in the submodel. It is important for gauging whether the contributions of those variables are statistically significant.

TABLE A.1
Analysis of Variance for a Multiple Regression

Source	SS	df	MS
Full-model	$SSR_1 \equiv \sum(Y_i - \bar{Y})^2 - RSS_1$	s	SSR_1/s
Sub-model	$SSR_0 \equiv \sum(Y_i - \bar{Y})^2 - RSS_0$	$s - q$	$SSR_0/(s - q)$
Difference	$RSS_0 - RSS_1$	q	$(RSS_0 - RSS_1)/q$
Residuals	RSS_1	$n - s - 1$	$RSS_1/(n - s - 1)$
Total	$SST = \sum(y_i - \bar{y})^2$	$n - 1$	

A.3.4 An illustration

In this section, we illustrate the concepts of linear regression via analysis of an environmental data set.

Example 7 (continued) Uniform design $U_{17}(17^6)$ (see Chapter 3 for the basics of uniform design) was used to conduct experiments. The actual design is displayed in the left panel of Table A.2, extracted from Fang (1994). For each combination in the left panel of Table A.2, three experiments were conducted, and the outputs (mortality) are depicted in the right panel of Table A.2. The mortality in the last row, corresponding to high levels of all metals, is higher than the other ones. This implies that the contents of metals may affect the mortality.

TABLE A.2
Environmental Data

Cd	Cu	Zn	Ni	Cr	Pb	Y_1	Y_2	Y_3
0.01	0.2	0.8	5.0	14.0	16.0	19.95	17.6	18.22
0.05	2.0	10.0	0.1	8.0	12.0	22.09	22.85	22.62
0.1	10.0	0.01	12.0	2.0	8.0	31.74	32.79	32.87
0.2	18.0	1.0	0.8	0.4	4.0	39.37	40.65	37.87
0.4	0.1	12.0	18.0	0.05	1.0	31.90	31.18	33.75
0.8	1.0	0.05	4.0	18.0	0.4	31.14	30.66	31.18
1.0	8.0	2.0	0.05	12.0	0.1	39.81	39.61	40.80
2.0	16.0	14.0	10.0	5.0	0.01	42.48	41.86	43.79
4.0	0.05	0.1	0.4	1.0	18.0	24.97	24.65	25.05
5.0	0.8	4.0	16.0	0.2	14.0	50.29	51.22	50.54
8.0	5.0	16.0	2.0	0.01	10.0	60.71	60.43	59.69
10.0	14.0	0.2	0.01	16.0	5.0	67.01	71.99	67.12
12.0	0.01	5.0	8.0	10.0	2.0	32.77	30.86	33.70
14.0	0.4	18.0	0.2	4.0	0.8	29.94	28.68	30.66
16.0	4.0	0.4	14.0	0.8	0.2	67.87	69.25	67.04
18.0	12.0	8.0	1.0	0.1	0.05	55.56	55.28	56.52
20.0	20.0	20.0	20.0	20.0	20.0	79.57	79.43	78.48

Note that the ratio of maximum to minimum levels of the six metals is 2000. To stabilize numerical computation, it is recommended to standardize all six variables. Denote $x_1, ..., x_6$ for the standardized variables of Cd, Cu, Zn, Ni, Cr and Pb, respectively. That is, for example, $x_1 = (Cd - 6.5624)/7.0656$, where 6.5624 and 7.0656 are the sample mean and standard deviation of 17 Cd contents, respectively. As an illustration, we take the average of the three outcomes as the response variable. The data were fitted by a linear regression model

$$y = \beta_0 + \beta_1 x_1 + \cdots, \beta_6 x_6 + \varepsilon.$$

The estimates, standard errors, and P-values of the regression coefficients are depicted in Table A.3, from which it can be seen that only Cd and Cu are very significant. Other metals seem not to be statistically significant, under the assumption that the model under consideration is appropriate.

TABLE A.3
Estimates, Standard Errors and P-Value

X-variable	Estimate	Standard Error	P-value
Intercept	42.8639	2.9313	0.0000
Cd	10.7108	3.3745	0.0099
Cu	8.1125	3.1694	0.0284
Zn	−2.1168	3.3745	0.5445
Ni	3.8514	3.1694	0.2522
Cr	−0.3443	3.1410	0.9149
Pb	0.3579	3.1410	0.9115

From Table A.3, it is of interest to test whether the effects of Zn, Ni, Cr, and Pb are significant. This leads us to consider the following null hypothesis:

$$H_0 : \beta_3 = \beta_4 = \beta_5 = \beta_6 = 0 \quad \text{versus} \quad H_1 : \text{not all } \beta_3, \beta_4, \beta_5 \text{ and } \beta_6 \text{ equal } 0.$$

Hence we take $\mathbf{C} = (\mathbf{0}, \mathbf{0}, \mathbf{0}, \mathbf{0}, \mathbf{I}_4)$, where I_4 is the identity matrix of order 4, and $\mathbf{h} = \mathbf{0}$. The ANOVA table for the hypothesis is given in Table A.4, from which we can easily calculate the F-statistic:

$$F = 64.45/146.07 = 0.4412$$

with P-value=0.7765. This indicates in favor of H_0. Here we consider this example only for the purpose of illustration. One may further consider interaction effects and quadratic effects of the six metals. The model becomes

$$y = \beta_0 + \sum_{i=1}^{6} \beta_i x_i + \sum_{1=i \leq j \leq 6} \beta_i x_i x_j + \varepsilon,$$

which involves 22 items, only a few of which we believe are significant. Thus, selection of variables is an important issue.

TABLE A.4
ANOVA Table for Example 7

Source	SS	df	MS
Sub-model	3420.7	2	1710.3
Difference	257.8	4	64.45
Residuals	1460.7	10	146.07
Total	5139.2	16	

A.4   Variable Selection for Linear Regression Models

As discussed in Section 5.1, variable selection can be viewed as a type of regularization and may be useful for choosing a parsimonious metamodel. There is a considerable literature on the topic of variable selection. Miller (2002) gives a systematic account of this research area. In this section, we briefly review some recent developments of this topic.

Suppose that $\{(x_{i1}, \cdots, x_{is}, y_i), i = 1, \cdots, n\}$ is a random sample from the following linear regression model:

$$y_i = \sum_{j=0}^{s} x_{ij}\beta_j + \varepsilon_i,$$

where $x_{i0} \equiv 1$, β_0 is the intercept, $E(\varepsilon_i) = 0$, and $\mathrm{Var}(\varepsilon_i) = \sigma^2$. Then a penalized least squares is defined as

$$Q(\boldsymbol{\beta}) = \frac{1}{2}\sum_{i=1}^{n}(y_i - \sum_{j=0}^{s} x_{ij}\beta_j)^2 + n\sum_{j=0}^{s} p_\lambda(|\beta_j|),$$

where $p_\lambda(\cdot)$ is a pre-specified non-negative penalty function, and λ is a regularization parameter. The factor "1/2" in the definition of the penalized least squares comes from the likelihood function with normal distribution for the random errors.

Using matrix notation defined in the last section, the penalized least squares can be rewritten as

$$Q(\boldsymbol{\beta}) = \frac{1}{2}\|\mathbf{y} - \mathbf{X}\boldsymbol{\beta}\|^2 + n\sum_{j=0}^{s} p_\lambda(|\beta_j|). \tag{A.12}$$

Many classical variable selection criteria in linear regression analysis can be derived from the penalized least squares. Take the penalty function in (A.12) to be the L_0 penalty:

$$p_\lambda(|\beta|) = \frac{1}{2}\lambda^2 I(|\beta| \neq 0),$$

where $I(\cdot)$ is the indicator function. Note that $\sum_{j=1}^{s} I(|\beta_j| \neq 0)$ equals the number of nonzero regression coefficients in the model. Hence many popular variable selection criteria can be derived from (A.12) with the L_0 penalty by choosing different values of λ. For instance, the C_p (Mallows (1973)), AIC (Akaike (1974)), BIC (Schwarz (1978)), ϕ-criterion (Hannan and Quinn (1979), and Shibata (1984)), and RIC (Foster and George (1994)) correspond to $\lambda = \sqrt{2}(\sigma/\sqrt{n})$, $\sqrt{2}(\sigma/\sqrt{n})$, $\sqrt{\log n}(\sigma/\sqrt{n})$, $\sqrt{\log\log n}(\sigma/\sqrt{n})$ and $\sqrt{2\log(s+1)}(\sigma/\sqrt{n})$, respectively, although these criteria were motivated by different principles.

Since the L_0 penalty is discontinuous, it requires an exhaustive search over all 2^d possible subsets of x-variables to find the solution. This approach is very expensive in computational cost. Therefore, people usually employ some less expensive algorithms, such as forward subset selection and stepwise regression (see, for example, Miller (2002)), to select a good subset of variables. Furthermore, subset variable selection suffers from other drawbacks, the most severe of which is its lack of stability as analyzed, for instance, by Breiman (1996).

A.4.1 Nonconvex penalized least squares

To avoid the drawbacks of the subset selection, heavy computational load, and the lack of stability, Tibshirani (1996) proposed the LASSO, which can be viewed as the solution of (A.12) with the L_1 penalty defined below:

$$p_\lambda(|\beta|) = \lambda|\beta|.$$

LASSO retains the virtues of both best subset selection and ridge regression which can be viewed as the solution of penalized least squares with the L_2 penalty:

$$p_\lambda(|\beta|) = \lambda\beta^2.$$

Frank and Friedman (1993) considered the L_q penalty:

$$p_\lambda(|\beta|) = \lambda|\beta|^q, \qquad q > 0,$$

which yields a "bridge regression." The issue of selection penalty function has been studied in depth by various authors, for instance, Antoniadis and Fan (2001). Fan and Li (2001) suggested the use of the smoothly clipped absolute deviation (SCAD) penalty, defined by

$$p'_\lambda(\beta) = \lambda\{I(\beta \le \lambda) + \frac{(a\lambda - \beta)_+}{(a-1)\lambda}I(\beta > \lambda)\} \text{ for some } a > 2 \quad \text{and } \beta > 0,$$

with $p_\lambda(0) = 0$. This penalty function involves two unknown parameters, λ and $a = 3.7$ (cf. Fan and Li (2001)). Figure A.1 depicts the plots of the SCAD, $L_{0.5}$, and L_1 penalty functions.

As shown in Figure A.1, the three penalty functions all are singular at the origin. This is a necessary condition for sparsity in variable selection: the resulting estimator automatically sets small coefficients to be zero. Furthermore, the SCAD and $L_{0.5}$ penalties are nonconvex over $(0, +\infty)$ in order to reduce estimation bias. We refer to penalized least squares with the nonconvex penalties over $(0, \infty)$ as *nonconvex penalized least squares* in order to distinguish from the L_2 penalty. The L_1 penalty is convex over $(-\infty, \infty)$, but it may be viewed as a concave function over $(0, +\infty)$. The SCAD is an improvement over the L_0 penalty in two aspects: saving computational cost and resulting

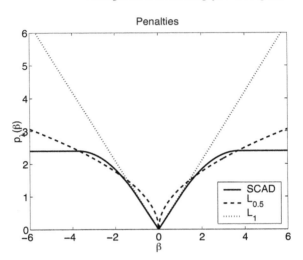

FIGURE A.1
Plot of penalty functions

in a continuous solution to avoid unnecessary modeling variation. Furthermore, the SCAD improves bridge regression by reducing modeling variation in model prediction. Although similar in spirit to the L_1 penalty, the SCAD may also improve the L_1 penalty by avoiding excessive estimation bias because the solution of the L_1 penalty could shrink all regression coefficients by a constant, for instance, the soft thresholding rule (Donoho and Johnstone (1994) and Tibshirani (1996)).

A.4.2   Iteratively ridge regression algorithm

The L_q for $0 < q < 1$ and the SCAD penalty functions are singular at the origin, and they do not have continuous second-order derivatives. This poses a challenge in searching for the solution of the penalized least squares. Fan and Li (2001) proposed using a quadratic function to locally approximate the penalty functions. Their procedure can be summarized as follows. As usual, we set the ordinary least squares estimate as the initial value of β. Suppose that we are given an initial value β_0 that is close to the minimizer of (A.12). If β_{j0} is very close to 0, then set $\hat{\beta}_j = 0$. Otherwise the penalty functions can be locally approximated by a quadratic function as

$$[p_\lambda(|\beta_j|)]' = p'_\lambda(|\beta_j|)\mathrm{sgn}(\beta_j) \approx \{p'_\lambda(|\beta_{j0}|)/|\beta_{j0}|\}\beta_j,$$

when $\beta_j \neq 0$. In other words,

$$p_\lambda(|\beta_j|) \approx p_\lambda(|\beta_{j0}|) + \frac{1}{2}\{p'_\lambda(|\beta_{j0}|)/|\beta_{j0}|\}(\beta_j^2 - \beta_{j0}^2), \quad \text{for } \beta_j \approx \beta_{j0}. \quad (A.13)$$

Set the least squares estimate as an initial value for β. With the local quadratic approximation, the solution for the penalized least squares can be found by iteratively computing the following ridge regression:

$$\beta_{k+1} = \{\mathbf{X}'\mathbf{X} + n\Sigma_\lambda(\beta_k)\}^{-1}\mathbf{X}'\mathbf{y} \qquad (A.14)$$

for $k = 0, 1, \cdots$, with β_0 being the least squares estimate, where

$$\Sigma_\lambda(\beta_k) = \mathrm{diag}\{p'_\lambda(|\beta_{1k}|)/|\beta_{1k}|, \cdots, p'_\lambda(|\beta_{sk}|)/|\beta_{sk}|\}.$$

In the course of iterations, if β_{jk} is set to be zero, the variable X_j is deleted from the model. We call this algorithm *iteratively ridge regression algorithm*. Hunter and Li (2005) studied the convergence of this algorithm.

Choice of the regularization parameter is important. It is clear that the cross validation method (5.1) can be directly used to select the λ. Alternatively, we may define the effective number of parameters using the expression (A.14), and further define generalized cross validation (GCV) scores. Detailed description of GCV scores has been given in Fan and Li (2001). Thus, we can use the GCV procedure presented in Section 7.3 to select the λ.

We summarize the variable selection procedure via nonconvex penalized least squares as the following algorithm:

Algorithm for Nonconvex Penalized Least Squares

Step 1. Choose a grid point set, $(\lambda_1, \cdots, \lambda_S)$ say, and let $i = 1$ and $\lambda = \lambda_i$.

Step 2. With λ_i, compute the $\hat{\beta}$ using the iterative ridge regression algorithm.

Step 3. Compute the CV or GCV score with $\lambda = \lambda_i$. Let $i = i + 1$.

Step 4. Repeat steps 2 and 3 until all S grid points are exhausted.

Step 5. The final estimator for β is the one that has the lowest GCV.

A.4.3 An illustration

We next illustrate the variable selection procedure introduced in previous sections by application to the data in Example 7. For stepwise regression with F_{in} and F_{out}, we follow the traditional regression analysis by setting the F-values to be the 95% percentile of the corresponding F distributions. For the AIC, BIC, RIC, and ϕ-criterion, the subset is also selected by using the stepwise regression algorithm. For the LASSO and the SCAD, the tuning parameter λ is chosen by minimizing the GCV statistic. The corresponding values of λ equal 0.5863 and 0.5877 for the SCAD and the LASSO, respectively. The resulting estimate and the corresponding residual sum of squares (RSS) after variable selection is depicted in Table A.5, from which we can see that all procedures select exactly the same significant variables. We are surprised that all variable selection procedures except the LASSO yield exactly the same final model. LASSO slightly shrinks the regression coefficients and yields a slightly larger residual sum of squares.

TABLE A.5
Estimates and RSS after Variable Selection

Variable	F-Values	AIC	BIC	RIC	ϕ	SCAD	LASSO
Intercept	42.864	42.864	42.864	42.864	42.864	42.864	42.276
Cd	10.528	10.528	10.528	10.528	10.528	10.528	10.020
Cu	8.014	8.014	8.014	8.014	8.014	8.014	7.506
Zn	–	–	–	–	–	0	0
Ni	–	–	–	–	–	0	0
Cr	–	–	–	–	–	0	0
Pb	–	–	–	–	–	0	0
RSS	101.087	101.087	101.087	101.087	101.087	101.087	102.029

Acronyms

AIC	Akaike information criterion		regression splines
ANOVA	Analysis of variance	MLHS	Midpoint Latin hypercube
BIBD	Balanced incomplete block		sampling
	design	MLP	Multi-layer perceptron
BIC	Bayesian information criterion	Mm	Maximin distance design
BLUE	Best linear unbiased predictor	MMSE	Maximum mean squared
BLUP	Linear unbiased predictor		error
BPH	Balance-pursuit heuristic	MS	Mean squares
CD	Centered L_2-discrepancy	MSE	Mean square error
cdf	Cumulative distribution	NVH	Noise, vibration and
	function		harshness
CFD	Computational fluid dynamics	OA	Orthogonal array
CR	Correction ratio		OA-based
CV	Cross validation	LHD	Orthogonal array-based
DACE	Design and analysis of		Latin hypercube design
	computer experiments	OCLHD	Orthogonal column Latin
df	Degrees of freedom		designs
EFAST	Extended Fourier amplitude	PCC	Partial correlation
	sensitivity test		coefficient
ESE	Enhanced stochastic	pdf	Probability density
	evolutionary		function
FAST	Fourier amplitude sensitivity	PE	Prediction error
	test	PRCC	Partial rank correlation
FFD	Fractional factorial design		coefficient
FLM	Functional linear model	PRESS	Predicted error of sum of
GCV	Generalized cross validation		squares
glp	Good lattice point	RBD	Reliability-based design
IMSE	Integrated mean squares error	RBF	Radial basis function
LASSO	Least absolute shrinkage and	RBIBD	Resolvable balanced
	selection operator		incomplete block design
LHD	Latin hypercube design	REML	Restricted maximum
LHS	Latin hypercube sampling		likelihood
LHSA	Latin hypercube sampling	RIC	Risk information criterion
	algorithm	RPM	Rotation per minute
LS	Local search	RSM	Response surface
MAR	Median of absolute residuals		methodology
MARS	Multivariate adaptive	RSS	Residual sum of squares

SA	Simulated annealing or sensitivity analysis
SCAD	Smoothly clipped absolute deviation
SE	Stochastic evolutionary
SLHD	Symmetric Latin hypercube design
SLSE	Simple least squares estimate
SRC	Standardized regression coefficient
SRRC	Standardized rank regression coefficient
SS	Sum of squares
SSE	Sum of squares of error
SSR	Sum of squares of regression
SST	Sum of squares of total
TA	Threshold accepting
UD	Uniform design
URBD	Uniformly resolvable block design
VCE	Variance conditional expectation
WD	Wrap-around L_2-discrepancy
WMSE	Weighted mean square error

References

Abel, R. J. R., Ge, G., Greig, M. and Zhu, L. (2001), Resolvable balanced incomplete block designs with a block size of 5, *J. Statist. Plann. Inference* **95**, 49–65.

Akaike, H. (1973), Maximum likelihood identification of gaussian autoregressive moving average models, *Biometrika* **60**, 255–265.

Akaike, H. (1974), A new look at the statistical model identification, *IEEE Trans. on Autom. Contr.* **19**, 716–723.

Allen, D. M. (1974), The relationship between variable selection and data augmentation and a method for prediction, *Technometrics* **16**, 125–127.

An, J. and Owen, A. B. (2001), Quasi-regression, *J. Complexity* **17**, 588–607.

Antoniadis, A. and Fan, J. (2001), Regularization of wavelets approximations (with discussions), *J. Amer. Statist. Assoc.* **96**, 939–967.

Antoniadis, A. and Oppenheim, G. (1995), *Wavelets and Statistics*, Springer-Verlag, New York.

Atkinson, A., Bogacka, B. and Bagacki, M. (1998), D- and T-optimum designs for the kinetics of a reversible chemical reaction, *Chemom. Intell. Lab. Systems* **43**, 185–198.

Atkinson, A. and Donev, A. (1992), *Optimum Experimental Designs*, Oxford Science Publications, Oxford.

Bates, R. A., Buck, R. J., Riccomagno, E. and Wynn, H. P. (1996), Experimental design and observation for large systems (with discussion), *J. Royal Statist. Soc. Ser. B* **58**, 77–94.

Bates, R. A., Giglio, B. and Wynn, H. P. (2003), A global selection procedure for polynomial interpolators, *Technometrics* **45**, 246–255.

Bates, R. A., Riccomagno, E., Schwabe, R. and Wynn, H. P. (1995), Lattices and dual lattices in experimental design for Fourier model, *in Workshop on Quasi-Monte Carlo Methods and Their Applications,* pp. 1–14, Hong Kong Baptist University, Hong Kong.

Beckman, R. J., Conover, W. J. and McKay, M. D. (1979), A comparision of three methods for selecting values of input variables in the analysis of output from a computer code, *Technometrics* **21**, 239–245.

Bellman, R. E. (1961), *Adaptive Control Processes*, Princeton University Press, Princeton.

Berger, J. O. and Pericchi, L. R. (1996), The intrinsic Bayes factor for model selection and prediction, *J. Amer. Statist. Assoc.* **91**, 109–122.

Bickel, P. (1975), One-step Huber estimates in linear models, *J. Amer. Statist. Assoc.* **70**, 428–433.

Bishop, C. (1995), *Neural Networks for Pattern Recognition*, Oxford University Press, Oxford.

Booker, A. J., Dennis, J. J., Frank, P., Serafini, D., Torczon, V. and Trosset, M. (1999), A rigorous framework for optimization of expensive function by surrogates, *Structural Optimization* **17**, 1–13.

Box, G. E. P., Hunter, W. G. and Hunter, J. S. (1978), *Statistics for Experimenters, An Introduction to Design, Data Analysis, and Model Building*, Wiley, New York.

Breiman, L. (1995), Better subset regression using the nonnegative garrote, *Technometrics* **37**, 373–384.

Breiman, L. (1996), Heuristics of instability and stabilization in model selection, *Ann. Statist.* **24**, 2350–2383.

Breitung, K. (1984), Asymptotic approximations for multinomial integrals, *J. Engineer. Meth. Div.* **110**, 357–367.

Bundschuh, P. and Zhu, Y. C. (1993), A method for exact calculation of the discrepancy of low-dimensional finite point sets (I), *in, Abhandlungen aus Math. Seminar* pp. 115–133, Univ. Hamburg, Bd. 63.

Carroll, R. J., Chen, R., Li, T. H., Newton, H. J., Schmiediche, H., Wang, N. and George, E. I. (1997), Modeling ozone exposure in Harris County, Texas, *J. Amer. Statist. Assoc.* **92**, 392–413.

Chan, K., Saltelli, A. and Tarantola, S. (1997), Sensitivity analysis of model output: variance-based methods make the difference, *in* S. Andradóttir, K. J. Healy, D. H. Withers and B.L. Nelson eds, *Proceedings of the 1997 Winter Simulation Conference* pp. 261–268.

Chan, L. Y., Fang, K. T. and Mukerjee, R. (2001), A characterization for orthogonal arrays of strength two via a regression model, *Statist. & Probab. Letters* **54**, 189–192.

Chang, Y. J. and Notz, W. I. (1996), Model robust designs, *in* S. Ghosh and C. R. Rao, eds, *Handbook of Statistics* **13**, 1055–1098, Elsevier Science B. V., Amsterdam.

Chatterjee, K., Fang, K. T. and Qin, H. (2004), A lower bound for centered L_2-discrepancy on asymmetric factorials and its application, *J. Statist. Plann. Inference* **128**, 593–607.

Chen, T. Y., Zwick, J., Tripathy, B. and Novak, G. (2002), 3D engine analysis and mls cylinder head gaskets design, Society of Automotive Engineers, SAE paper 2002-01-0663.

Chen, W., Allen, J., Mistree, F. and Tsui, K.-L. (1996), A procedure for robust design: Minimizing variations caused by noise factors and control factors, *ASME Journal of Mechanical Design* **18**, 478–485.

Cheng, C. S. (1980), Orthogonal arrays with variable numbers of symbols, *Ann. Statist.* **8**, 447–453.

Cheng, C. S. (1997), $E(s^2)$-optimal superaturated designs, *Statistica Sinica* **7**, 929–939.

Chiang, C. T., Rice, J. A. and Wu, C. O. (2001), Smoothing spline estimation for varying coefficient models with repeatedly measured dependent variables, *J. Amer. Statist. Assoc.* **96**, 605–619.

Chui, C. K. (1992), *Wavelets: A Tutorial in Theory and Applications*, Academic Press Inc., Boston.

Colbourn, C. J. and Dinita, J. H. (1996), *CRC Handbook of Combinatorial Designs*, CRC Press, New York.

Craven, P. and Wahba, G. (1979), Smoothing noisy data with spline functions: estimating the correct degree of smoothing by the method of generalized cross-validation, *Numer. Math.* **31**, 377–403.

Cressie, N. A. (1993), *Statistics for Spatial Data*, Wiley, New York.

Cukier, R. I., Levine, H. B. and Schuler, K. E. (1978), Nonlinear sensitivity analysis of multiparameter model systems, *J. Computing Physics* **26**, 1–42.

Currin, C., Mitchell, T., Morris, M. and Ylvisaker, D. (1991), Bayesian prediction of deterministic functions, with applications to the design and analysis of computer experiments, *J. Amer. Statist. Assoc.* **86**, 953–963.

Cybenko, G. (1989), Approximation by superposition of a sigmoidal function, *Math. Contr. Signal Systems* **2**, 303–314.

Darken, C. and Moody, J. (1992), Toward faster stochastic gradient search, *in* J. E. Moody, S. J. Hanson, and R. P. Lippmann eds, *Advances in Neural Information Processing System* **3**, 1009–1016, Morgan Kaufmann, San Matco, CA.

Daubechies, I. (1992), *Ten Lectures on Wavelets*, CBMS-NSF, SIAM, Philadelphia.

De Boor, C. (1978), *A Practical Guide to Splines*, Springer-Verlag, New York.

de Luca, J. C. and Gerges, S. N. Y. (1996), Piston slap excitation: literature review, Society of Automative Engineering, SAE paper 962396.

Dey, A. and Mukerjee, R. (1999), *Fractional Factorial Plans*, John Wiley, New York.

Diaconis, P. (1988), Bayesian numerical analysis, *in* S. S. Gupta and J. O. Berger, eds, *Statistical Decision Theory and Related Topics* **IV**, 163–175, Springer–Verlag, New York.

Donoho, D. L. (2004*a*), For most large underdetermined systems of linear equations, the minimal l_1-norm near-solution approximates the sparsest near-solution, Manuscript.

Donoho, D. L. (2004*b*), For most large underdetermined systems of linear equations, the minimal l_1-norm solution is also the sparsest solution, Manuscript.

Donoho, D. L. and Johnstone, I. M. (1994), Ideal spatial adaptation by wavelet shrinkage, *Biometrika* **81**, 425–455.

Draper, N. R. and Smith, H. (1981), *Applied Regression Analysis,* 2nd ed., Wiley, New York.

Du, X. and Chen, W. (2004), Sequential optimization and reliability assessment method for efficient probabilistic design, *ASME Journal of Mechnical Design* **126**, 225–233.

Du, X. and Sudjianto, A. (2004), The first order saddlepoint approximation for reliability analysis, *AIAA J.* **42**, 1199–1207.

Du, X., Sudjianto, A. and Chen, W. (2004), An integrated framework for optimization under uncertainty using inverse reliability strategy, *ASME J. Meth. Design* **126**, 1–9.

Dueck, G. and Scheuer, T. (1991), Threshold accepting: A general purpose algorithm appearing superior to simulated annealing, *J. Computational Physics* **90**, 161–175.

Eilers, P. H. C. and Marx, B. D. (1996), Flexible smoothing with *b*-splines and penalties, *Statist. Sci.* **11**, 89–121.

Engle, R. F., Granger, C. W. J., Rice, J. and Weiss, A. (1986), Semiparametric estimates of the relation between weather and electricity sales, *J. Amer. Statist. Assoc.* **81**, 310–320.

Fan, J. (1992), Design-adaptive nonparametric regression, *J. Amer. Statist. Assoc.* **87**, 99–104.

Fan, J. and Gijbels, I. (1996), *Local Polynomial Modelling and Its Applica-*

tions, Chapman & Hall, London.

Fan, J. and Huang, T. (2005), Profile likelihood inferences on semiparametric varying-coefficient partially linear models, *Bernoulli* to appear.

Fan, J. and Li, R. (2001), Variable selection via nonconcave penalized likelihood and its oracle properties, *J. Amer. Statist. Assoc.* **96**, 1348–1360.

Fan, J. and Li, R. (2004), New estimation and model selection procedures for semiparametric modeling in longitudinal data analysis, *J. Amer. Statist. Assoc.* **99**, 710–723.

Fan, J. and Zhang, J. (2000), Two-step estimation of functional linear models with applications to longitudinal data, *J. Royal Statist. Soc. Ser. B* **62**, 303–322.

Fang, K. T. (1980), The uniform design: application of number-theoretic methods in experimental design, *Acta Math. Appl. Sinica* **3**, 363–372.

Fang, K. T. (1994), *Uniform Design and Uniform Design Tables*, Science Press, Beijing. (in Chinese).

Fang, K. T. (2001), Some applications of quasi-Monte Carlo methods in statistics, *in* K. T. Fang, F. J. Hickernell and H. Niederreiter, *Monte Carlo and Quasi-Monte Carlo Methods 2000* pp. 10–26, Springer, New York.

Fang, K. T. (2002), Theory, method and applications of the uniform design, *Inter. J. Reliability, Quality, and Safety Engineering* **9**, 305–315.

Fang, K. T. and Ge, G. N. (2004), An efficient algorithm for the classification of Hadamard matrices, *Math. Comp.* **73**, 843–851.

Fang, K. T., Ge, G. N. and Liu, M. Q. (2002), Construction of $e(f_{NOD})$-optimal supersaturated designs via room squares, *in* A. Chaudhuri and M. Ghosh, *Calcutta Statistical Association Bulletin* **52**, 71–84.

Fang, K. T., Ge, G. N. and Liu, M. Q. (2003), Construction of optimal supersaturated designs by the packing method, *Science in China (Series A)* **33**, 446–458.

Fang, K. T., Ge, G. N., Liu, M. Q. and Qin, H. (2002), Construction on minimum generalized aberration designs, *Metrika* **57**, 37–50.

Fang, K. T., Ge, G. N., Liu, M. Q. and Qin, H. (2004), Construction of uniform designs via super-simple resolvable t-designs, *Utilitas Mathematica* **66**, 15–32.

Fang, K. T. and Hickernell, F. J. (1995), The uniform design and its applications, in *Bulletin of The International Statistical Institute* pp. 339–349, Beijing.

Fang, K. T. and Hickernell, F. J. (1996), Discussion of the papers by Atkinson, and Bates et al., *J. Royal Statist. Soc. Ser. B* **58**, 103.

Fang, K. T. and Li, J. K. (1994), Some new results on uniform design, *Chinese Science Bulletin* **40**, 268–272.

Fang, K. T. and Lin, D. K. J. (2003), Uniform designs and their application in industry, *in* R. Khattree and C.R. Rao, *Handbook on Statistics in Industry* pp. 131–170, Elsevier, North–Holland, Amsterdam.

Fang, K. T., Lin, D. K. J. and Liu, M. Q. (2003), Optimal mixed-level supersaturated design and computer experiment, *Metrika* **58**, 279–291.

Fang, K. T., Lin, D. K. J. and Ma, C. X. (2000), On the construction of multi-level supersaturated designs, *J. Statist. Plann. Inference* **86**, 239–252.

Fang, K. T., Lin, D. K. J., Winker, P. and Zhang, Y. (2000), Uniform design: Theory and applications, *Technometrics* **42**, 237–248.

Fang, K. T., Lu, X., Tang, Y. and Yin, J. (2003), Constructions of uniform designs by using resolvable packings and coverings, *Discrete Mathematics* **19**, 692–711.

Fang, K. T., Lu, X. and Winker, P. (2003), Lower bounds for centered and wrap-around l_2-discrepancies and construction of uniform, *J. Complexity* **20**, 268–272.

Fang, K. T. and Ma, C. (2001a), *Orthogonal and Uniform Experimental Designs*, Science Press, Beijing.

Fang, K. T. and Ma, C. X. (2001b), Wrap-around L_2-discrepancy of random sampling, Latin hypercube and uniform designs, *J. Complexity* **17**, 608–624.

Fang, K. T. and Ma, C. X. (2002), Relationship between uniformity, aberration and correlation in regular fractions 3^{s-1}, *in* K. T. Fang and F. J. Hickernell and H. Niederreiter, *Monte Carlo and Quasi-Monte Carlo Methods 2000* pp. 213–231, Springer, New York.

Fang, K. T., Ma, C. X. and Mukerjee, R. (2001), Uniformity in fractional factorials, *in* K. T. Fang, F. J. Hickernell and H. Niederreiter, *Monte Carlo and Quasi-Monte Carlo Methods 2000* pp. 232–241, Springer, New York.

Fang, K. T., Ma, C. X. and Winker, P. (2000), Centered L_2-discrepancy of random sampling and latin hypercube design, and construction of uniform designs, *Math. Comp.* **71**, 275–296.

Fang, K. T., Ma, C. X. and Winker, P. (2002), Centered L_2-discrepancy of random sampling and Latin Hypercube design, and construction of uniform designs, *Math. Comp.* **71**, 275–296.

Fang, K. T., Maringer, D., Tang, Y. and Winker, P. (2005), Lower bounds and stochastic optimization algorithms for uniform designs with three or four levels, *Math. Comp.* **76**, to appear.

Fang, K. T. and Mukerjee, R. (2000), A connection between uniformity and aberration in regular fractions of two-level factorials, *Biometrika* **87**, 1993–198.

Fang, K. T. and Qin, H. (2002), A note on construction of nearly uniform designs with large number of runs, *Statist. & Probab. Letters* **61**, 215–224.

Fang, K. T. and Qin, H. (2004a), Discrete discrepancy in factorial designs, *Metrika* **60**, 59–72.

Fang, K. T. and Qin, H. (2004b), Uniformity pattern and related criteria for two-level factorials, *Science in China Ser. A. Math.* **47**, 1–12.

Fang, K. T., Shiu, W. C. and Pan, J. X. (1999), Uniform designs based on Latin squares, *Statistica Sinica* **9**, 905–912.

Fang, K. T., Tang, Y. and Yin, J. (2005), Lower bounds for wrap-around l_2-discrepancy and constructions of symmetrical uniform designs, *J. Complexity* **21**, to appear.

Fang, K. T. and Wang, Y. (1994), *Number-Theoretic Methods in Statistics*, Chapman & Hall, London.

Fang, K. T. and Winker, P. (1998), Uniformity and orthogonality, Technical Report MATH-175, Hong Kong Baptist University.

Fang, K. T. and Yang, Z. H. (1999), On uniform design of experiments with restricted mixtures and generation of uniform distribution on some domains, *Statist. & Probab. Letters* **46**, 113–120.

Fang, Y. (1995), Relationships between uniform design and orthogonal design, *in Proceedings of The 3rd International Chinese Statistical Association Statistical Conference* Beijing.

Faraway, J. (1997), Regression analysis for functional response, *Technometrics* **39**, 254–261.

Foster, D. P. and George, E. I. (1994), The risk inflation criterion for multiple regression, *Ann. Statist.* **22**, 1947–1975.

Frank, I. E. and Friedman, J. H. (1993), A statistical view of some chemometrics regression tools, *Technometrics* **35**, 109–148.

Friedman, J. H. (1991), Multivariate adaptive regression splines, *Ann. Statist.* **19**, 1–141.

George, E. I. (2000), The variable selection problem, *J. Amer. Statist. Assoc.* **95**, 1304–1308.

George, E. I. and McCulloch, R. E. (1993), Variable selection via gibbs sampling, *J. Amer. Statist. Assoc.* **88**, 991–889.

Gérard, F., Tournour, M., El Masri, N., Cremers, L., Felice, M. and Selmane, A. (2001), Numerical modeling of engine noise radiation through the use of acoustic transfer vectors - a case study, Society of Automotive Engineer, SAE Paper 2001-01-1514.

Giglio, B., Riccomagno, E. and Wynn, H. P. (2000), Gröbner bases strategies in regression, *J. Appl. Statist.* **27**, 923–938.

Hagan, M. T., Demuth, H. B. and Beale, M. H. (1996), *Neural Network Design*, PWS Publishing, Boston.

Haldar, A. and Mahadevan, S. (2000), *Reliability Assessment Using Stochastic Finite Element Analysis*, John Wiley & Sons, New York.

Hannan, E. J. and Quinn, B. G. (1979), The determination of the order of autoregression, *J. Royal Statist. Soc. Ser. B* **41**, 190–195.

Hasofer, A. M. and Lind, N. C. (1974), Exact and invariant second-moment code format, *J. Engineer. Meth.* **100**, 111–121.

Hassoun, M. H. (1995), *Fundamentals of Artificial Neural Networks*, The MIT Press, Cambridge, Massachusetts.

Hastie, T. and Tibshirani, R. (1993), Varying-coefficient models (with discussion), *J. Royal Statist. Soc. Ser. B* **55**, 757–796.

Haykin, S. S. (1998), *Neural Networks: A Comprehensive Foundation*, Prentice Hall, Upper Saddle River, NJ.

Hazime, R. M., Dropps, S. H., Anderson, D. H. and Ali, M. Y. (2003), Transient non-linear fea and tmf life estimates of cast exhaust manifolds, Society of Automotive Engineers, SAE 2003-01-0918.

Heckman, N. (1986), Spline smoothing in partly linear models, *J. Royal Statist. Soc. Ser. B* **48**, 244–248.

Hedayat, A. S., Sloane, N. J. A. and Stufken, J. (1999), *Orthogonal Arrays: Theory and Applications*, Springer, New York.

Hickernell, F. J. (1998*a*), A generalized discrepancy and quadrature error bound, *Math. Comp.* **67**, 299–322.

Hickernell, F. J. (1998*b*), Lattice rules: How well do they measure up?, *in* P. Hellekalek and G. Larcher, *Random and Quasi-Random Point Sets* pp. 106–166, Springer–Verlag, New York.

Hickernell, F. J. (1999), Goodness-of-fit statistics, discrepancies and robust designs, *Statist. & Probab. Letters* **44**, 73–78.

Hickernell, F. J. and Liu, M. Q. (2002), Uniform designs limit aliasing, *Bio-*

metrika **89**, 893–904.

Ho, W. M. (2001), Case studies in computer experiments, applications of uniform design and modern modeling techniques, Master's thesis, Hong Kong Baptist University.

Ho, W. M. and Xu, Z. Q. (2000), Applications of uniform design to computer experiments, *J. Chinese Statist. Assoc.* **38**, 395–410.

Hoerl, A. E. and Kennard, R. W. (1970), Ridge regression: Biased estimation for non-orthogonal problems, *Technometrics* **12**, 55–67.

Hoffman, R. M., Sudjianto, A., Du, X. and Stout, J. (2003), Robust piston design and optimization using piston secondary motion analysis, Society of Automotive Engineers, SAE Paper 2003-01-0148.

Homma, T. and Saltelli, A. (1996), Importance measures in global sensitivity analysis of nonlinear models, *Reliability Engineering and System Safety* **52**, 1–17.

Hoover, D. R., Rice, J. A., Wu, C. O. and Yang, L. P. (1998), Nonparametric smoothing estimates of time-varying coefficient models with longitudinal data, *Biometrika* **85**, 809–822.

Hora, S. C. and Iman, R. L. (1989), A comparison of maximum/boulding and Bayesian/Monte Carlo for fault tree uncertainty analysis, Technical Report SAND85-2839, Sandia National Laboratories.

Høst, G. (1999), Kriging by local polynomials, *Comp. Statist. & Data Anal.* **29**, 295–312.

Hua, L. K. and Wang, Y. (1981), *Applications of Number Theory to Numerical Analysis*, Springer and Science Press, Berlin and Beijing.

Huang, J. Z., Wu, C. O. and Zhou, L. (2002), Varying-coefficient models and basis function approximations for the analysis of repeated measurements, *Biometrika* **89**, 111–128.

Hunter, D. R. and Li, R. (2005), Variable selection using mm algorithms, *Ann. Statist.* **33**, 1617–1642.

Iman, R. L. and Conover, W. J. (1982), A distribution-free approach to inducing rank correlation among input variables, *Comm. Statist., Simulation and Computation* **11**, 311–334.

Iman, R. L. and Helton, J. C. (1988), An investigation of uncertainty and sensitivity analysis techniques for computer models, *Risk Analysis* **8**, 71–90.

Iman, R. L. and Hora, S. C. (1990), A robust measure of uncertainty importance for use in 44 fault tree system analysis, *Risk Analysis* **10**, 401–406.

Jin, R., Chen, W. and Sudjianto, A. (2004), Analytical metamodel-based global sensitivity analysis and uncertainty propagation for robust design, Society of Automotive Engineer, SAE Paper 2004-01-0429.

Jin, R., Chen, W. and Sudjianto, A. (2005), An efficient algorithm for constructing optimal design of computer experiments, *J. Statist. Plann. Inference* **134**, 268–287.

Jin, R., Du, X. and Chen, W. (2003), The use of metamodeling techniques for design under uncertainty, *Structural and Multidisciplinary Optimization* **25**, 99–116.

Joe, S. and Sloan, I. H. (1994), *Lattice Methods for Multiple Integration*, Clarendon Press, Oxford.

Johnson, M. E., Moore, L. M. and Ylvisaker, D. (1990), Minimax and maxmin distance design, *J. Statist. Plann. Inference* **26**, 131–148.

Jones, M. C., Marron, J. S. and Sheather, S. J. (1996a), A brief survey of bandwidth selection for density estimation, *J. Amer. Statist. Assoc.* **91**, 401–407.

Jones, M. C., Marron, J. S. and Sheather, S. J. (1996b), Progress in data-based bandwidth selection for kernel density estimation, *Computat. Statist.* **11**, 337–381.

Kalagnanam, J. and Diwekar, U. (1997), An efficient sampling techniques for off-line quality control, *Technometrics* **39**, 308–319.

Kendall, M. and Stuart, A. (1979), *The Advanced Theory of Statistics,* 4th ed, MacMillan Publishing Co., New York.

Kiefer, J. (1959), Optimum experimental design, *J. Royal Statist. Soc. Ser. B* **21**, 272–297.

Kimeldorf, G. S. and Wahba, G. (1970), A correspondence between Bayesian estimation on stochastic processes and smoothing by splines, *Ann. Math. Statist.* **41**, 495–502.

Kirkpatrick, S., Gelett, C. and Vecchi, M. (1983), Optimization by simulated annealing, *Science* **220**, 621–630.

Kleijnen, J. P. C. (1987), *Statistical Tools for Simulation Practitioners*, Marcel Decker, New York.

Koehler, J. R. and Owen, A. B. (1996), Computer experiments, *in* S. Ghosh and C. R. Rao, eds, *Handbook of Statistics* **13**, 261–308, Elsevier Science B.V., Amsterdam.

Korobov, N. M. (1959a), The approximate computation of multiple integrals, *Dokl. Akad. Nauk. SSSR* **124**, 1207–1210.

Korobov, N. M. (1959b), Computation of multiple integrals by the method of optimal coefficients, *Vestnik Moskow Univ. Sec. Math. Astr. Fiz. Him.* **4**, 19–25.

Krige, D. G. (1951), A statistical approach to some mine valuations and allied problems at the Witwatersrand, Master's thesis, University of Witwatersrand.

Krzykacz, B. (1990), Samos: a computer program for the derivation of empirical sensitivity measures of results from large computer models, Technical Report GRS-A-1700, Gesellschaft fur Reaktorsicherheit (GRS) mbH, Garching, Germany.

Le Cun, Y., Simard, P. and Pearlmutter, B. (1993), Automatic learning rate maximization by on-line estimation of Hessian's eigenvector, *in* eds. S. J. Hanson, J. D. Cowan, and C. L. Giles, *Advances in Neural Information Processing System* 5, 156–163, Morgan Kaufmann, San Mateo, CA.

Lee, A. W. M., Chan, W. F., Yuen, F. S. Y., Tse, P. K., Liang, Y. Z. and Fang, K. T. (1997), An example of a sequential uniform design: Application in capillary electrophoresis, *Chemom. Intell. Lab. Systems* **39**, 11–18.

Li, K. C. (1987), Asymptotic optimality for C_p, C_l, cross-validation and generalized cross validation: discrete index set, *Ann. Statist.* **15**, 958–975.

Li, R. (2002), Model selection for analysis of uniform design and computer experiment, *Inter. J. Reliability, Quality, and Safety Engineering* **9**, 305–315.

Li, R. and Sudjianto, A. (2005), Analysis of computer experiments using penalized likelihood in Gaussian Kriging models, *Technometrics* **47**, 111–120.

Li, R., Sudjianto, A. and Zhang, Z. (2005), Modeling computer experiments with functional response, Society of Automative Engineers, SAE paper 2005-01-1397.

Li, W. W. and Wu, C. F. J. (1997), Columnwise-pairwise algorithms with applications to the construction of supersaturated designs, *Technometrics* **39**, 171–179.

Liang, Y. Z., Fang, K. T. and Xu, Q. S. (2001), Uniform design and its applications in chemistry and chemical engineering, *Chemom. Intell. Lab. Systems* **58**, 43–57.

Lin, D. K. J. (1993), A new class of supersaturated designs, *Technometrics* **35**, 28–31.

Lin, D. K. J. (1995), Generating systematic supersaturated designs, *Technometrics* **37**, 213–225.

Lin, D. K. J. (1999), Recent developments in supersaturated designs, *in* S. H.

Park and G. G. Vining, eds, *Statistical Process Monitoring and Optimization,* Chapter 18, Marcel Dekker, New York.

Lin, D. Y. and Ying, Z. (2001), Semiparametric and nonparametric regression analysis of longitudinal data (with discussions), *J. Amer. Statist. Assoc.* **96**, 103–126.

Liu, M. Q. and Hickernell, F. J. (2002), $E(s^2)$-optimality and minimum discrepancy in 2-level supersaturated designs, *Statistica Sinica* **12**, 931–939.

Lo, Y. K., Zhang, W. J. and Han, M. X. (2000), Applications of the uniform design to quality engineering, *J. Chinese Statist. Assoc.* **38**, 411–428.

Loibnegger, B., Rainer, G. P., Bernard, L., Micelli, D. and Turino, G. (1997), An integrated numerical tool for engine noise and vibration simulation, Society of Automative Engineers, SAE paper 971992.

Lu, X. and Meng, Y. (2000), A new method in the construction of two-level supersaturated designs, *J. Statist. Plann. Inference* **86**, 229–238.

Lu, X. and Sun, Y. (2001), Supersaturated designs with more than two levels, *Chinese Ann. Math., Ser. B* **22**, 69–73.

Ma, C., Fang, K. T. and Lin, D. K. J. (2001), On isomorphism of factorial designs, *J. Complexity* **17**, 86–97.

Ma, C. X. and Fang, K. T. (2001), A note on generalized aberration in factorial designs, *Metrika* **53**, 85–93.

Ma, C. X. and Fang, K. T. (2004), A new approach to construction of nearly uniform designs, *Inter. J. Materials and Product Technology* **20**, 115–126.

Ma, C. X., Fang, K. T. and Lin, D. K. J. (2002), A note on uniformity and orthogonality, *J. Statist. Plann. Inference* **113**, 323–334.

MacKay, D. J. C. (1998), Introduction to gaussian processes, *in* C. M. Bishop, ed., *Neural Networks and Machine Learning,* NATO ASI Series pp. 133–166, Kluwer, Dordrecht.

Mallows, C. L. (1973), Some comments on C_p, *Technometrics* **15**, 661–675.

Marron, J. S. and Nolan, D. (1988), Canonical kernels for density estimation, *Statist. & Probab. Letters* **7**, 195–199.

Matheron, G. (1963), Principles of geostatistics, *Econm. Geol.* **58**, 1246–1266.

McKay, M. D. (1995), Evaluating prediction uncertainty, Technical Report NU REG/CR-6311, Nuclear Regulatory Commission and Los Alamos National Laboratory.

McKay, M. D., Beckman, R. J. and Conover, W. J. (1979), A comparison of three methods for selecting values of input variables in the analysis of output from a computer code, *Technometrics* **21**, 239–245.

Melchers, R. (1999), *Structural Reliability Analysis and Prediction,* John Wiley & Sons, Chichester, England.

Micchelli, C. A. and Wahba, G. (1981), Design problems for optimal surface interpolation, *in* Z. Ziegler, eds, *Approximation Theory and Applications* Academic Press, New York.

Miller, A. (2002), *Subset Selection in Regression,* 2nd ed, Chapman & Hall/CRC, London.

Miller, D. and Frenklach, M. (1983), Sensitivity analysis and parameter estimation in dynamic modeling of chemical kinetics, *Inter. J. Chem. Kinetics* **15**, 677–696.

Mitchell, T. J. and Beauchamp, J. J. (1988), Bayesian variable selection in linear regression (with discussion), *J. Amer. Statist. Assoc.* **83**, 1023–1036.

Montgomery, D. C. (2001), *Design and Analysis of Experiments,* John Wiley & Sons, New York.

Moody, J. and Darken, C. (1989), Fast learning in networks of locally-tuned processing units, *Neural Comput.* **1**, 281–294.

Morris, M. D. and Mitchell, T. J. (1995), Exploratory design for computational experiments, *J. Statist. Plann. Inference* **43**, 381–402.

Morris, M. D., Mitchell, T. J. and Ylvisaker, D. (1993), Bayesian design and analysis of computer experiments: Use of derivatives in surface prediction, *Technometrics* **35**, 243–255.

Myers, R. H. (1990), *Classical and Modern Regression with Applications,* 2nd ed, Duxbury Press, Belmont, CA.

Myers, R. H. and Montgomery, D. C. (1995), *Response Surface Methodology: Process and Product Optimization Using Designed Experiments,* Wiley, New York.

Nadaraya, E. A. (1964), On estimating regression, *Theory Prob. Appl.* **9**, 141–142.

Nair, V. J., Taam, W. and Ye, K. Q. (2002), Analysis of functional responses from robust design studies, *J. Qual. Contr.* **34**, 355–369.

Neter, J., Kutner, M. H., Nachtsheim, C. J. and Wasserman, W. (1996), *Applied Linear Statistical Models,* 4th ed, Irwin, Chicago.

Niederreiter, H. (1992), *Random Number Generation and Quasi-Monte Carlo Methods,* Applied Mathematics, SIAM CBMS-NSF Regional Conference, Philadelphia.

Owen, A. B. (1992*a*), A central limit theorem for Latin hypercube sampling, *J. Royal Statist. Soc. Ser. B* **54**, 541–551.

Owen, A. B. (1992*b*), Randomly orthogonal arrays for computer experiments, integration and visualization, *Statistica Sinica* **2**, 439–452.

Owen, A. B. (1994*a*), Controlling correlations in Latin hypercube samples, *J. Amer. Statist. Assoc.* **89**, 1517–1522.

Owen, A. B. (1994*b*), Lattice sampling revisited: Monte Carlo variance of means over randomized orthogonal array, *Ann. Statist.* **22**, 930–945.

Owen, A. B. (1997), Monte Carlo variance of scrambled net quadrature, *SIAM J. Numer. Anal.* **34**, 1884–1910.

Pan, J. X. and Fang, K. T. (2002), *Growth Curve Models and Statistical Diagnostics*, Springer, New York.

Park, J.-S. (1994), Optimal Latin-hypercube designs for computer experiments, *J. Statist. Plann. Inference* **39**, 95–111.

Parkinson, A., Sorensen, C. and Pourhassan, N. (1993), A general approach for robust optimal design, *ASME J. Mech. Design* **115**, 74–80.

Patterson, H. D. and Thompson, R. (1971), Recovery of inter-block information when block sizes are unequal, *Biometrika* **58**, 545–554.

Phadke, M. (1989), *Quality Engineering Using Robust Design*, Prentice Hall, Englewood Cliffs, NJ.

Philips, P. J., Schamel, A. R. and Meyer, J. (1989), An efficient model for valvetrain and spring dynamics, Society of Automotive Engineers, SAE paper 890619.

Powell, M. J. D. (1987), Radial basis functions for multivariable interpolation: a review, *in* J. C. Mason and M. G. Cox, eds, *Algorithm for Approximation* pp. 143–167, Clarendon Press, Oxford.

Press, W. H., Teukolsky, S. A., Vetterling, W. T. and Flannery, B. P. (1992), *Numerical Recipes in C: The Art of Scientific Computing,* 2nd ed, Cambridge University Press, Cambridge, UK.

Pukelsheim, F. (1993), *Optimum Design of Experiments*, Wiley, New York.

Qin, H. (2002), Construction of uniform designs and usefulness of uniformity in fractional factorial designs, Ph.D thesis, Hong Kong Baptist University.

Qin, H. and Fang, K. T. (2004), Discrete discrepancy in factorial designs, *Metrika* **60**, 59–72.

Ramsay, J. O. and Silverman, B. W. (1997), *Functional Data Analysis*, Springer, New York.

Rao, C. R. (1973), *Linear Statistical Inference and its Applications*, Wiley, New York.

Reedijk, C. I. (2000), Sensitivity analysis of model output: Performance of various local and global sensitivity measures on reliability problems, Delft University of Technology, Master's thesis.

Rees, R. and Stinson, D. (1992), Frames with block size four, *Canad. J. Math.* **44**, 1030.

Riccomango, E., Schwabe, R. and Wynn, H. P. (1997), Lattice-based D-optimal design for Fourier regression, **25**, 2313–2317.

Rumelhart, D., Hinton, G. and Williams, R. (1986), Learning internal representations by error propagation, *in* D. E. Rumelhart, J.L. Mcclelland, and the PDP Research Group, eds, *Parallel Distributed Processing: Explorations in the Microstructure of Cognition*, MIT Press, Cambridge, MA.

Ruppert, D. (2002), Selecting the number of knots for penalized splines, *J. Comput. Graph. Statist.* **11**, 735–757.

Ruppert, D. and Carroll, R. J. (2000), Spatially-adaptive penalties for spline fitting, *Austral. & New Zealand J. Statist.* **42**, 205–223.

Saab, Y. G. and Rao, Y. B. (1991), Combinatorial optimization by stochastic evolution, *IEEE Trans. on Computer-aided Design* **10**, 525–535.

Sack, J. and Ylvisaker, D. (1985), Model robust design in regression: Bayes theory, *in* L. M. lecam and R. A. Olshen, *Proceedings of the Berkeley Conference in Honor of Jerzy Neyman and Jack Kiefer, Vol. II* pp. 667–679, Wadsworth, Belmont, CA.

Sacks, J., Schiller, S. B. and Welch, W. J. (1989), Designs for computer experiments, *Technometrics* **31**, 41–47.

Sacks, J., Welch, W. J., Mitchell, T. J. and Wynn, H. P. (1989), Design and analysis of computer experiments (with discussion), *Statistica Sinica* **4**, 409–435.

Sacks, J. and Ylvisaker, D. (1984), Some model robust designs in regression, *Ann. Statist.* **12**, 1324–1348.

Saitoh, S. (1988), *Theory of Reproducing Kernels and Its Applications*, Longman Scientific and Technical, Essex, England.

Saltelli, A., Chan, K. and Scott, E. M. (2000), *Sensitivity Analysis*, Wiley, New York.

Santner, T. J., Williams, B. J. and Notz, W. I. (2003), *The Design and Analysis of Computer Experiments*, Springer, New York.

Satoh, K., Kawai, T., Ishikawa, M. and Matsuoka, T. (2000), Development of method for predicting efficiency of oil mist separators, Society of Automobile Engineering, SAE Paper 2000-01-1234.

Schwarz, G. (1978), Estimating the dimension of a model, *Ann. Statist.* **6**, 461–464.

Seber, G. A. F. and Wild, C. J. (1989), *Nonlinear Regression*, Wiley, New York.

Shaw, J. E. H. (1988), A quasirandom approach to integration in Bayesian statistics, *Ann. Statist.* **16**, 859–914.

Shewry, M. C. and Wynn, H. P. (1987), Maximum entropy sampling, *J. Appl. Statist.* **14**, 165–170.

Shibata, R. (1984), Approximation efficiency of a selection procedure for the number of regression variables, *Biometrika* **71**, 43–49.

Silverman, B. W. (1984), Spline smoothing: the equivalent variable kernel method, *Ann. Statist.* **12**, 898–916.

Simpson, T. W., Booker, A. J., Ghosh, D., Giunta, A. A., Koch, P. N. and Yang, R.-J. (2000), Approximation methods in multidisciplinary analysis and optimization: A panel discussion, *in AIAA/ISSMO Symposium on Multi-disciplinary Analysis and Optimization* **9,** Atlanta.

Simpson, T. W., Peplinski, J. D., Koch, P. N. and Allen, J. K. (2001), Metamodels for computer-based engineering design: survey and recommendations, *Engineering with Computers* **17**, 129–150.

Smola, A. J. and Schölkopf, B. (1998), A tutorial on support vector regression, Manuscript.

Sobol', I. M. (1993), Sensitivity analysis for nonlinear mathematical models, *Mathematical Modeling and Computational Experiment* **1**, 407–414.

Sobol', I. M. (2001), Global sensitivity indices for nonlinear mathematical models and their Monte Carlo estimates, *Math. and Comp. in Simulation* **55**, 271–280.

Sobol', I. M. (2003), Theorems and examples on high dimensional model representation, *Reliability Engneering & System Safety* **79**, 187–193.

Speckman, P. (1988), Kernel smoothing in partial linear models, *J. Royal Statist. Soc. Ser. B* **50**, 413–436.

Stein, M. (1987), Large sample properties of simulations using Latin hypercube sampling, *Technometrics* **29**, 143–151.

Stone, C. J. (1985), Additive regression and other nonparametric models, *Ann. Statist.* **13**, 689–705.

Stone, C. J., Hansen, M. H., Kooperberg, C. and Truong, Y. K. (1997), Polynomial splines and their tensor products in extended linear modeling, *Ann. Statist.* **25**, 1371–1470.

Sudjianto, A., Juneja, L., Agrawal, H. and Vora, M. (1998), Computer-aided reliability and robustness assessment, *Inter. J. Quality, Reliability, and Safety* **5**, 181–193.

Taguchi, G. (1986), *Introduction to Quality Engineering*, Asian Production Organization, Tokyo.

Taguchi, G. (1993), *Taguchi on Robust Technology Development: Bringing Quality Engineering Upstream*, ASME Press, New York.

Tang, B. (1993), Orthogonal array-based Latin hypercubes, *J. Amer. Statist. Assoc.* **88**, 1392–1397.

Tang, Y. (2005), Combinatorial properties of uniform designs and their applications in the constructions of low-discrepancy designs, Ph.D thesis, Hong Kong Baptist University.

Tibshirani, R. (1996), Regression shrinkage and selection via the lasso, *J. Royal Statist. Soc. Ser. B* **58**, 267–288.

Tu, J. (2003), Cross-validated multivariate metamodeling methods for physics-based computer simulations, *in Proceedings of the IMAC-XXI: A Conference on Structural Dynamics* Feb. 3–6, 2003, Kissimmee Florida.

Tu, J., Choi, K. K. and Young, H. P. (1999), A new study on reliability-based design optimization, *ASME J. Mech. Engineering* **121**, 557–564.

Tu, J. and Jones, D. R. (2003), Variable screening in metamodel design by cross-validated moving least squares method, *in The 44th AIAA/ASME/ASCE/AHS Structural Dynamics, and Material Conference* AIAA2003-1669, April 7–10, 2003, Norfolk, Virginia.

Wahba, G. (1978), Improper priors, spline smoothing and the problem of guarding against model errors in regression, *J. Royal Statist. Soc. Ser. B* **40**, 364–372.

Wahba, G. (1990), *Spline Models for Observational Data*, SIAM, Philadelphia.

Wang, L., Grandhi, R. V. and Hopkins, D. A. (1995), Structural reliability optimization using an efficient safety index calculation procedure, *Inter. J. Numer. Methods in Engineering* **38**, 1721–1738.

Wang, X. and Fang, K. T. (2005), Extensible uniform designs for computer experiments, Hong Kong Baptist University, Technical Report Math-404.

Wang, Y. and Fang, K. T. (1981), A note on uniform distribution and experimental design, *KeXue TongBao* **26**, 485–489.

Wang, Y. and Fang, K. T. (1996), Uniform design of experiments with mixtures, *Science in China (Series A)* **39**, 264–275.

Wang, Y. J., Lin, D. K. J. and Fang, K. T. (1995), Designing outer array points, *J. Qual. Tech.* **27**, 226–241.

Warnock, T. T. (1972), Computational investigations of low discrepancy point sets, *in* S. K. Zaremba, *Applications of Number Theory to Numerical Analysis* pp. 319–343, Academic Press, New York.

Watson, G. S. (1963), Smooth regression analysis, *Sankhyā Ser. A* **26**, 359–372.

Weisberg, S. (1985), *Applied Linear Regression*, 2 edn, Wiley & Sons, New York.

Weyl, H. (1916), Über die Gleichverteilung der Zahlem mod Eins, *Math. Ann.* **77**, 313–352.

Weyl, H. (1938), Mean motion, *Am. J. Math.* **60**, 889–896.

Wiens, D. (1991), Designs for approximately linear regression: Two optimality properties of uniform designs, *Statist. & Probab. Letters* **12**, 217–221.

Winker, P. (2001), *Optimization Heuristics in Econometrics: Applications of Threshold Accepting*, Wiley, Chichester.

Winker, P. and Fang, K. T. (1997), Application of threshold accepting to the evaluation of the discrepancy of a set of points, *SIAM Numer. Analysis* **34**, 2038–2042.

Winker, P. and Fang, K. T. (1998), Optimal *u*-type design, *in* H. Niederreiter, P. Zinterhof and P. Hellekalek, *Monte Carlo and Quasi-Monte Carlo Methods 1996* pp. 436–448, Springer, New York.

Worley, B. A. (1987), Deterministic uncertainty analysis, *in* 'ORNL—0628', National Technical Information Service, 5285 Port Royal Road, Springfield, VA 22161, USA.

Wu, C. F. J. (1993), Construction of supersaturated designs through partially aliased interactions, *Biometrika* **80**, 661–669.

Wu, C. F. J. and Hamada, M. (2000), *Experiments: Planning, Analysis, and Parameter Design Optimization*, John Wiley & Sons, New York.

Wu, C. O., Chiang, C. T. and Hoover, D. R. (1998), Asymptotic confidence regions for kernel smoothing of a varying-coefficient model with longitudinal data, *J. Amer. Statist. Assoc.* **93**, 1388–1403.

Wu, Y.-T. and Wang, W. (1998), Efficient probabilistic design by converting reliability constraints to approximately equivalent deterministic constraints, *J. Integrated Design and Process Sciences* **2**, 13–21.

Xie, M. Y. and Fang, K. T. (2000), Admissibility and minimaxity of the uniform design in nonparametric regression model, *J. Statist. Plann. Inference* **83**, 101–111.

Xu, Q. S., Liang, Y. Z. and Fang, K. T. (2000), The effects of different experimental designs on parameter estimation in the kinetics of a reversible chemical reaction, *Chemom. Intell. Lab. Systems* **52**, 155–166.

Yamada, S. and Lin, D. K. J. (1999), 3-level supersaturated designs, *Statist. & Probab. Letters* **45**, 31–39.

Yang, R.-J., Gu, L., Tho, C. H. and Sobieszczanski-Sobieski, J. (2001), Multidisciplinary design optimization of a full vehicle with high performance computing, American Institute of Aeronautics and Astronautics, AIAA-2001-1273, Seattle, WA., April 16-19, 2001.

Ye, Q. (1998), Orthogonal column latin hypercubes and their application in computer experiments, *J. Amer. Statist. Assoc.* **93**, 1430–1439.

Ye, Q., Li, W. and Sudjianto, A. (2003), Algorithmic construction of optimal symmetric latin hypercube designs, *J. Statist. Plann. Inference* **90**, 145–159.

Yue, R. X. (2001), A comparison of random and quasirandom points for nonparametric response surface design,, *Statist. & Probab. Letters* **52**, 129–142.

Yue, R. X. and Hickernell, F. J. (1999), Robust optimal designs for fitting approximately liner models, *Statistica Sinica* **9**, 1053–1069.

Zhang, A. J., Fang, K. T., Li, R. and Sudjianto, A. (2005), Majorization framework for balanced lattice designs, *Ann. Statist.* **33**, in press.

Zou, T., Mahadevan, S., Mourelatos, Z. and Meernik, P. (2002), Reliability analysis of automotive body-door subsystem, *Reliability Engineering and System Safety* **78**, 315–324.

Index

Author Index